Annals of Mathematics Studies

Number 136

Hyperfunctions on Hypo-Analytic Manifolds

by

Paulo D. Cordaro and François Treves

PRINCETON UNIVERSITY PRESS

PRINCETON, NEW JERSEY

1994

Princeton University Press books are printed on acid-free paper and meet the guidelines for permanence and durability of the Committee on Production Guidelines for Book Longevity of the Council on Library Resources

Printed in the United States of America

10 9 8 7 6 5 4 3 2 1

Library of Congress Cataloging-in-Publication Data

Cordaro, Paulo D.
Hyperfunctions on hypo-analytic manifolds / by Paulo D. Cordaro and François Treves.
p. cm. — (Annals of mathematics studies ; no. 136)
Includes bibliographical references and index.
ISBN 0-691-02993-8 (cloth) ISBN 0-691-02992-X (pbk.)
1. Hyperfunctions. 2. Submanifolds. I. Treves, François, 1930–. II. Title. III. Series.
QA324.C67 1994
515'.782—dc20 94-29391

The publisher would like to acknowledge the authors of this volume for providing the camera-ready copy from which this book was printed

To André Martineau, a friend, *in memoriam*

CONTENTS

CHAPTER III
HYPERFUNCTION SOLUTIONS IN A HYPO—ANALYTIC MANIFOLD

CHAPTER IV
TRANSVERSAL SMOOTHNESS OF HYPERFUNCTION SOLUTIONS

PREFACE

The material in the present monograph represents, in a sense, the next logical step in the development of the local (and microlocal) theory of *hypo–analytic structures* described in the book [TREVES, 1992]. Roughly speaking, a hypo—analytic structure on a $\mathcal{C}^\infty$ manifold is what makes possible the holomorphic extension of functions, distributions, and more generally, as we show in the pages that follow, hyperfunctions. The theory developed here relies on concepts, methods and results found in the book cited above. However, we have made a rule of referring to it only for technical assistance, and of repeating every definition needed for the comprehension of the present text.

Distributions are the objects of study in an *involutive structure*: as solutions, right—hand sides, initial data, etc. An involutive structure on a $\mathcal{C}^\infty$ manifold $\mathcal{M}$ is the datum of a vector subbundle $\mathcal{V}$ of the complexified tangent bundle $\mathbb{C}T\mathcal{M}$ that satisfies the Frobenius condition $[\mathcal{V},\mathcal{V}] \subset \mathcal{V}$ (ie., the commutation bracket of two smooth sections is a section). For a distribution u in an open subset Ω of $\mathcal{M}$ to be a solution means that $\mathrm{L}u = 0$ for all $\mathcal{C}^\infty$ sections L of $\mathcal{V}$ over Ω, ie., $\mathrm{L} \in \mathcal{C}^\infty(\Omega;\mathcal{V})$.

An important subclass of involutive structures are the *locally integrable structures*: the structure on $\mathcal{M}$ defined by the vector bundle $\mathcal{V}$ (whose rank is equal to n) is locally integrable if every point $\wp$ has an open neighborhood $\mathrm{U}_\wp$ in which there are $\mathrm{m} = \dim \mathcal{M} - \mathrm{n}$ "*first integrals*", ie., $\mathcal{C}^\infty$ solutions $Z_1,\ldots,Z_\mathrm{m}$ such that $\mathrm{d}Z_1\wedge\cdots\wedge\mathrm{d}Z_\mathrm{m} \neq 0$ at $\wp$. Real structures (ie., $\mathcal{V}$ is locally spanned by real vector fields) and real—analytic structures (ie., $\mathcal{M}$ and $\mathcal{V}$ are of class $\mathcal{C}^\omega$), complex structures (ie., $\mathcal{V}\oplus\overline{\mathcal{V}} = \mathbb{C}T\mathcal{M}$) and more generally elliptic structures (ie., $\mathcal{V}+\overline{\mathcal{V}} = \mathbb{C}T\mathcal{M}$) are locally integrable; and so are strongly pseudoconvex CR structures of the hyper—surface type if $\dim \mathcal{M} \geq 7$. However there exist involutive structures that are not locally integrable. On this and on related topics we refer the reader to Chapters VI, VII of [TREVES, 1992].

Dealing with a fixed set of first integrals $Z_1,\ldots,Z_m$ in an open subset Ω of $\mathcal{M}$ leads to novel questions. The approximation formula in [BAOUENDI–TREVES, 1981] (see also Section II.2, [TREVES, 1992]) asserts that each point $\wp \in \Omega$ has a neighborhood $U_\wp$ in which every C^0 solution h is constant on the pre–image $Z^{-1}(z_0)$ of every point $z_0 \in \mathbb{C}^m$ in the range of $Z = (Z_1,\ldots,Z_m)$. This allows us to pushforward the solution h under the map Z, to get a kind of CR function $\tilde{h}$ on the, generally highly singular, set $Z(U_\wp)$. In turn this leads to the question of the possibility, or impossi–bility, of extending the pushforward $\tilde{h}$ as a holomorphic function in a do–main of $\mathbb{C}^m$ larger than the union of the ranges $Z(U_\wp)$. Such an extension, however, would become meaningless if we were to change our choice of the system of first integrals $Z_1,\ldots,Z_m$ — unless, of course, we were to replace it by one of its biholomorphic "substitutions." This is the observation that leads to the concept of *hypo–analytic structure.*

On the manifold $\mathcal{M}$ a hypo–analytic structure of rank $m \in \mathbb{Z}_+$ is defined by an atlas of hypo–analytic charts (U^α, Z^α): the domains U^α make up an open covering of $\mathcal{M}$; Z^α is a C^∞ map $U^\alpha \to \mathbb{C}^m$ whose components Z^α_i have linearly independent differentials. On the overlap $U^\alpha \cap U^\beta$, if it is nonempty, Z^α and Z^β are biholomorphically equivalent: there is a biholomorphism $F_\alpha{}^\beta$ of an open neighborhood of $Z^\alpha(U^\alpha \cap U^\beta)$ onto one of $Z^\beta(U^\alpha \cap U^\beta)$ such that $Z^\beta = F_\alpha{}^\beta \circ Z^\alpha$.

Hypo–analytic structures are the natural framework for the study of hyperfunctions. Suppose the hypo–analytic structure of $\mathcal{M}$ is *maximal*, ie., $m = \dim \mathcal{M}$. Then Z^α realizes an immersion of U^α as a totally real submanifold $\mathcal{X}$ of $\mathbb{C}^m$ of real dimension m (what will be always called, in this text, a *maximally real submanifold*). The concept of a hyperfunction on a totally real submanifold of $\mathbb{C}^m$ has been around for a while. Chapter I of this book provides what we believe is the most succint, and elementary, definition of that concept. At the end of Chapter I we have inserted a section that duplicates the definition from a slightly more sophisticated viewpoint, based on polynomial convexity and Serre duality. Both approaches follow the original prescription of A. Martineau and are based on the use of analytic functionals (introduced and rapidly studied in sections I.1, I.2).

Now the sheaf of hyperfunctions on $Z^\alpha(U^\alpha)$ can be pulled back as a sheaf $\mathcal{B}^\alpha$ to U^α. The biholomorphic equivalence of Z^α and Z^β on $U^\alpha \cap U^\beta$ ensures that $\mathcal{B}^\alpha$ and $\mathcal{B}^\beta$ are equal there, and thus the local sheaves $\mathcal{B}^\alpha$ can be patched up to define a sheaf $\mathcal{B}_{\mathcal{M}}$ on the whole of $\mathcal{M}$. The sheaf of hyperfunctions in $\mathcal{M}$, $\mathcal{B}_{\mathcal{M}}$, will only depend on the hypo—analytic structure of $\mathcal{M}$. It contains the sheaf $\mathcal{D}_{\mathcal{M}}'$ of the germs of distributions in $\mathcal{M}$. When the manifold $\mathcal{M}$ is real analytic the hyperfunctions on $\mathcal{M}$ defined following our recipe are the same as those introduced in [SATO, 1960].

Boundary values of holomorphic functions in wedges with edge an open subset U of $\mathcal{X}$, whose absolute values grow arbitrarily "at the edge" U, are hyperfunctions in U; and conversely any hyperfunction in $\mathcal{X}$ is locally the sum of boundary values of holomorphic functions in wedges of that kind (sections II.1, II.3). We are therefore in the ideal situation to study holomorphic extension, without the need to limit the rate of its "blow up" as the edge is approached. This leads to a very general Edge of the Wedge Theorem (section II.5). Our tool of choice for the analysis that leads to it, and to the related concept of the *hypo–analytic wave–front set* of a hyperfunction, is the Fourier—Brós—Iagolnitzer (FBI) transform (section II.2).

The contents of Chapters I and II are more or less well charted territory. The authors might claim credit for a modest amount of stream—lining and simplification (thanks, often, to the use of Gaussian approxi—mation and of singular integrals of the Cauchy—Fantappiè kind); and for extending the range of certain theorems, as with the Edge of the Wedge alluded to above. Since its cohomological inception in the works of M. Sato in the late Fifties, and its analytic reformulation by A. Martineau in the Sixties, a consensus has developed on how hyperfunction theory should look — at least on Euclidean space $\mathbb{R}^m$ or on a totally real submanifold of $\mathbb{C}^m$.

It is when attempting to define hyperfunctions in a hypo—analytic structure of codimension n (= dim $\mathcal{M}$ — m) $\geq$ 1 that we enter new terri—tory. Locally one can select first integrals $Z_1,\ldots,Z_m$ and real coordinates $t_1,\ldots,t_n$ that are transverse to one another: $dZ_1\wedge\cdots\wedge dZ_m\wedge dt_1\wedge\cdots\wedge dt_n \neq 0$. It is easy enough to define hyperfunctions in z—space, on the image of the map Z for constant t, that depend smoothly on t (see sections IV.1, IV.2). The problem is that the definition of these hyperfunctions is not going to be independent of the choice of the transverse coordinates t_j; it is not going to be *hypo—analytically invariant.*

To go around this obstacle one limits one's ambitions to defining *hyperfunction solutions.* Even this is not automatic: it suffices to note that the coefficients of the vector fields $L_1,\ldots,L_n$ that span (locally) the tangent structure bundle $\mathcal{V}$ are generally not hypo—analytic; per se they will not act multiplicatively on hyperfunctions. Instead we avail ourselves of the fact that our base manifold $\mathcal{M}$ can be locally embedded into $\mathbb{C}^{m+n}$ (m+n = dim $\mathcal{M}$) by means of local maps $\wp \to (Z(\wp),Z'(\wp))$ whose first m com—ponents $Z_1,\ldots,Z_m$ are hypo—analytic whereas the last n ones, $Z'_1,\ldots,Z'_n$, are arbitrary (but smooth). A standard way of doing this is by introducing (as done above) coordinates $t_1,\ldots,t_n$ transverse to $Z_1,\ldots,Z_m$ and putting $Z'(\wp)$

$= t(\wp)$. But in any case, whether the $Z'_1,\dots,Z'_n$ are real or not, the image of the map (Z,Z') is a maximally real submanifold Σ of $\mathbb{C}^{m+n}$ — on which hyperfunctions are well defined (in Chapter I). At this stage we observe that the hypo—analytic functions on Σ for the hypo—analytic structure pushed forward from $\mathcal{M}$ are equal to the restrictions of the holomorphic functions of $(z,z') \in \mathbb{C}^{m+n}$ *that are locally constant with respect to* z'. Whence the definition of a hyperfunction solution in the domain $\mathcal{M}'$ of the map (Z,Z'): it is the "pullback" of a hyperfunction u in Σ such that $\partial_{z'} u = 0$ (holomorphic vector fields, such as $\partial/\partial z'_j$, do act on hyperfunctions in Σ).

The first thing to do is to check that such a definition encompasses that of a distribution solution: it does (Example III.1.2). The second thing to do, equally important, is to check that it is hypo—analytically invariant. It is, as proved by the Main Theorem of Chapter III, which states that the space of hyperfunction solutions in an appropriately small open subset U of Σ, is naturally isomorphic to the relative cohomology space $H_U^{m,m+n}(\mathbb{C}^{m+n}\backslash\partial U)$ for the Dolbeault—De Rham complex $\bar\partial_z + d_{z'}$, — precisely the differential complex attached to the hypo—analytic structure in $\mathbb{C}^{m+n}$ that induces on Σ the hypo—analytic structure pushed forward from $\mathcal{M}$. We use the notation ∂U to denote the boundary of U relative to Σ; $H_U^{m,m+n}(\mathbb{C}^{m+n}\backslash\partial U)$ is the cohomology space relative to the subset U, which is closed in $\mathbb{C}^{m+n}\backslash\partial U$.

When $n = 0$ and $\Sigma = \mathbb{R}^m$ this is exactly the definition of a hyper—function in U according to Sato.

Actually the Main Theorem of Chapter III identifies the cohomo—logy classes in $H_U^{m,m+n+q}(\mathbb{C}^{m+n}\backslash\partial U)$ to the cohomology classes in the differential complex $\partial_{z'}$ on Σ — for any $q = 0,1,\dots,n$. For $q \geq 1$ the nonvanishing of $H_U^{m,m+n+q}(\mathbb{C}^{m+n}\backslash\partial U)$ signals the nonsolvability in U of the equation $\partial_{z'}u = f$ for some f such that $\partial_{z'}f = 0$. Here $f = \sum_{|I|=q} f_I dz'_I$ is a "current" whose coefficients are hyperfunctions in U, u is required to be a similar current but of degree q—1, and of course

$$\partial_{z'}f = \sum_{|I|=q} \sum_{j=1}^{n} (\partial f_I/\partial z'_j)dz'_j \wedge dz'_I.$$

When specialized to true currents f, ie., currents whose coefficients are distributions on Σ, the equation $\partial_{z'}u = f$ is equivalent to the standard equation $Lu = f$, where now L is the tangential differential operator on Σ induced by the operator $\bar{\partial}_z + d_{z'}$ in $\mathbb{C}^{m+n}$. The reader will notice that the nonsolvability of the equation $Lu = f$ is thus invariantly defined although the hyperfunction currents f on Σ are not (and the operation Lf is not at all defined).

The same argument can be made for special embeddings, e.g., the "coarse local embedding" $\wp \to (Z(\wp),t(\wp))$ now viewed as valued in $\mathbb{C}^m \times \mathbb{R}^n$. The hypo—analytic structure on its range Σ transferred from $\mathcal{M}$ is the same as that inherited from the elliptic structure on $\mathbb{C}^m \times \mathbb{R}^n$ defined by the complex coordinates $z_1,...,z_m$; Σ is a "generic" submanifold of $\mathbb{C}^m \times \mathbb{R}^n$ in the sense that the pullbacks to Σ of $dz_1,...,dz_m$ are linearly independent. The differential operator on $\mathbb{C}^m \times \mathbb{R}^n$ associated with the elliptic structure is the Dolbeault—De Rham derivative $\bar{\partial}_z + d_t$; Σ inherits a tangential differ— ential operator, made up of the linear combinations of the partial deriva— tives $\partial/\partial t_j$ (j = 1,...,n) and of the Cauchy—Riemann operators $\partial/\partial \bar{z}_i$ (i = 1,...,m) tangent to Σ. Each "slice" Σ_{t_0} of Σ by an affine subspace $t = t_0$ is a maximally real submanifold of $\mathbb{C}^m$ in the usual sense.

As mentioned earlier it is not difficult to define the space $\mathcal{B}(U, C_t^\infty)$

of hyperfunctions in $U \subset \Sigma$ that depend smoothly on t (U is open and suitably small). We define the space of *hyperfunctions solutions in* U *depending smoothly on t*, $\mathfrak{Sol}(U,\mathcal{C}^\infty_t)$, as the subspace of $\mathcal{B}(U,\mathcal{C}^\infty_t)$ consisting of those hyperfunctions u such that $d_t u \equiv 0$. We also define the cohomology spaces $\mathfrak{Sol}^{(q)}(U,\mathcal{C}^\infty_t)$ of the differential complex d_t: $\mathcal{B}(U,\mathcal{C}^\infty_t;\Lambda^{q-1}) \to \mathcal{B}(U,\mathcal{C}^\infty_t;\Lambda^q)$ ($q = 1,2,\ldots,n$). A typical element of $\mathcal{B}(U,\mathcal{C}^\infty_t;\Lambda^q)$ is a "current" $f = \sum_{|I|=q} f_I dt_I$ whose coefficients are hyperfunctions in U depending smoothly on t, ie., elements of $\mathcal{B}(U,\mathcal{C}^\infty_t)$. The Main Theorem of Chapter III has its counterpart here: $\mathfrak{Sol}^{(q)}(U,\mathcal{C}^\infty_t) \cong H^{m,m+q}_U(\mathbb{C}^m\times\mathbb{R}^n\backslash\partial U)$ (section IV.3).

Now, still within the framework of the coarse embedding, every distribution solution in U is a $\mathcal{C}^\infty$ function of t valued in the spaces of distributions on the slices Σ_t ([TREVES, 1992], Proposition I.4.3). The central result of Chapter IV is that the analogue is valid for hyperfunctions. This is established by proving that

$$H^{m,m+n+q}_U(\mathbb{C}^m\times\mathbb{C}^n\backslash\partial U) \cong H^{m,m+q}_U(\mathbb{C}^m\times\mathbb{R}^n\backslash\partial U)$$

Recalling that Σ, and thus its subset U, are contained in the subspace $\mathcal{Im}\, z' = 0$, the preceding isomorphism is constructed by fibre integration in the fibration $\mathbb{C}^m\times\mathbb{C}^n\backslash\partial U \to \mathbb{C}^m\times\mathbb{R}^n\backslash\partial U$ (section IV.4).

The transversal smoothness of all hyperfunction solutions has some important consequences: for instance the fact that the trace of a hyperfunction solution in $\mathcal{M}$, on any maximally real submanifold $\mathcal{X}$ of $\mathcal{M}$ is well–defined (Theorem IV.5.1); the uniqueness in the Cauchy problem with data on $\mathcal{X}$ (Theorem IV.5.2); the fact that if an orbit of the involutive structure of $\mathcal{M}$ ([TREVES, 1992], Definition I.11.1) intersects the support of a hyperfunction solution it is entirely contained in it (Theorem

IV.5.3). Another noteworthy consequence is that, if a hyperfunction $h \in \mathcal{B}(\mathrm{U},\mathcal{C}^{\infty}_{t})$ is the boundary value of a $\mathcal{C}^{\infty}$ function $\tilde{h}$ in a wedge in (z,t)–space $\mathbb{C}^{m}\times\mathbb{R}^{n}$ with edge on U, holomorphic with respect to z, in order for h to be a solution it is necessary and sufficient that $\tilde{h}$ be locally constant with respect to t (Theorem IV.5.5).

Section IV.6 extends the FBI transform to hyperfunctions which depend smoothly on t — simply by treating t as a parameter. The FBI transform commutes with differentiation with respect to t, and therefore the FBI transform of a hyperfunction solution is locally constant with respect to t. Using the FBI transform one can decompose any hyperfunction u depending smoothly on t into a sum of boundary values of $\mathcal{C}^{\infty}$ functions in wedges in (z,t)–space $\mathbb{C}^{m}\times\mathbb{R}^{n}$ that are holomorphic with respect to z (a counter–example of Trépreau shows that it is not always possible to decompose a hyperfunction solution into a sum of boundary values that are also solutions, ie., boundary values of functions holomorphic with respect to z and locally constant with respect to t). There is a version of the classical Sato's theorem on singularities of solutions of linear PDEs that fits our situation: the hypo–analytic wave front set of a hyperfunction solution is contained in the *characteristic set* of the locally integrable structure of $\mathcal{M}$ (the characteristic set is the set of common zeros of the symbols of all the vector fields sections of $\mathcal{V}$; see Theorem IV.6.6 and following remark).

The preceding constructions can be duplicated (see Section IV.7) for any *fine* local embedding. An embedding of this type originates with a choice of local coordinates (x_i,y_j,s_k,t_l) $(1 \leq i,\ j \leq \nu,\ 1 \leq k \leq d = m-\nu,\ 1 \leq l \leq n-\nu)$ such that $z_j = x_j+iy_j$ $(j = 1,\dots,\nu)$ and $w_k = s_k+i\phi_k(z,s,t)$ $(k = 1,\dots,d)$ are hypo–analytic in their domain of definition $\mathcal{M}'$ and thus make up a system of first integrals (since $m = \nu+d$); the functions ϕ_k are real–valued and smooth in $\mathcal{M}'$. The number d is equal to the dimension of the characteristic set at a central point O of $\mathcal{M}'$, which is equivalent to saying that the differentials $d\phi_k$ vanish at O for all k = 1,...,d. The map $(z,s,t) \to (z,w,t)$ realizes an embedding of $\mathcal{M}'$ as a submanifold Σ of $\mathbb{C}^\nu \times \mathbb{C}^d \times \mathbb{R}^{n-\nu}$ and leads to dealing with hyperfunctions that depend smoothly on (z,t). For such a hyperfunction to be a solution it is necessary and sufficient that it be annihilated by $\bar{\partial}_z + d_t$, ie., that it be *holomorphic with respect to z and locally constant with respect to t.* Extending the results of section IV.3 and IV.4 is routine, and one concludes that every hyperfunction solution is smooth with respect to (z,t).

The fine embedding is especially well suited to the study of CR manifolds; in this case n = ν, ie., there are no variables t. A hyperfunction that depends smoothly on z is a CR hyperfunction if and only if it is holomorphic with respect to z. And if a hyperfunction $h \in \mathcal{B}(U, \mathcal{C}^\infty_z)$ is the boundary value of a $\mathcal{C}^\infty$ function $\tilde{h}$ in a wedge in (z,w)—space $\mathbb{C}^m$ with edge on U, holomorphic with respect to w, in order for h to be a CR hyperfunction it is necessary and sufficient that $\tilde{h}$ be holomorphic also with respect to z (Theorem IV.7.3).

The analogue of the FBI transform in a fine local embedding is the FBI *minitransform.* It is particularly well suited to analysis on a hypo—analytic manifold $\mathcal{M}$ of the *hypersurface type,* meaning that the number d of variables s (or w) is equal to one (the characteristic set at O has rank one). Thus $(z,s,t) \to (z,w,t)$ realizes a local embedding of $\mathcal{M}$ as a hypersurface in $\mathbb{C}^\nu \times \mathbb{C} \times \mathbb{R}^{n-\nu}$. Then the inversion formula for the FBI mini—transform enables one to prove that, locally, every hyperfunction solution on $\mathcal{M}$ is equal to the difference between the boundary value, itself a solution, of a holomorphic function of (z,w) constant with respect to t, on one side of $\mathcal{M}$, and the boundary value of a similar function on the opposite side (Theorem IV.8.2).

We close Chapter IV by showing that a number of nonsolvabity results valid for distributions, which include the classical ones for the Lewy and Mizohata vector fields, remain valid for hyperfunctions.

The main definitions and results about sheaves and sheaf cohomology needed in the text are gathered in two short introductory sections, 0.1 and 0.2. We have omitted most of the (mostly semitrivial) proofs.

We have thought it proper to add a postscript, briefly surveying the rather rich history of hyperfunction theory and highlighting the contributions of various authors to the presentation in this monograph.

Much of the research and the writing for the monograph was done while Cordaro was a visiting member at the Institute for Advanced Study in Princeton, N. J. The visit of Cordaro was funded in part by Grant No 92/1402—7 from FAPESP (São Paulo, Brazil) and in part by a grant CNPq/NSF under the US—Brazil Cooperative Research program. The work of F. Treves was supported by NSF Grant DMS—9201980, as well as by NSF Grant INT—9103833 (US—Brazil Cooperative Research).

Paulo D. Cordaro, Instituto de Matemática e Estatística, Universidade de São Paulo, C.P. 20570, 01452—990 São Paulo, S.P. (Brazil).

Francois Treves, Department of Mathematics, Rutgers University, New Brunswick, N. J. 08903 (USA).

Hyperfunctions on Hypo-Analytic Manifolds

0.1 BACKGROUND ON SHEAVES OF VECTOR SPACES OVER A MANIFOLD

Like most presentations of hyperfunction theory, ours makes much use of the language of sheaves and of sheaf cohomology. In these preliminary sections 0.1 and 0.2 we recall some of the basic statements of that language. For the proofs that are missing we refer the reader to the textbooks on sheaf theory (e.g. [CARTAN, 1950/51], [GODEMENT, 1958], [BREDON, 1968]).

We shall be dealing exclusively with sheaves of complex vector spaces over a $\mathcal{C}^\infty$ manifold $\mathcal{M}$, countable at infinity (and therefore para-compact). For us a sheaf $\mathcal{F}$ will be primarily defined through a presheaf: we are given a family $\mathcal{U}$ of open subsets U of $\mathcal{M}$ which is a basis of the topology of $\mathcal{M}$ and: for each $U \in \mathcal{U}$, a vector space F(U); for each pair of elements U, V of $\mathcal{U}$ such that $V \subset U$, a linear map r_V^U: $F(U) \to F(V)$ (thought of as a restriction mapping), with the standard transitivity property: if $W \subset V \subset U$, $r_W^V \circ r_V^U = r_W^U$. We list right away some of the main examples of presheaves $(F(U), r_V^U)$ we shall be dealing with. For the sake of simplicity, in every one of these cases we take the family $\mathcal{U}$ to comprise all open subsets of $\mathcal{M}$ (in the text we shall sometimes be forced to select smaller families). In every one of these examples the maps r_V^U are truly defined by restriction from U to $V \subset U$. We content ourselves with indicating the selection of the linear space F(U).

Basic Examples:

1) $\mathcal{C}^k(U)$, the space of all $\mathcal{C}^k$ functions in U $(0 \leq k \leq +\infty)$;

2) $\mathcal{D}'(U)$, the space of distributions in U;

3) when $\mathcal{M}$ is real analytic, $\mathcal{C}^\omega(U)$, the space of real analytic

functions in U;

4) when $\mathcal{M} = \mathbb{C}^m$, $\mathcal{O}(U)$, the space of holomorphic functions in U. □

Presheaves of spaces of differential forms: 5) Returning to a real manifold $\mathcal{M}$, we can also select $F(U) = \mathcal{C}^\infty(U;\Lambda^q)$, the space of smooth differential q–forms in U. In the domain of local coordinates $x_1,..,x_N$ ($N = \dim \mathcal{M}$) such a form has an expression

$$f = \sum_{|I|=q} f_I \, dx_I. \tag{0.1.1}$$

We use multi–index notation: $I = \{i_1,..,i_q\}$ with integers i_ν such that $1 \leq i_1 < \cdots < i_q \leq N$, $dx_I = dx_{i_1} \wedge \cdots \wedge dx_{i_q}$; and $f_I \in \mathcal{C}^\infty(U) = \mathcal{C}^\infty(U;\Lambda^0)$.

With these spaces one can define the *De Rham differential complex*, ie. the sequence of differential operators

$$d\colon \mathcal{C}^\infty(U;\Lambda^q) \to \mathcal{C}^\infty(U;\Lambda^{q+1}) \quad q = 0,1,..., \tag{0.1.2}$$

where the *exterior derivative* d acts in the usual fashion on a form (0.1.1):

$$df = \sum_{|I|=q} \sum_{j=1}^{N} (\partial f_I/\partial x_j) dx_j \wedge dx_I; \tag{0.1.3}$$

of course, $d^2 = 0$. We shall use the standard terminology: f is said to be *closed* if $df \equiv 0$, *exact* if there is $u \in \mathcal{C}^\infty(U;\Lambda^{q-1})$ such that $du = f$.

6) Now, once again, suppose $\mathcal{M} = \mathbb{C}^m$ (where the coordinates are denoted by $z_1,..,z_m$) and let p be an integer ≥ 1. We can take $F(U) = \mathcal{O}^{(p)}(U)$, the space of *holomorphic differential* p*–forms* in the open

subset U of $\mathbb{C}^m$,

(0.1.4) $$f = \sum_{|I|=p} f_I \, dz_I$$

where $f_I \in \mathcal{O}(U)$ for every multi–index I of length p. This leads to the differential complex

(0.1.5) $$\partial\colon \mathcal{O}^{(p)}(U) \to \mathcal{O}^{(p+1)}(U),\ p = 0,1,\ldots,$$

where ∂ stands for the *holomorphic exterior derivative*:

(0.1.6) $$\partial f = \sum_{|I|=p} \sum_{j=1}^{m} (\partial f_I/\partial z_j) dz_j \wedge dz_I.$$

Of course, $\partial^2 = 0$ and $\mathcal{O}^{(0)}(U) = \mathcal{O}(U)$, $\mathcal{O}^{(p)}(U) = 0$ if $p > m$.

7) Another very important example is that of $F(U) = C^\infty(U;\Lambda^{p,q})$, the space of forms of *bidegree* (or *type*) (p,q) in $U \subset \mathbb{C}^m$,

(0.1.7) $$f = \sum_{|I|=p} \sum_{|J|=q} f_{I,J} \, dz_I \wedge d\bar{z}_J,$$

with coefficients $f_{I,J} \in C^\infty(U)$. The Cauchy–Riemann operator $\bar{\partial}$ acts of forms (0.1.7):

(0.1.8) $$\bar{\partial} f = \sum_{|I|=p} \sum_{|J|=q} \sum_{k=1}^{m} (\partial f_{I,J}/\partial \bar{z}_k) d\bar{z}_k \wedge dz_I \wedge d\bar{z}_J,$$

and defines the *Cauchy–Riemann,* or *Dolbeault* complexes (for each p = 0,1,...)

$$(0.1.9) \qquad \bar{\partial}\colon \mathcal{C}^{\infty}(U;\Lambda^{p,q}) \to \mathcal{C}^{\infty}(U;\Lambda^{p,q+1}),\ q = 0,1,\ldots.$$

Notice that $\mathcal{O}^{(p)}(U) = \{\, f \in \mathcal{C}^{\infty}(U;\Lambda^{p,0});\ \bar{\partial}f \equiv 0 \,\}$.

□

Presheaves of spaces of currents: The examples 5) and 7) of presheaves of spaces of differential forms have their "current" counterparts: in coordinates, we allow the coefficients in the "forms" (0.1.1) and (0.1.7) to be distributions in U. The corresponding spaces will be denoted by $\mathcal{D}'(U;\Lambda^{p})$ and $\mathcal{D}'(U;\Lambda^{p,q})$ respectively. □

We return to the general case. For every $\wp \in \mathcal{M}$ select the subfamily $\mathcal{U}_{\wp}$ of the family $\mathcal{U}$ consisting of the open sets $U \in \mathcal{U}$ containing $\wp$; $\mathcal{U}_{\wp}$ is a basis of the filter of neighborhoods of $\wp$ in $\mathcal{M}$, and we can define the *inductive limit* $F_{\wp}$ of the spaces F(U), $U \in \mathcal{U}_{\wp}$: it is the quotient of the disjoint union of the spaces F(U) for the following equivalence relation: $\sigma_U \approx \sigma_V$, where $\sigma_U \in F(U)$, $\sigma_V \in F(V)$, $U, V \in \mathcal{U}_{\wp}$, if there is $W \in \mathcal{U}_{\wp}$, $W \subset U \cap V$, such that $r^{U}_{W}\sigma_U = r^{V}_{W}\sigma_V$. [This is the abstract version of the standard definition of the germ of a function, a distribution, etc., at a point.] If $U \in \mathcal{U}_{\wp}$ there is a natural linear map $r^{U}_{\wp}\colon F(U) \to F_{\wp}$. It assigns to every element $\sigma_U \in F(U)$ its equivalence class modulo the "germ relation" at $\wp$.

By *the sheaf over* $\mathcal{M}$ *associated to the presheaf* $(F(U), r^{U}_{V})$ we shall mean the disjoint union $\mathcal{F}$ of the vector spaces $F_{\wp}$ (then $F_{\wp}$ is called the *stalk* of $\mathcal{F}$ at $\wp$). Included in the definition of $\mathcal{F}$ is the specification of its topology, or, perhaps more importantly, that of the continuous sections of $\mathcal{F}$: let A be an arbitrary subset of $\mathcal{M}$; a section of $\mathcal{F}$ over A is a map $s\colon A \to$

$\mathcal{F}$ such that, for every $\wp \in \Omega$, $s(\wp) \in F_{\wp}$. It is said to be *continuous at* $\wp_0$ if there are $U \in \mathcal{U}_{\wp_0}$ and an element $\sigma \in F(U)$ such that $r_{\wp}^{U}\sigma = s(\wp)$ for every $\wp \in A \cap U$; it is said to be continuous in (or over) A if it is continuous at every point of A. Two remarks are in order at this juncture:

(0.1.10) *If a continuous section s over* A *vanishes at a point* $\wp_0$ *then it vanishes identically in a full neighborhood of* $\wp_0$ *in* A.

Indeed, in the above notation, $r_{\wp_0}^{U}\sigma = 0 \Leftrightarrow \exists\, V \in \mathcal{U}_{\wp_0}$, $V \subset U$, $r_V^U\sigma = 0$, and the latter entails $r_{\wp}^{U}\sigma = s(\wp) = 0$ for all $\wp \in V \cap A$.

(0.1.11) *Every continuous section s over a closed subset* A *of* $\mathcal{M}$ *extends as a continuous section over a neighborhood of* A *in* $\mathcal{M}$.

Indeed, to every $\wp_0 \in A$ there is an open neighborhood U of $\wp_0$ in $\mathcal{M}$ and an element $\sigma_U \in F(U)$ such that $s(\wp) = r_{\wp}^{U}\sigma_U$ for all $\wp \in A \cap U$. We can therefore form a *locally finite* covering of A by open sets U of this kind. Let $U_1,..,U_\nu$ be ν elements of that covering such that $A \cap U_1 \cap \cdots \cap U_\nu \neq \emptyset$. By (0.1.10) the sections $\wp \to r_{\wp}^{U_i}\sigma_{U_i}$, $i = 1,..,\nu$, are equal in a neighborhood of $A \cap U_1 \cap \cdots \cap U_\nu$, whence easily our claim.

The *support* of a continuous section s over $A \subset \mathcal{M}$ is the relatively closed subset of A, which we denote by supp s, consisting of the points $\wp$ such that $s(\wp) \neq 0$.

This said, the topology of $\mathcal{F}$ is defined as the coarsest topology for which the so–called continuous sections are indeed continuous. In practi–

cal terms, this means that the images of the open subsets of $\mathcal{M}$ under continuous sections form a basis of the topology of $\mathcal{F}$. It is traditional to denote by $\Gamma(A,\mathcal{F})$ the complex vector space of all the continuous sections of $\mathcal{F}$ over A. If $B \subset A$ we have a true restriction mapping ρ_B^A: $\Gamma(A,\mathcal{F}) \to \Gamma(B,\mathcal{F})$, and we see that the collection of spaces and maps $(\Gamma(\Omega,\mathcal{F}),\rho_{\Omega'}^{\Omega})$, where (Ω,Ω') ranges over all possible pairs of open subsets of $\mathcal{M}$ such that $\Omega \supset \Omega'$, defines a presheaf over $\mathcal{M}$, whose associated sheaf is obviously $\mathcal{F}$. If $U \in \mathcal{U}$ there is a natural linear map ι_U: $F(U) \to \Gamma(U,\mathcal{F})$: to $\sigma \in F(U)$ it assigns the section $U \ni \wp \to r_\wp^U \sigma \in \mathcal{F}_\wp$. Sometimes one says that the presheaf $(F(U),r_V^U)$ is a sheaf if, for every $U \in \mathcal{U}$, the map ι_U is a bijection. In general, the map ι_U need not be surjective, nor injective, as we now show.

Let E be a complex vector space. For each open set U in $\mathcal{M}$ set $F(U) = E$. This defines the *constant sheaf* (with stalk E) over $\mathcal{M}$, $\mathcal{F}_E$. The continuous sections of $\mathcal{F}_E$ over an open subset Ω of $\mathcal{M}$ are the locally constant maps $\Omega \to E$. Clearly $F(\Omega)$ can be identified to a subspace of $\Gamma(\Omega,\mathcal{F}_E)$ but, unless Ω is connected or $E = \{0\}$, $F(\Omega) \neq \Gamma(\Omega,\mathcal{F}_E)$. Next take $E(U) = \Gamma(U,\mathcal{F}_E)/F(U)$; E(U) can be identified to the space of locally constant maps $U \to E$ which vanish identically in one and the same connected component of U. The restriction of E–valued functions defines a linear map r'^U_V: $E(U) \to E(V)$ if $V \subset U$. The presheaf $(E(U),r'^U_V)$ does not vanish identically, yet it defines the sheaf identically zero since the family $\mathcal{U}$ of *connected* open subsets of $\mathcal{M}$ is a basis of the topology of $\mathcal{M}$.

Sheaf homomorphisms, isomorphisms, subsheaves, quotient sheaves, products, direct sums, tensor products of sheaves, etc. have

evident definitions. The important point to keep in mind is that the base $\mathcal{M}$ is kept fixed. Thus, if $\mathcal{F}$ and $\mathcal{G}$ are two sheaves of complex vector spaces over $\mathcal{M}$, a *sheaf homomorphism* ϕ: $\mathcal{F} \to \mathcal{G}$ is a *continuous* map $\mathcal{F} \to \mathcal{G}$ such that, for every $\wp \in \mathcal{M}$, the restriction $\phi_\wp$ of ϕ to $\mathcal{F}_\wp$ induces a linear map $\mathcal{F}_\wp \to \mathcal{G}_\wp$. [The sheaf homomorphism ϕ is a sheaf isomorphism if it is bijective and if its inverse is a homomorphism.] If $\mathcal{F}'$ is a subset of $\mathcal{F}$ and A a subset of $\mathcal{M}$, denote by $\Gamma(A,\mathcal{F}')$ the subset of $\Gamma(A,\mathcal{F})$ consisting of those sections s such that $s(\wp) \in \mathcal{F}'$ for every $\wp \in A$. We say that $\mathcal{F}'$ is a *subsheaf* of $\mathcal{F}$ if $\mathcal{F}'$ is an open subset of $\mathcal{F}$ and if $\mathcal{F}'_\wp$ is a linear subspace of $\mathcal{F}_\wp$ for every $\wp \in \mathcal{M}$. Note that, as a consequence, for any pair $\Omega \supset \Omega'$ of open subsets of $\mathcal{M}$, the restriction mapping $r^{\Omega}_{\Omega'}$: $\Gamma(\Omega,\mathcal{F}) \to \Gamma(\Omega',\mathcal{F})$ induces a linear mapping $r'^{\Omega}_{\Omega'}$: $\Gamma(\Omega,\mathcal{F}') \to \Gamma(\Omega',\mathcal{F}')$. We can then form the quotient sheaf $\mathcal{F}/\mathcal{F}'$; it is associated to the presheaf defined by the quotient vector spaces $\Gamma(\Omega,\mathcal{F})/\Gamma(\Omega,\mathcal{F}')$ and by the obvious restriction mappings.

Let f be a continuous map of a manifold $\mathcal{M}'$ into $\mathcal{M}$. By definition, the *pullback* $f^*\mathcal{F}$ of the sheaf $\mathcal{F}$ to $\mathcal{M}'$ is the sheaf over $\mathcal{M}'$ whose stalk at an arbitrary point $\wp' \in \mathcal{M}'$ is equal to $\mathcal{F}_{f(\wp')}$. For a section s' of $f^*\mathcal{F}$ over a set $A' \subset \mathcal{M}'$ to be continuous at a point $\wp'_0 \in A$ it is necessary and sufficient that there be an open neighborhood U of $f(\wp'_0)$ in $\mathcal{M}$, a continuous section s of $\mathcal{F}$ over $U \cap f(A')$ and an open neighborhood $U' \subset \overset{-1}{f}(U)$ of $\wp'_0$ such that $s'(\wp') = s(f(\wp'))$ for all $\wp' \in A' \cap U'$.

Going back to the basic examples we shall use the following notation for the corresponding sheaves:

1) $\mathscr{C}^k_{\mathcal{M}}$: sheaf of germs of C^k functions in $\mathcal{M}$ $(0 \le k \le +\infty)$;

2) $\mathscr{D}'_{\mathcal{M}}$: sheaf of germs of distributions in $\mathcal{M}$;

3) $\mathscr{C}^{\omega}$: sheaf of germs of real analytic functions in the real analytic manifold $\mathcal{M}$;

4) ${}_m\mathcal{O}$: sheaf of germs of holomorphic functions in $\mathbb{C}^m$;

5) $\mathscr{C}_{\mathcal{M}}^{\infty}\Lambda^p$: sheaf of germs of C^{∞} p–forms in $\mathcal{M}$;

6) ${}_m\mathcal{O}^{(p)}$: sheaf of germs of holomorphic p–forms in $\mathbb{C}^m$;

7) $\mathscr{C}^{\infty}\Lambda^{p,q}$: sheaf of germs of C^{∞} (p,q)–forms in $\mathbb{C}^m$;

8) $\mathscr{D}_{\mathcal{M}}'\Lambda^p$: sheaf of germs of p–currents in $\mathcal{M}$;

9) $\mathscr{D}'\Lambda^{p,q}$: sheaf of germs of (p,q)–currents in $\mathbb{C}^m$.

There are three classes of sheaves that are important in the applications: the soft, the flabby and the fine sheaves. A sheaf $\mathcal{F}$ over $\mathcal{M}$ is called *soft* (resp., *flabby*) if every continuous section over any closed (resp., open) subset of $\mathcal{M}$ extends as a continuous section over the whole of $\mathcal{M}$.

The sheaf $\mathcal{F}$ is said to be *fine* if to every pair of closed subsets A and B of $\mathcal{M}$ such that $A \cap B = \emptyset$, there is a sheaf homomorphism $\mathcal{F} \to \mathcal{F}$ which is equal to zero in a neighborhood of A and to the identity map I in a neighborhood of B.

PROPOSITION 0.1.1.— *If the sheaf $\mathcal{F}$ of complex vector spaces over the manifold $\mathcal{M}$ is flabby (resp., fine) then $\mathcal{F}$ is soft.*

Proof: According to (0.1.11) if A is a closed subset of $\mathcal{M}$ and $s \in \Gamma(A,\mathcal{F})$ there are a closed subset B of $\mathcal{M}$ such that $A \cap B = \emptyset$ and a section $\tilde{s} \in \Gamma(\mathcal{M}\setminus B,\mathcal{F})$ such that $\tilde{s}\big|_A = s$. When $\mathcal{F}$ is flabby $\tilde{s}$ extends as a continuous section of $\mathcal{F}$ over $\mathcal{M}$. When $\mathcal{F}$ is fine there is a sheaf homomorphism ϕ: $\mathcal{F} \to \mathcal{F}$, $\phi = 0$ in a neighborhood V of B and $\phi = I$ in a neighborhood U of A. We can extend $\phi \circ \tilde{s}$ as a continuous section $\hat{s}$ over $\mathcal{M}$ by setting it to be zero in B. We have $\hat{s} = \tilde{s}$ in U. □

In the examples 1) to 9) above all the sheaves are fine except $\mathscr{C}^{\omega}$, ${}_m\mathcal{O}$, ${}_m\mathcal{O}^{(p)}$. None of the sheaves is flabby.

0.2 BACKGROUND ON SHEAF COHOMOLOGY

As before we consider a sheaf $\mathcal{F}$ of vector spaces over the C^∞ manifold $\mathcal{M}$. Let $\mathcal{F}^{(0)}$ denote the sheaf of germs of sections, *not necessarily continuous*, of $\mathcal{F}$. If U is an open subset of $\mathcal{M}$ there is a natural isomorphism between $\Gamma(U,\mathcal{F}^{(0)})$ and the space of (not necessarily continuous) sections of $\mathcal{F}$ over U. Since a section of the latter kind can be extended to $\mathcal{M}$ by setting it equal to zero in $\mathcal{M}\backslash U$ we see that the restriction map $\Gamma(\mathcal{M},\mathcal{F}^{(0)}) \to \Gamma(U,\mathcal{F}^{(0)})$ is surjective: $\mathcal{F}^{(0)}$ is flabby. Identifying $\mathcal{F}$ to a subsheaf of $\mathcal{F}^{(0)}$ we form the quotient sheaf $\mathcal{F}^{(0)}/\mathcal{F}$ [defined by the quotient linear spaces $\Gamma(U,\mathcal{F}^{(0)})/\Gamma(U,\mathcal{F})$ and the obvious restriction maps]: it is the sheaf of germs of discontinuous sections of $\mathcal{F}$ congruent modulo germs of continuous sections. Call then $\mathcal{F}^{(1)}$ the sheaf of germs of sections, not necessarily continuous, of $\mathcal{Z}^{(1)} = \mathcal{F}^{(0)}/\mathcal{F}$. Using induction on $\nu \geq 1$ we define $\mathcal{F}^{(\nu+1)}$ as the sheaf of germs of not necessarily continuous sections of $\mathcal{Z}^{(\nu+1)} = \mathcal{F}^{(\nu)}/\mathcal{Z}^{(\nu)}$. We have the exact sequences of sheaf homomorphisms

$$0 \to \mathcal{F} \to \mathcal{F}^{(0)} \to \mathcal{Z}^{(1)} \to 0,$$

$$0 \to \mathcal{Z}^{(\nu)} \to \mathcal{F}^{(\nu)} \to \mathcal{Z}^{(\nu+1)} \to 0, \quad \nu = 1,2,..$$

Out of these short exact sequences we can form the long exact sequence

$$0 \to \mathcal{F} \xrightarrow{i} \mathcal{F}^{(0)} \xrightarrow{d_0} \cdots \xrightarrow{d_{\nu-1}} \mathcal{F}^{(\nu)} \xrightarrow{d_\nu} \mathcal{F}^{(\nu+1)} \to \cdots, \tag{0.2.1}$$

where the map d_ν is the compose $\mathcal{F}^{(\nu)} \to \mathcal{Z}^{(\nu+1)} \to \mathcal{F}^{(\nu+1)}$. We shall refer to (0.2.1) as the *canonical resolution* of $\mathcal{F}$. To it there corresponds the sequence of linear maps

$$0 \to \Gamma(\mathcal{M},\mathcal{F}) \xrightarrow{i} \Gamma(\mathcal{M},\mathcal{F}^{(0)}) \xrightarrow{d_0} \cdots \xrightarrow{d_{\nu-1}} \Gamma(\mathcal{M},\mathcal{F}^{(\nu)}) \xrightarrow{d_\nu} \tag{0.2.2}$$

$$\Gamma(\mathcal{M},\mathcal{F}^{(\nu+1)}) \overset{d_{\nu+1}}{\rightarrow} \cdots,$$

which in general is not exact (it is always exact when $\mathcal{F}$ is soft). Note that the sequence

$$0 \rightarrow \Gamma(\mathcal{M},\mathcal{F}) \overset{i}{\rightarrow} \Gamma(\mathcal{M},\mathcal{F}^{(0)}) \overset{d_0}{\rightarrow} \Gamma(\mathcal{M},\mathcal{F}^{(1)})$$

is exact since, if $s \in \Gamma(\mathcal{M},\mathcal{F}^{(0)})$, $d_0 s = 0$ means that $s(\mathfrak{p}) \in \mathcal{F}_{\mathfrak{p}}$ whatever $\mathfrak{p} \in \mathcal{M}$. At this point we can define the *cohomology spaces with values in* $\mathcal{F}$:

(0.2.3) $\quad H^0(\mathcal{M},\mathcal{F}) = \Gamma(\mathcal{M},\mathcal{F})$; if $q \geq 1$, $H^q(\mathcal{M},\mathcal{F}) = \operatorname{Ker} d_q/\operatorname{Im} d_{q-1}$.

It can be shown that $H^q(\mathcal{M},\mathcal{F}) = 0$ for all $q \geq 1$, whenever $\mathcal{F}$ is soft, therefore whenever $\mathcal{F}$ is flabby or fine (cf. Proposition 0.2.2 below). In particular the cohomology with values in the sheaves 1), 2), 5), 7), 8), 9) in our list of basic examples (see Section 0.1) is always trivial in dimensions ≥ 1.

We shall also make use of the *cohomology with supports*. To define it one introduces *a family of supports* Φ: Φ is a family of *closed* subsets of $\mathcal{M}$ satisfying the following conditions: A, B $\in \Phi \Rightarrow A \cup B \in \Phi$; A $\in \Phi$, B $\subset$ A, B closed, $\Rightarrow$ B $\in \Phi$. Two frequently used special cases are the family of all compact subsets of $\mathcal{M}$ and the family of all closed subsets of a given closed subset of $\mathcal{M}$. Given a family of supports Φ we denote by $\Gamma_\Phi(\mathcal{M},\mathcal{F})$ the space of continuous sections of $\mathcal{F}$ over $\mathcal{M}$ whose supports are elements of Φ. We have the analogues of (0.2.2) and (0.2.3):

(0.2.4) $\quad 0 \rightarrow \Gamma_\Phi(\mathcal{M},\mathcal{F}) \overset{i}{\rightarrow} \Gamma_\Phi(\mathcal{M},\mathcal{F}^{(0)}) \overset{d_0}{\rightarrow} \cdots \overset{d_{\nu-1}}{\rightarrow} \Gamma_\Phi(\mathcal{M},\mathcal{F}^{(\nu)}) \overset{d_\nu}{\rightarrow} \cdots,$

(0.2.5) $\quad H^0_\Phi(\mathcal{M},\mathcal{F}) = \Gamma_\Phi(\mathcal{M},\mathcal{F})$; if $q \geq 1$, $H^q_\Phi(\mathcal{M},\mathcal{F}) = \operatorname{Ker} d_q/\operatorname{Im} d_{q-1}$.

Let us also point out that if $\Phi \subset \Phi'$ are two families of supports there is a natural "inclusion" map

$$i_{\Phi,\Phi'} : H^q_\Phi(\mathcal{M},\mathcal{F}) \to H^q_{\Phi'}(\Omega,\mathcal{F}),\ q = 0,1,\ldots,$$

whose definition follows at once from the fact that supp $d_\nu s \subset$ supp s if $s \in \Gamma(\mathcal{M},\mathcal{F}^{(\nu)})$.

In practice the canonical resolution (0.2.1) is seldom used. Whatever the exact sequence (0.2.1) one uses, provided the sheaves $\mathcal{F}^{(\nu)}$ are flabby the cohomology spaces arrived at will be naturally isomorphic to the spaces $H^*_\Phi(\mathcal{M},\mathcal{F})$ originating with the canonical resolution. Greater care, however, must be taken when the sheaves $\mathcal{F}^{(\nu)}$ in the resolution are not flabby but are fine. The conditions on the family of supports Φ must be strengthened; Φ must also satisfy

(0.2.6) $$\forall\, A \in \Phi,\ \exists\, B \in \Phi,\ A \subset \textit{Interior of}\ B.$$

This is expressed by saying that the family Φ is *paracompactifying*. Keep in mind that here the base $\mathcal{M}$ is assumed to be a C^∞ manifold countable at infinity. When the family Φ is paracompactifying, the qth cohomology space defined by means of a fine resolution of $\mathcal{F}$ is naturally isomorphic to $H^q_\Phi(\mathcal{M},\mathcal{F})$ (with the latter defined by means of the canonical resolution). All the natural isomorphisms alluded to follow easily from the axiomatic characterization of the cohomology spaces $H^*_\Phi(\mathcal{M},\mathcal{F})$.

Here are a few examples of often used *fine* resolutions. First of all, the *De Rham resolution(s)*:

(0.2.7) $$0 \to \mathcal{F}_{\mathbb{C}} \xrightarrow{i} \mathcal{C}^\infty_{\mathcal{M}} \xrightarrow{d} \mathcal{C}^\infty_{\mathcal{M}}\Lambda^1 \xrightarrow{d} \cdots \xrightarrow{d} \mathcal{C}^\infty_{\mathcal{M}}\Lambda^q \xrightarrow{d} \cdots,$$

$$(0.2.8)\qquad 0 \to \mathcal{F}_{\mathbb{C}} \xrightarrow{i} \mathcal{D}_{\mathcal{M}}' \xrightarrow{d} \mathcal{D}_{\mathcal{M}}'\Lambda^1 \xrightarrow{d} \cdots \xrightarrow{d} \mathcal{D}_{\mathcal{M}}'\Lambda^q \xrightarrow{d} \cdots,$$

where $\mathcal{F}_{\mathbb{C}}$ is the constant sheaf (with stalk $\mathbb{C}$) over $\mathcal{M}$ and d is the exterior derivative. The fact that the sequences of sheaf homomorphisms (0.2.7) and (0.2.8) are exact follows from the Poincaré lemma (locally, every closed form is exact). A continuous section s of $\mathcal{C}_{\mathcal{M}}^{\infty}\Lambda^q$ can be identified to a C^{∞} differential q–form f; and supp s = supp f. Thus $\Gamma(\mathcal{M}, \mathcal{C}_{\mathcal{M}}^{\infty}\Lambda^q) \cong C^{\infty}(\mathcal{M};\Lambda^q)$; likewise $\Gamma(\mathcal{M}, \mathcal{D}_{\mathcal{M}}'\Lambda^q) \cong \mathcal{D}'(\mathcal{M};\Lambda^q)$. If Φ is taken to be the family of all closed subsets of $\mathcal{M}$, the resulting cohomology spaces are the *De Rham cohomology spaces* $H^q(\mathcal{M})$, q = 0,1,.... If Φ is the family of all com– pact subsets of $\mathcal{M}$, $\Gamma_{\Phi}(\mathcal{M}, \mathcal{C}_{\mathcal{M}}^{\infty}\Lambda^q) \cong C_c^{\infty}(\mathcal{M};\Lambda^q)$, the space of compactly supported smooth q–forms, and $\Gamma_{\Phi}(\mathcal{M}, \mathcal{D}_{\mathcal{M}}'\Lambda^q) \cong \mathcal{E}'(\mathcal{M};\Lambda^q)$, the space of compactly supported q–currents in $\mathcal{M}$. [Of course, in this case $\Gamma_{\Phi}(\mathcal{M},\mathcal{F}_{\mathbb{C}}) = \{0\}$ unless $\mathcal{M}$ has compact connected components.] The resulting space $H_c^q(\mathcal{M})$ is naturally isomorphic, and therefore identified, to the *singular homology space* in the complementary dimension, N–q, $H_{N-q}(\mathcal{M})$.

In complex variable theory (when $\mathcal{M}$ is a domain in $\mathbb{C}^m$ or more generally any complex analytic manifold countable at infinity) the De Rham resolution is replaced by the *Dolbeault resolution(s)*:

$$(0.2.9)\qquad 0 \to \mathcal{O}^{(p)} \xrightarrow{i} \mathcal{C}^{\infty}\Lambda^{p,0} \xrightarrow{\bar\partial} \mathcal{C}^{\infty}\Lambda^{p,1} \xrightarrow{\bar\partial} \cdots \xrightarrow{\bar\partial} \mathcal{C}^{\infty}\Lambda^{p,q} \xrightarrow{\bar\partial} \cdots,$$

$$(0.2.10)\qquad 0 \to \mathcal{O}^{(p)} \xrightarrow{i} \mathcal{D}'\Lambda^{p,0} \xrightarrow{\bar\partial} \mathcal{D}'\Lambda^{p,1} \xrightarrow{\bar\partial} \cdots \xrightarrow{\bar\partial} \mathcal{D}'\Lambda^{p,q} \xrightarrow{\bar\partial} \cdots,$$

where $\bar\partial$ is the Cauchy–Riemann operator. When p = 0 these are the

Dolbeault resolution(s) of the sheaf ${}_m\mathcal{O}$ of germs of holomorphic functions of m complex variables. That these sequences of sheaf homomorphisms are exact is a consequence of the analogue of the Poincaré lemma for $\bar{\partial}$, ie., the Dolbeault (or Dolbeault—Grothendieck) lemma.

Taking Φ to be the family of all closed subsets of $\mathcal{M}$ the sequences (0.2.9) and (0.2.10) establishes the isomorphism between the cohomology space $H^q(\mathcal{M},\mathcal{O}^{(p)})$ with values in the sheaf $\mathcal{O}^{(p)}$ and the *Dolbeault co—homology space* $H^{p,q}(\mathcal{M})$, ie. the qth cohomology space of the differential complex (0.1.9). We have $\Gamma(\mathcal{M}, \mathscr{C}_{\mathcal{M}}^{\infty}\Lambda^{p,q}) \cong \mathcal{C}^{\infty}(\mathcal{M};\Lambda^{p,q})$, the space of $\mathcal{C}^{\infty}$ (p,q)—forms, and $\Gamma(\mathcal{M},\mathscr{D}'\Lambda^{p,q}) \cong \mathcal{D}'(\mathcal{M};\Lambda^{p,q})$, the space of (p,q)—currents in $\mathcal{M}$. If Φ is the family of all compact subsets of $\mathcal{M}$, $\Gamma_{\Phi}(\mathcal{M}, \mathscr{C}_{\mathcal{M}}^{\infty}\Lambda^{p,q}) \cong \mathcal{C}_c^{\infty}(\mathcal{M};\Lambda^{p,q})$, the space of compactly supported smooth (p,q)—forms, and $\Gamma_{\Phi}(\mathcal{M},\mathscr{D}'\Lambda^{p,q}) \cong \mathcal{E}'(\mathcal{M};\Lambda^{p,q})$, the space of compactly supported (p,q)—currents in $\mathcal{M}$. In this case we obtain the cohomology spaces $H_c^*(\mathcal{M},\mathcal{O}^{(p)})$ with compact support; they are isomorphic to the cohomology spaces of the differential complex

(0.2.11) $$\bar{\partial}\colon \mathcal{C}_c^{\infty}(\mathcal{M},\Lambda^{p,q}) \to \mathcal{C}_c^{\infty}(\mathcal{M},\Lambda^{p,q+1}),\ q = 0,1,\ldots,$$

as well as to those of the complex

(0.2.12) $$\bar{\partial}\colon \mathcal{E}'(\mathcal{M},\Lambda^{p,q}) \to \mathcal{E}'(\mathcal{M},\Lambda^{p,q+1}),\ q = 0,1,\ldots.$$

In Chapters III and IV of this book we shall make use of a com—bination of the De Rham and the Dolbeault resolutions: the base manifold will be an open subset Ω of $\mathbb{C}^m\times\mathbb{R}^n$ (where the variables are denoted by z and t respectively), the differential operator will be $\bar{\partial}_z + d_t$, and the basic sheaf will be the sheaf of germs of functions that are holomorphic with respect to z and locally constant with respect to t. Details will be provided at the appropriate time, but the scheme followed is the same as the one just described.

Of course, what counts are the functorial properties of the cohomology. We recall a few of them here, without their (easy) proofs.

First note that, if $\mathcal{F}$ and $\mathcal{G}$ are sheaves of vector spaces over the manifold $\mathcal{M}$, to each sheaf homomorphism $\phi\colon \mathcal{F} \to \mathcal{G}$ there correspond homomorphisms $\phi^{(\nu)}\colon \mathcal{F}^{(\nu)} \to \mathcal{G}^{(\nu)}$ between the sheaves in the canonical resolutions of $\mathcal{F}$ and $\mathcal{G}$ [if s is a section of $\mathcal{F}$, not necessarily continuous, $\phi \circ s$ is one of $\mathcal{G}$]. This leads to a homomorphism of the canonical resolutions, ie. to the commutative diagram

$$\begin{array}{ccccccccccccc} 0 & \to & \mathcal{F} & \overset{i}{\to} & \mathcal{F}^{(0)} & \overset{d_0}{\to} & \cdots & \overset{d_{\nu-1}}{\to} & \mathcal{F}^{(\nu)} & \overset{d_\nu}{\to} & \mathcal{F}^{(\nu+1)} & \to \cdots, \\ & & \phi\downarrow & i' & \downarrow\phi^{(0)} & d_0' & & d_{\nu-1}' & \downarrow\phi^{(\nu)} & d_\nu' & \downarrow\phi^{(\nu+1)} & \\ 0 & \to & \mathcal{G} & \to & \mathcal{G}^{(0)} & \to & \cdots & \to & \mathcal{G}^{(\nu)} & \to & \mathcal{G}^{(\nu+1)} & \to \cdots. \end{array} \tag{0.2.13}$$

In turn this leads to the corresponding commutative diagram for the spaces of continuous sections $\Gamma(\mathcal{M},\cdot)$. Finally this defines a linear map

$$\phi^{*(q)}\colon \mathrm{H}^q(\mathcal{M},\mathcal{F}) \to \mathrm{H}^q(\mathcal{M},\mathcal{G}),\ q = 0,1,.... \tag{0.2.14}$$

For $q = 0$, it is the obvious map $\Gamma(\mathcal{M},\mathcal{F}) \ni s \to \phi \circ s \in \Gamma(\mathcal{M},\mathcal{G})$. When there is no risk of confusion we shall often write ϕ^* rather than $\phi^{*(q)}$.

PROPOSITION 0.2.1.— *Let Φ be a family of supports. To each short exact sequence of sheaf homomorphisms* (over the manifold $\mathcal{M}$)

(0.2.15)
$$0 \to \mathcal{F}' \overset{\phi}{\to} \mathcal{F} \overset{\psi}{\to} \mathcal{F}'' \to 0$$

there correspond the long exact sequence of cohomology

(0.2.16)
$$0 \to \Gamma_\Phi(\mathcal{M},\mathcal{F}') \overset{\phi^*}{\to} \Gamma_\Phi(\mathcal{M},\mathcal{F}) \overset{\psi^*}{\to} \Gamma_\Phi(\mathcal{M},\mathcal{F}'') \overset{\delta}{\to} H^1_\Phi(\mathcal{M},\mathcal{F}') \overset{\phi^*}{\to} \cdots$$
$$\to H^q_\Phi(\mathcal{M},\mathcal{F}) \overset{\psi^*}{\to} H^q_\Phi(\mathcal{M},\mathcal{F}'') \overset{\delta}{\to} H^{q+1}_\Phi(\mathcal{M},\mathcal{F}') \overset{\phi^*}{\to} \cdots.$$

Moreover, if we have a commutative diagram of exact sequences

(0.2.17)
$$\begin{array}{ccccccccc} 0 & \to & \mathcal{F}' & \to & \mathcal{F} & \to & \mathcal{F}'' & \to & 0 \\ & & \downarrow & & \downarrow & & \downarrow & & \\ 0 & \to & \mathcal{G}' & \to & \mathcal{G} & \to & \mathcal{G}'' & \to & 0, \end{array}$$

the corresponding diagrams

(0.2.18)
$$\begin{array}{ccc} H^q_\Phi(\mathcal{M},\mathcal{F}'') & \overset{\delta}{\to} & H^{q+1}_\Phi(\mathcal{M},\mathcal{F}') \\ \downarrow & & \downarrow \\ H^q_\Phi(\mathcal{M},\mathcal{G}'') & \overset{\delta}{\to} & H^{q+1}_\Phi(\mathcal{M},\mathcal{G}') \end{array}$$

are commutative.

The "connecting" homomorphism δ is defined as follows. It is not difficult to see that (0.2.15) implies, for each $q \geq 0$, the exactness of corresponding sequences
$$0 \to \mathcal{F}'^{(q)} \overset{\phi^{(q)}}{\to} \mathcal{F}^{(q)} \overset{\psi^{(q)}}{\to} \mathcal{F}''^{(q)} \to 0,$$
and therefore, given any family of supports Φ, also the exactness of the sequence
$$0 \to \Gamma_\Phi(\mathcal{M},\mathcal{F}'^{(q)}) \overset{\phi^{(q)*}}{\to} \Gamma_\Phi(\mathcal{M},\mathcal{F}^{(q)}) \overset{\psi^{(q)*}}{\to} \Gamma_\Phi(\mathcal{M},\mathcal{F}''^{(q)}) \to 0.$$

[The only aspect of this assertion to be checked is the surjectivity of the map $\psi^{(q)*}$: it is proved by identifying a continuous section of $\mathcal{F}''^{(q)}$ to a not necessarily continuous section of $\mathcal{Z}''^{(q)}$.] If $s'' \in \Gamma_\Phi(\mathcal{M},\mathcal{F}''^{(q)})$, $d''_q s'' = 0$, there is $s \in \Gamma_\Phi(\mathcal{M},\mathcal{F}^{(q)})$ such that $\psi^{(q)*}s = s''$. The latter implies $\psi^{(q+1)*}(d_q s) = d''_q(\psi^{(q)*}s) = 0$, hence there exists $s' \in \Gamma_\Phi(\mathcal{M},\mathcal{F}^{(q+1)})$ such that $d_q s = \phi^{(q+1)*}(s')$, and

$$d_{q+1}\phi^{(q+1)*}(s') = \phi^{(q+2)*}(d'_{q+1}s') = 0 \Rightarrow d'_{q+1}s' = 0.$$

It is readily checked that the class $[s'] \in H^{q+1}_\Phi(\mathcal{M},\mathcal{F}')$ only depends on the class $[s''] \in H^q_\Phi(\mathcal{M},\mathcal{F}'')$. We set $\delta[s''] = [s']$.

COROLLARY 0.2.1.— *Given the exact sequence* (0.2.15) *the sequence*

$$0 \to \Gamma_\Phi(\mathcal{M},\mathcal{F}') \xrightarrow{\phi^*} \Gamma_\Phi(\mathcal{M},\mathcal{F}) \xrightarrow{\psi^*} \Gamma_\Phi(\mathcal{M},\mathcal{F}'') \to 0 \tag{0.2.19}$$

is exact if $H^1_\Phi(\mathcal{M},\mathcal{F}') = 0$.

PROPOSITION 0.2.2.— *Let $\mathcal{F}$ be a sheaf of vector spaces over the manifold $\mathcal{M}$, Φ a family of supports. Suppose either that $\mathcal{F}$ is flabby or that the family Φ is paracompactifying and $\mathcal{F}$ is soft. Then* $H^q_\Phi(\mathcal{M},\mathcal{F}) = 0$ *for all* $q \geq 1$.

Other properties concern changes in the base manifold. We shall content ourselves with a few words about the restriction r_Ω from $\mathcal{M}$ to an open subset Ω of $\mathcal{M}$. If the vertical arrows stand for r_Ω we get the commutative diagram

$$\begin{array}{ccccccccccc} 0 \to \Gamma(\mathcal{M},\mathcal{F}) & \xrightarrow{i} & \Gamma(\mathcal{M},\mathcal{F}^{(0)}) & \xrightarrow{d_0} & \cdots & \xrightarrow{d_{\nu-1}} & \Gamma(\mathcal{M},\mathcal{F}^{(\nu)}) & \xrightarrow{d_\nu} & \Gamma(\mathcal{M},\mathcal{F}^{(\nu+1)}) & \to \cdots, \\ \downarrow & & \downarrow & & & & \downarrow & & \downarrow & \\ 0 \to \Gamma(\Omega,\mathcal{F}) & \xrightarrow{i} & \Gamma(\Omega,\mathcal{F}^{(0)}) & \xrightarrow{d_0} & \cdots & \xrightarrow{d_{\nu-1}} & \Gamma(\Omega,\mathcal{F}^{(\nu)}) & \xrightarrow{d_\nu} & \Gamma(\Omega,\mathcal{F}^{(\nu+1)}) & \to \cdots. \end{array}$$

This leads straight away to linear maps

(0.2.20) $$r_\Omega^{*(q)} : H^q(\mathcal{M},\mathcal{F}) \to H^q(\Omega,\mathcal{F}),\ q = 0,1,...,$$

to which we shall often refer as the restriction to Ω. Greater care must be exercised when dealing with a family of supports, as is clear when dealing with the family of all compact subsets of $\mathcal{M}$.

We close this section with some remarks about *relative cohomology*. In what follows A will be a *closed* subset of $\mathcal{M}$. We shall denote by $H_A^*(\mathcal{M},\mathcal{F})$ the spaces $H_\Phi^*(\mathcal{M},\mathcal{F})$ where Φ is the family of all closed subsets of A. Except when A is equal to the union of connected components of $\mathcal{M}$ the family Φ is *not* paracompactifying. In this connection we have no choice but to use flabby resolutions of the sheaf $\mathcal{F}$.

We are going to avail ourselves repeatedly of the long exact sequence associated with the cohomology relative to A, which we now recall. It is the exact sequence

(0.2.21) $$0 \to \Gamma_A(\mathcal{M},\mathcal{F}) \xrightarrow{\iota_0} \Gamma(\mathcal{M},\mathcal{F}) \xrightarrow{r_0} \Gamma(\mathcal{M}\backslash A,\mathcal{F}) \xrightarrow{\delta_0} H_A^1(\mathcal{M},\mathcal{F}) \xrightarrow{\iota_1} \cdots$$
$$\to H^q(\mathcal{M},\mathcal{F}) \xrightarrow{r_q} H^q(\mathcal{M}\backslash A,\mathcal{F}) \xrightarrow{\delta_q} H_A^{q+1}(\mathcal{M},\mathcal{F}) \xrightarrow{\iota_{q+1}} \cdots.$$

The maps ι_q stand for the natural "inclusion maps" and the maps r_q for the natural "restriction maps" from $\mathcal{M}$ to $\mathcal{M}\backslash A$ [see (0.2.20)]. It remains to explain the connecting maps δ_q in (0.2.21). First consider the case $q = 0$

and $s \in \Gamma(\mathcal{M}\backslash A,\mathcal{F})$; let $\tilde{s}$ be an arbitrary extension of s to $\mathcal{M}$. Since $\mathcal{F}$ is not assumed to be flabby, in general $\tilde{s}$ will not be continuous. However, it can be identified to a continuous section of $\mathcal{F}^{(0)}$; its image in $\mathcal{Z}^{(1)} = \mathcal{F}^{(0)}/\mathcal{F}$ under the quotient map obviously vanishes identically in $\mathcal{M}\backslash A$; in other words, supp $d_0\tilde{s} \subset A$, and therefore $d_0\tilde{s} \in \Gamma_A(\mathcal{M},\mathcal{F}^{(1)})$. But clearly $d_1(d_0\tilde{s}) \equiv 0$, hence $d_0\tilde{s}$ defines a class in $H^1_A(\mathcal{M},\mathcal{F})$ which we set to be $\delta_0[s]$. When $q \geq 1$ we consider a section $s \in \Gamma(\mathcal{M}\backslash A,\mathcal{F}^{(q)})$ such that $d_q s \equiv 0$. ie., $s(\wp) \in \mathcal{Z}^{(q)}_{\wp}$ for all $\mathfrak{p} \in \mathcal{M}\backslash A$. Let $\tilde{s}$ be a continuous section of $\mathcal{F}^{(q)}$ over the whole of $\mathcal{M}$ extending s. Then $d_q\tilde{s}$ is a continuous section of $\mathcal{F}^{(q+1)}$ such that supp $d_q\tilde{s} \subset A$ and $d_{q+1}(d_q\tilde{s}) \equiv 0$. The cohomology class of $d_q\tilde{s}$ is set to be $\delta_q[s]$. This construction shows the advantage of using flabby sheaves [here the sheaves $\mathcal{F}^{(\nu)}$].

CHAPTER I

HYPERFUNCTIONS IN A MAXIMAL HYPO–ANALYTIC STRUCTURE

INTRODUCTION

The first two sections of this chapter provide an introduction to the theory of analytic functionals in $\mathbb{C}^m$ with special emphasis on the concept of *carrier*. Much of the effort in the subsequent section will consist in extending to analytic functionals carried by a sufficiently small open subset Ω of a maximally real submanifold $\mathcal{X}$ of $\mathbb{C}^m$ the results, principally Theorem I.2.3, valid for a Runge compact subset of the plane, as well as the method of the Cauchy transform, as described in the last part of section I.2. The complex coordinates $z_1,\ldots,z_m$ can be chosen in such a way that near one of its points (taken to be the origin of $\mathbb{C}^m$) the submanifold $\mathcal{X}$ is tangent to real space $\mathbb{R}^m$, and quite close to it. Then if the open neighborhood Ω of 0 in $\mathcal{X}$ is sufficiently small, every compact subset K of Ω will have the Runge property (and in fact be polynomially convex, cf. Lemma I.6.1). We introduce, in (I.3.27), the Cauchy–Fantappiè transform $\Gamma_{\mathcal{X}}\mu$ of an arbitrary analytic functional μ carried by K. We select a geometrically simple open subset of $\mathbb{C}^m$ containing K; in this text we take a "biball" $B_{01} = B_0+iB_1$. And we define $\Gamma_{\mathcal{X}}\mu$ to be a smooth, *closed* differential form in $B_{01}\backslash K$, of type (m,m–1), endowed with the following property: if χ is any compactly supported $\mathcal{C}^\infty$ function in B_{01} such that $\chi \equiv 1$ in a neighborhood of K, then (Theorem I.3.1), given any entire holomorphic function h in $\mathbb{C}^m$,

$$\langle\mu,h\rangle = -\int h\bar{\partial}\chi\wedge\Gamma_{\mathcal{X}}\mu.$$

Being $\bar{\partial}$–closed $\Gamma_{\mathcal{X}}\mu$ defines a class $[\Gamma_{\mathcal{X}}\mu] \in H^{m,m-1}(B_{01}\backslash K)$ in the cohomology space of the Dolbeault complex, and the map $\mu \to [\Gamma_{\mathcal{X}}\mu]$ is an isomorphism of the space $\mathcal{O}'(K)$ of analytic functionals carried by K onto $H^{m,m-1}(B_{01}\backslash K)$ (Theorem I.4.2). Actually this remains valid however large is the biball B_{01} and leads to a natural isomorphism

$$\mathcal{O}'(K) \cong H^{m,m-1}(\mathbb{C}^m \backslash K).$$

This provides us with all we need, namely the following: Let K_1 and K_2 be two compact subsets of Ω ($0 \in \Omega \subset \mathcal{X}$, Ω open in $\mathcal{X}$ and diam Ω suitably small). *If an analytic functional* μ *is carried by* K_1 *and by* K_2 μ *is carried by* $K_1 \cap K_2$ (Proposition I.4.1). *If* μ *is carried by* $K_1 \cup K_2$ *there are analytic functionals* μ_1 *and* μ_2, *respectively carried by* K_1 *and* K_2, *such that* $\mu = \mu_1 + \mu_2$ (Proposition I.4.2).

Let now U be a relatively compact open subset of Ω. The space of hyperfunctions in U, $\mathcal{B}(U)$, is the quotient linear space $\mathcal{O}'(\mathscr{C}\ell U)/\mathcal{O}'(\partial U)$ (Definition I.5.1). If $V \subset U$ is also open in $\mathcal{X}$ there is a natural restriction map ρ_V^U: $\mathcal{B}(U) \to \mathcal{B}(V)$. This defines a presheaf on $\mathcal{X}$; the associated sheaf is the sheaf of (germs of) hyperfunctions in $\mathcal{X}$, $\mathscr{B}$. It is a flabby sheaf: every continuous section of $\mathscr{B}$ over any open subset of $\mathcal{X}$ extends as a continuous section to the whole of $\mathcal{X}$. The sheaf of germs of distributions in $\mathcal{X}$, $\mathscr{D}'$, is a subsheaf of $\mathscr{B}$

It is pure routine to go from the definition of the sheaf of hyperfunctions on a maximally real submanifold of $\mathbb{C}^m$ to that of the analogous sheaf on a manifold equipped with a maximal hypo-analytic structure. It suffices to define local sheaves by pullback from $\mathbb{C}^m$ via the local hypo-analytic embeddings (cf. Preface).

Section I.6 is entirely devoted to a repetition of the construction, but without recourse to integral kernels (such as the one that defines the Cauchy-Fantappiè transform). The argument exploits fully the fact that every compact subset of Ω is polynomially convex and uses sheaf cohomology and Functional Analysis, mainly Serre's duality, to establish the isomorphism $\mathcal{O}'(K) \cong H^{m,m-1}(\mathbb{C}^m \backslash K)$. It is hoped that this change of viewpoint will give the reader a better understanding of the underpinnings of the theory. Furthermore, parts of the argument are needed in section IV.5, in the proof of the uniqueness in the Cauchy problem for hyperfunction solutions on a general hypo—analytic manifold.

I.1 ANALYTIC FUNCTIONALS IN OPEN SUBSETS OF $\mathbb{C}^m$

In this section U will denote an open subset of $\mathbb{C}^m$ and $\mathcal{O}(U)$ the space of holomorphic functions in U, equipped with the topology of uniform convergence on the compact subsets of U. The topology of $\mathcal{O}(U)$ is defined by the seminorms

$$h \to \operatorname*{Max}_K |h|,$$

where K is any compact subset of U. [Note that in general the above seminorm is not a norm; it is a norm when the interior of K, $\mathcal{Int}\,K$, intersects every connected component of U.] To define the topology of $\mathcal{O}(U)$ it suffices to let K range over an exhausting sequence $\{K_\nu\}_{\nu=1,2,\dots}$ of compact subsets of U: U is equal to the union of the sets K_ν; and $K_\nu \subset \mathcal{Int}\,K_{\nu+1}$. This shows that the space $\mathcal{O}(U)$ is locally convex and metrizable (ie., its topology can be defined by means of a metric). It is an elementary fact that it is also complete: it is a Fréchet space. Moreover, Montel's theorem states that every closed and bounded subset of $\mathcal{O}(U)$ is compact or, which amounts to the same, every bounded sequence of elements of $\mathcal{O}(U)$ contains a convergent subsequence: $\mathcal{O}(U)$ is a Montel space.

DEFINITION I.1.1.— *Any element of the dual* $\mathcal{O}'(U)$ *is called an* ***analytic functional*** *in* U.

Let V denote an open subset of U, and $\rho_V^U\colon \mathcal{O}(U) \to \mathcal{O}(V)$ the restriction map. In general the transpose of ρ_V^U, ${}^t\rho_V^U\colon \mathcal{O}'(V) \to \mathcal{O}'(U)$, need not be injective:

EXAMPLE I.1.1.— Let $\mathbb{C}\backslash\{0\}$ denote the complement of the origin in the complex plane $\mathbb{C}$. Define $\mu \in \mathcal{O}'(\mathbb{C}\backslash\{0\})$ by the formula

$$(I.1.1) \qquad <\mu,h> = (2\imath\pi)^{-1} \oint_{|z|=1} h(z)dz.$$

We have $<\mu,h> = 0$ whenever $h \in \mathcal{O}(\mathbb{C})$ (ie., h is an entire function). On the other hand, $1/z \in \mathcal{O}(\mathbb{C}\backslash\{0\})$ and $<\mu,z^{-1}> = 1$. □

It follows from the Hahn—Banach theorem that the injectivity of ${}^t\rho_V^U$: $\mathcal{O}'(V) \to \mathcal{O}'(U)$ is equivalent to the density of $\rho_V^U(\mathcal{O}(U))$ in $\mathcal{O}(V)$. We recall that V is said to have the *Runge property* with respect to U, or to be a *Runge subset* of U, if $\rho_V^U(\mathcal{O}(U))$ is dense in $\mathcal{O}(V)$. Barring exceptional circumstances, when $U = \mathbb{C}^m$ and the open subset V of $\mathbb{C}^m$ has the Runge property we shall omit mentioning U and simply say that V has the Runge property, or that V is a *Runge open set.*

DEFINITION I.1.2.— *An analytic functional* μ *in* U *is said to be* ***carried by*** V *if it belongs to the range of* ${}^t\rho_V^U$. *The analytic functional* μ *is said to be carried by a compact set* $K \subset U$ *if it is carried by every open set* $V \subset U$ *such that* $K \subset V$.

We shall denote by $\mathcal{O}'(U,K)$ the space of analytic functionals in U carried by the compact subset K of U. We shall simply write $\mathcal{O}'(K)$ when $U = \mathbb{C}^m$.

In any open set U there exist analytic functionals that are carried by *disjoint* compact subsets. We content ourselves with showing this when m = 1 and leave the higher dimensional case as an exercise to the reader.

EXAMPLE I.1.2.— The analytic functional defined by (I.1.1) is obviously carried by every circle $|z| = r > 0$, since $\oint_{|z|=1} h(z)dz = \oint_{|z|=r} h(z)dz$, for all $h \in \mathcal{O}(\mathbb{C}\backslash\{0\})$. By moving the origin to an arbitrary point of U and taking r sufficiently small, the circles $|z| = r$ can be brought inside U. □

Proposition I.1.1 below shows that an analytic functional in U is carried by an open subset V of U if, and only if, it is carried by some compact subset of V. This allows us to extend the terminology of Definition I.1.2 and to say that an analytic functional in U is carried by an arbitrary subset S of U if it is carried by some compact subset of U contained in S.

PROPOSITION I.1.1.— *For* $\mu \in \mathcal{O}'(\mathrm{U})$ *to be carried by the open subset* V *of* U *it is necessary and sufficient that there be a compact subset* K *of* V *and a constant* $C > 0$ *such that, for all* $h \in \mathcal{O}(\mathrm{U})$:

$$|\langle\mu,h\rangle| \leq C \operatorname{Max}_{\mathrm{K}} |h|. \tag{I.1.2}$$

Proof: Suppose there is $\tilde{\mu} \in \mathcal{O}'(\mathrm{V})$ such that $\mu = {}^t\rho_{\mathrm{V}}^{\mathrm{U}}\tilde{\mu}$. Then, given any $h \in \mathcal{O}(\mathrm{U})$, $\langle\mu,h\rangle = \langle\tilde{\mu},\rho_{\mathrm{V}}^{\mathrm{U}}h\rangle$. There is a compact set $\mathrm{K} \subset \mathrm{V}$ and a constant $C > 0$ such that $|\langle\tilde{\mu},g\rangle| \leq C \operatorname{Max}_{\mathrm{K}} |g|$, $\forall\, g \in \mathcal{O}(\mathrm{V})$, whence (I.1.2). Conversely suppose (I.1.2) holds, with $\mathrm{K} \subset\subset \mathrm{V}$. First we note that $\rho_{\mathrm{V}}^{\mathrm{U}}h \to \langle\mu,h\rangle$ is a well defined linear functional on the subspace $\rho_{\mathrm{V}}^{\mathrm{U}}\mathcal{O}(\mathrm{U})$ of $\mathcal{O}(\mathrm{V})$: by (I.1.2), $\rho_{\mathrm{V}}^{\mathrm{U}}h = 0 \Rightarrow \langle\mu,h\rangle = 0$. The right—hand side of (I.1.2) defines a continuous seminorm not only on $\rho_{\mathrm{V}}^{\mathrm{U}}\mathcal{O}(\mathrm{U})$ but on $\mathcal{O}(\mathrm{V})$. By the Hahn—Banach theorem, the linear functional $\rho_{\mathrm{V}}^{\mathrm{U}}h \to \langle\mu,h\rangle$ extends from

$\rho_V^U \mathcal{O}(U)$ to $\mathcal{O}(V)$ as a linear functional $\tilde{\mu}$ that satisfies (I.1.2). Clearly $\mu = {}^t\rho_V^U \tilde{\mu}$. □

COROLLARY I.1.1.— *For* $\mu \in \mathcal{O}'(U)$ *to be carried by the compact subset* K *of* U *it is necessary and sufficient that to each compact subset* K′ *of* U *whose interior contains* K *there be a constant* $C > 0$ *such that, for all* $h \in \mathcal{O}(U)$:

$$|\langle\mu,h\rangle| \leq C \operatorname*{Max}_{K'} |h|. \tag{I.1.3}$$

If $K \subset \mathcal{I}nt K' \subset K' \subset U$ the infimum of the constants C that can be used in (I.1.3) defines a norm $\|\mu\|_{K'}$. The space $\mathcal{O}'(U,K)$ of the analytic functionals in U carried by K can be equipped with the locally convex topology defined by all these norms $\mu \to \|\mu\|_{K'}$. Since it suffices to let K′ range over a countable basis of compact neighborhoods of K this topology is metrizable. The strong dual $\mathcal{O}'(U)$ of $\mathcal{O}(U)$ is complete. Since the subset of $\mathcal{O}'(U)$ consisting of the analytic functionals μ that verify (I.1.3) for some $C > 0$ (depending on μ) is a closed linear subspace of $\mathcal{O}'(U)$ we conclude that $\mathcal{O}'(U,K)$ is complete: it is a Fréchet space.

Let U be an open subset of $\mathbb{C}^m$ and let $\mathcal{C}(U)$ denote the space of continuous functions in U equipped with the topology of uniform convergence on the compact subsets of U. By the definition of the topology of $\mathcal{O}(U)$ the latter can be regarded as a linear subspace of $\mathcal{C}(U)$, also in the topological sense; $\mathcal{O}(U)$ is a *closed* subspace of $\mathcal{C}(U)$, since it is complete.

We recall that the dual of $\mathcal{C}(U)$ is the space of the compactly supported *Radon measures* on U, $\mathcal{M}_c(U)$. The transpose of the natural injection $\mathcal{O}(U) \to \mathcal{C}(U)$ (which is a homeomorphism onto its image) yields a *surjection* $\mathcal{M}_c(U) \to \mathcal{O}'(U)$. As a matter of fact, $\mathcal{O}'(U) \cong \mathcal{M}_c(U)/\mathcal{O}(U)^\perp$. The orthogonal of $\mathcal{O}(U)$ in $\mathcal{M}_c(U)$, $\mathcal{O}(U)^\perp$, is clearly not empty: it contains any Radon measure of the form $(\sum_{j=1}^{m} \partial u_j/\partial \bar{z}_j)dxdy$ with $u_j \in \mathcal{C}_c^1(U)$. Thus we may represent an arbitrary analytic functional μ in U by some compactly supported Radon measure μ^* in U:

$$\text{(I.1.4)} \qquad <\mu,h> = \int_U h d\mu^*.$$

It is obvious that if K is the support of the Radon measure μ^* then μ is carried by the compact set K.

REMARK I.1.1.— Actually the Radon measure μ^* in (I.1.4) can be taken to be a density $\phi(z)dxdy$ with $\phi \in \mathcal{C}_c^\infty(U)$. Indeed, $\mathcal{O}(U)$ can be regarded as a closed linear subspace of $\mathcal{D}'(U)$: closed since it is defined by the Cauchy–Riemann equations $\bar{\partial}h = 0$. The topology induced by $\mathcal{D}'(U)$ on $\mathcal{O}(U)$ is of course coarser than the one induced by $\mathcal{C}(U)$ (ie., the original topology of $\mathcal{O}(U)$). But actually these two topologies are the same because the closed graph theorem applies (see [SCHWARTZ, 1966], [DE WILDE, 1978]). It follows that the natural injection $\mathcal{O}(U) \to \mathcal{D}'(U)$ is also an isomorphism onto its image, and therefore its transpose is a surjection of the dual of $\mathcal{D}'(U)$, $\mathcal{C}_c^\infty(U)$, onto $\mathcal{O}'(U)$ (we identify $\phi \in \mathcal{C}_c^\infty(U)$ to the density $\phi(z)dxdy$).

For the readers uncomfortable with Functional Analysis let us recall a simple manner of associating a function $\phi \in \mathcal{C}_c^\infty(U)$ to the measure

μ^*. If K $\subset\subset$ U and $0 < r_j < \varepsilon$ ($j = 1,\dots,m$) for a suitably small $\varepsilon > 0$, then $z \in K \Rightarrow (z_1+r_1e^{i\theta_1},\dots,z_m+r_me^{i\theta_m}) \in U$ and as a consequence, we have, for all $h \in \mathcal{O}(U)$,

$$h(z) = (2\pi)^{-m}\int_0^{2\pi}\cdots\int_0^{2\pi} h(z_1+r_1e^{i\theta_1},\dots,z_m+r_me^{i\theta_m})d\theta_1\cdots d\theta_m.$$

Select $\psi_0 \in C^\infty(\mathbb{R})$ such that $\psi_0(r) = 0$ if $r < \frac{1}{4}\varepsilon$ or $r > \frac{1}{2}\varepsilon$ and $\int_0^{+\infty}\psi_0(r)r dr = 1$; and set $\psi(z) = \psi_0(|z_1|)\cdots\psi_0(|z_m|)$. Multiplying the preceding identity by $r_1\psi_0(r_1)\cdots r_m\psi_0(r_m)$ and integrating with respect to r_j from 0 to $+\infty$ for every $j = 1,\dots,n$, yields

$$h(z) = (2\pi)^{-m}\int_{\mathbb{R}^{2m}} h(z+z')\psi(z')dx'dy',$$

hence, if supp $\mu^* = K$,

$$\int_{\mathbb{R}^{2m}} hd\mu^* = (2\pi)^{-m}\int_{\mathbb{R}^{2m}}\left[\int_{\mathbb{R}^{2m}} h(z+z')\psi(z')dx'dy'\right]d\mu^*_z = \int_{\mathbb{R}^{2m}} h(z)\phi(z)dxdy,$$

where $\phi(z) = (2\pi)^{-m}\int_{\mathbb{R}^{2m}}\psi(z-w)d\mu^*_w \in C^\infty_c(U)$. □

Let $P(z,\zeta)$ be a polynomial with respect to $\zeta \in \mathbb{C}^m$ whose coefficients are holomorphic functions in U. We can view $P(z,\zeta)$ as the *symbol* of the *holomorphic differential operator* $P(z,\partial/\partial z)$ — by the rule that, if $f \in \mathcal{O}(U)$ then $f(z)\zeta_j$ is the symbol of the operator $f(z)\partial/\partial z_j$. Denote by ${}^tP(z,\partial/\partial z)$ its formal transpose [the formal transpose of the operator $u \to f(z)\partial u/\partial z_j$ is the operator $u \to -\partial/\partial z_j(f\,u)$]; ${}^tP(z,\partial/\partial z)$ acts continuously from $\mathcal{O}(U)$ to itself and defines, by transposition, the continuous linear operator $P(z,\partial/\partial z)$: $\mathcal{O}'(U) \to \mathcal{O}'(U)$. Thus

$$<P(z,\partial/\partial z)\mu,h> = <\mu,{}^tP(z,\partial/\partial z)h>,$$

The spaces $\mathcal{O}$(K) *and* $\mathcal{O}'$(K)

Let K be a compact subset of $\mathbb{C}^m$. We recall the definition of the space $\mathcal{O}$(K) of *germs at* K of holomorphic functions. It is the inductive limit of the spaces $\mathcal{O}$(U), with U open, U ⊃ K, ie., the quotient of the disjoint union of the spaces $\mathcal{O}$(U) modulo the following equivalence relation: if $h_j \in \mathcal{O}(U_j)$ (j = 1,2), $h_1 \approx h_2 \Leftrightarrow \exists$ V open, K ⊂ V ⊂ $U_1 \cap U_2$, such that $h_1 \equiv h_2$ in V. This defines a restriction map ρ_K^U: $\mathcal{O}(U) \to \mathcal{O}(K)$ for any open set U ⊃ K; to $h \in \mathcal{O}(U)$ it assigns the equivalence class $\rho_K^U h$ of h modulo the relation ≈. The kernel of the map ρ_K^U is a *closed* subspace of $\mathcal{O}$(U): it consists exactly of those holomorphic functions in U that vanish in some neighborhood of K in U, hence in every connected component of U which intersects K. Thus $\rho_K^U\mathcal{O}(U) \cong \mathcal{O}(U)/\text{Ker}\ \rho_K^U$ carries a natural Fréchet space structure; $\mathcal{O}$(K) is the union of all these Fréchet spaces, as U varies. We equip $\mathcal{O}$(K) with the locally convex inductive limit topology of the Fréchet spaces $\rho_K^U\mathcal{O}(U)$: a *convex* subset of $\mathcal{O}$(K) is open if its intersection with $\rho_K^U\mathcal{O}(U)$ is open [An arbitrary open set in $\mathcal{O}$(K) is the union of its convex open subsets. It is not true, in general, that a nonconvex subset of $\mathcal{O}$(K) will be open if its intersection with each subspace $\rho_K^U\mathcal{O}(U)$ is open.]

In practice, it suffices to use a countable basis of open neigh–borhoods U_j of K such that $\mathcal{C}\ell U_{j+1} \subset\subset U_j$ and such that every connected component of U_j intersects K (j = 1,2,...). The latter ensures that the maps $\rho_K^{U_j}$ are injective and therefore allows us to identify $\mathcal{O}(U_j)$ to $\rho_K^{U_j}\mathcal{O}(U_j) \subset \mathcal{O}(K)$ for every j. Furthermore, instead of the Fréchet spaces $\mathcal{O}(U_j)$ we can use the Banach spaces $\mathcal{O}_b(U_j)$ consisting of the bounded holomorphic functions in U_j (equipped with the sup norm). In this manner we see that $\mathcal{O}(K)$ is a countable inductive limit of *Banach spaces*, the spaces $\mathcal{O}_b(U_j)$ [identified to $\rho_K^{U_j}\mathcal{O}_b(U_j)$]. A convex subset of $\mathcal{O}(K)$ is a neighborhood of 0 if and only if, for each $j = 1,2,\ldots$, its intersection with $\mathcal{O}_b(U_j)$ contains an open ball centered at 0 in $\mathcal{O}_b(U_j)$. A linear map of $\mathcal{O}(K)$ into any locally convex topological vector space, in particular any linear functional on $\mathcal{O}(K)$, is continuous if and only if, whatever $j = 1,2,\ldots$, its restriction to $\mathcal{O}_b(U_j)$ is continuous.

Note, however, that the topology induced on $\mathcal{O}_b(U_j)$ by $\mathcal{O}_b(U_{j+1})$ and therefore the one induced on $\mathcal{O}_b(U_j)$ by $\mathcal{O}(K)$, is strictly coarser than the original Banach space topology. Indeed, the image of the closed unit ball in $\mathcal{O}_b(U_j)$ has a compact closure in $\mathcal{O}_b(U_{j+1})$ — by virtue of the Cauchy inequalities, and by the same argument that proves the Montel theorem. One says that $\mathcal{O}(K)$ is a *nonstrict* inductive limit of Banach (or of Fréchet) spaces. Nevertheless, a subset $\mathcal{A}$ of $\mathcal{O}(K)$ is bounded if and only if it is contained and bounded in $\mathcal{O}_b(U_j)$ for some j. The bounded set $\mathcal{A}$ will be closed, and therefore compact, in $\mathcal{O}(K)$ if and only if it is a compact subset of $\mathcal{O}_b(U_j)$ for some j. We conclude that every bounded and closed subset of $\mathcal{O}(K)$ is compact: $\mathcal{O}(K)$ is a Montel space (in particular, it is *reflexive*).

One says that K is a *Runge compact set* (or that K has the Runge property — in $\mathbb{C}^m$) if every holomorphic function h in an open set $U \supset K$ is the uniform limit of a sequence of entire functions, on a set $K_\epsilon = \{ z \in$

$\mathbb{C}^m$; dist$(z,K) \leq \epsilon \} \subset U$ (for a suitably small $\epsilon > 0$). This means precisely that the restriction to K maps $\mathcal{O}(\mathbb{C}^m)$ onto a dense subspace of $\mathcal{O}(K)$. The transpose of the restriction map is an isomorphism of the strong dual of $\mathcal{O}(K)$ onto the space of analytic functionals in $\mathbb{C}^m$ carried by K — what we have been calling $\mathcal{O}'(K)$. The Fréchet space structure of $\mathcal{O}'(K)$ is defined by the norms (cf. remark following Corollary I.1.1)

$$\mu \to \sup \left\{ |\langle\mu,h\rangle| \; ; h \in \mathcal{O}(\mathbb{C}^m), \; \underset{K_\epsilon}{\mathrm{Max}} \, |h| \leq 1 \right\} (\epsilon > 0).$$

I.2 ANALYTIC FUNCTIONALS IN $\mathbb{C}^m$

Analytic functionals in the whole space $\mathbb{C}^m$ (or, for that matter, in any open polydisk) have convenient representations. As the Taylor expansion of an arbitrary entire function h converges to h in $\mathcal{O}(\mathbb{C}^m)$ we get, for any analytic functional μ in $\mathbb{C}^m$,

$$<\mu,h> = \sum_{\alpha\in\mathbb{Z}_+^m} <\mu,z^\alpha> h^{(\alpha)}(0)/\alpha!. \tag{I.2.1}$$

[We make systematic use of the multi–index notation.] Using the derivatives of the Dirac distribution Formula (I.2.1) can be rewritten as

$$\mu = \sum_{\alpha\in\mathbb{Z}_+^m} (-1)^{|\alpha|} <\mu,z^\alpha> \delta^{(\alpha)}(0)/\alpha!. \tag{I.2.2}$$

This shows, among other things, that the finite linear combinations of the derivatives $\delta^{(\alpha)}(0)$ are dense in $\mathcal{O}'(\mathbb{C}^m)$.

It is traditional to introduce the Laplace–Borel transform

$$\hat{\mu}(\zeta) = <\mu_z, e^{z\cdot\zeta}>, \tag{I.2.3}$$

where $z\cdot\zeta = z_1\zeta_1+\cdots+z_m\zeta_m$. Formula (I.2.1) can then also be restated as

$$<\mu,h> = \sum_{\alpha\in\mathbb{Z}_+^m} \hat{\mu}^{(\alpha)}(0) h^{(\alpha)}(0)/\alpha!. \tag{I.2.4}$$

THEOREM I.2.1.— *The Laplace–Borel transform* $\mu \to \hat{\mu}$ *defines an isomorphism of* $\mathcal{O}'(\mathbb{C}^m)$ *onto the space* Exp$(\mathbb{C}^m)$ *of entire functions of exponential type in* $\mathbb{C}^m$.

Proof: Let $\mu \in \mathcal{O}'(\mathbb{C}^m)$. Then $\hat{\mu}(\zeta)$ is an entire function of ζ in $\mathbb{C}^m$ as one sees by letting the Cauchy–Riemann operator act "under the duality bracket" at the right in (I.2.2). Suppose the open polydisk $\Delta_r = \{ z \in \mathbb{C}^m; |z_j| < r_j \}$ with "multiradius" $r = (r_1,...,r_m)$, $r_j > 0$, carries $\mu \in \mathcal{O}'(\mathbb{C}^m)$. Then, for some constant $C > 0$,

$$|\hat{\mu}(\zeta)| \leq C \sup_{z \in \Delta_r} |e^{z \cdot \zeta}| = C \exp(\sum_{j=1}^{m} r_j |\zeta_j|).$$

This shows that $\hat{\mu} \in \mathrm{Exp}(\mathbb{C}^m)$.

Suppose $g \in \mathcal{O}(\mathbb{C}^m)$ satisfies $|g(\zeta)| \leq C \exp(\sum_{j=1}^{m} r_j |\zeta_j|)$ for some C, $r_j > 0$ $(1 \leq j \leq m)$ and all $\zeta \in \mathbb{C}^m$. It follows at once from the Cauchy inequalities in the polydisk Δ_ϱ that $|g^{(\alpha)}(0)| \leq C\left[\prod_{j=1}^{m} \alpha_j! \rho_j^{-\alpha_j}\right] e^{\rho \cdot r}$.

Selecting $\rho_j = \alpha_j / r_j$ yields

$$|g^{(\alpha)}(0)| \leq C \prod_{j=1}^{m} [\alpha_j!(r_j e/\alpha_j)^{\alpha_j}] \leq C \prod_{j=1}^{m} [(2\pi\alpha_j)^{\frac{1}{2}} r_j^{\alpha_j}]$$

by Stirling's formula. It follows that, whatever $h \in \mathcal{O}(\mathbb{C}^m)$ and $r'_j > r_j$ $(j = 1,...,m)$,

$$\left| \sum_{\alpha \in \mathbb{Z}_+^m} g^{(\alpha)}(0) h^{(\alpha)}(0)/\alpha! \right| \leq C' \sum_{\alpha \in \mathbb{Z}_+^m} r_1'^{\alpha_1} \cdots r_m'^{\alpha_m} |h^{(\alpha)}(0)|/\alpha!.$$

Applying once again Cauchy's inequalities, now in the polydisk $\Delta_{r''}$ with $r''_j > r'_j$ for each $j = 1,...,m$, gets us

$$r_1''^{\alpha_1} \cdots r_m''^{\alpha_m} |h^{(\alpha)}(0)|/\alpha! \leq \operatorname{Max}_{\Delta_{r''}} |h|,$$

whence

$$\left| \sum_{\alpha \in \mathbb{Z}_+^m} g^{(\alpha)}(0) h^{(\alpha)}(0)/\alpha! \right| \leq C \prod_{j=1}^{m} (1 - r_j'/r_j'')^{-1} \operatorname*{Max}_{\Delta_{r''}} |h|.$$

We conclude that $h \to \sum_{\alpha \in \mathbb{Z}_+^m} g^{(\alpha)}(0) h^{(\alpha)}(0)/\alpha!$ is an analytic functional carried by the closure of the polydisk $\Delta_{r'}$. Its Laplace—Borel transform is obviously equal to $g(\zeta)$. □

REMARK I.2.1.— Given any bounded subset S of $\mathbb{C}^m$ define

$$H_S(\zeta) = \sup_{z \in S} \mathcal{Re}(z \cdot \zeta).$$

It is obvious that if $\mu \in \mathcal{O}'(\mathbb{C}^m)$ is carried by a compact subset K of $\mathbb{C}^m$ then

(I.2.5) *to each* $\epsilon > 0$, *there is* $C_\epsilon > 0$ *such that*

$$|\hat{\mu}(\zeta)| \leq C_\epsilon \exp(H_K(\zeta) + \epsilon|\zeta|), \quad \forall\, \zeta \in \mathbb{C}^m.$$

Note that the function H_K does not distinguish between the set K and its convex hull. When K is a *convex* compact subset of $\mathbb{C}^m$ it is natural to ask whether (I.2.5) implies that μ is carried by K. Inspection of the proof of Theorem I.2.1 shows that this has been proved for any compact *polydisk*. Actually the result is general; but for an arbitrary convex compact set K its proof is more difficult; proofs can be found in [MARTINEAU, 1964, 1967], and in pp. 97—101 of [HÖRMANDER, 1966]. Granting this result one may state

THEOREM I.2.2.— *For* $\mu \in \mathcal{O}'(\mathbb{C}^m)$ *to be carried by a convex compact subset* K *of* $\mathbb{C}^m$ *it is necessary and sufficient that Property* (I.2.5) *hold.* □

We know, by the general considerations in section I.1, that any holomorphic differential operator $P(z,\partial/\partial z)$ whose coefficients are entire functions in $\mathbb{C}^m$ defines a bounded linear operator $\mathcal{O}'(\mathbb{C}^m) \to \mathcal{O}'(\mathbb{C}^m)$. Another operation that makes sense for any pair of analytic functionals in $\mathbb{C}^m$ is their *convolution*. If we regard provisionally w as the variable and z as an arbitrary point in $\mathbb{C}^m$, we see that the map $h(w) \to h(z-w)$ defines an automorphism of $\mathcal{O}(\mathbb{C}^m)$. We may therefore define the convolution

$$(\text{I.2.6}) \qquad (\mu \star h)(z) = <\mu_w, h(z-w)>,$$

where the subscript w indicates that the duality bracket is computed in w–space $\mathbb{C}^m$. As one sees by letting the operator $\bar{\partial}_z$ act inside the duality bracket, $(h,\mu) \to \mu\star h$ is a bilinear map $\mathcal{O}(\mathbb{C}^m)\times\mathcal{O}'(\mathbb{C}^m) \to \mathcal{O}(\mathbb{C}^m)$, easily seen to be separately continuous. This convolution is, in a sense, a *smoothing* operation.

We may also define the symmetry with respect to the origin, first of an entire function: $\check{h}(z) = h(-z)$, then of an analytic functional: $<\check{\mu},h> = <\mu,\check{h}>$. Finally, we may define the convolution of two analytic functionals in $\mathbb{C}^m$:

$$(\text{I.2.7}) \qquad <\mu\star\nu,h> = <\mu,\check{\nu}\star h>, \quad h \in \mathcal{O}(\mathbb{C}^m).$$

This convolution is a bilinear map $\mathcal{O}'(\mathbb{C}^m)\times\mathcal{O}'(\mathbb{C}^m) \to \mathcal{O}'(\mathbb{C}^m)$. This map is easily seen to be separately continuous.

REMARK I.2.2.— Actually the convolution of analytic functionals in $\mathbb{C}^m$ is a continuous bilinear map $\mathcal{O}'(\mathbb{C}^m)\times\mathcal{O}'(\mathbb{C}^m) \to \mathcal{O}'(\mathbb{C}^m)$. This is due to the fact that $\mathcal{O}'(\mathbb{C}^m)$ is the strong dual of a reflexive Fréchet space, $\mathcal{O}(\mathbb{C}^m)$ (reflexive because Montel); the latter property ensures that separate continuity of a bilinear map is equivalent to its full continuity (see [TREVES, 1967], Theorem 41.1). □

The space $\mathcal{O}'(\mathbb{C}^m)$ is a commutative (and associative!) algebra for convolution; it has a unit element, the Dirac distribution δ. In passing note that, if $\mu \in \mathcal{O}'(\mathbb{C}^m)$, $\alpha \in \mathbb{Z}_+^m$ and $z_0 \in \mathbb{C}^m$, and if we denote by τ_{z_0} the translation by z_0, then

$$\partial^\alpha \mu / \partial z^\alpha = \mu \star \delta^{(\alpha)}, \quad \tau_{z_0} \mu = \mu \star \tau_{z_0} \delta. \tag{I.2.8}$$

Also, it is easy to check that if a compact set K_j carries μ_j ($j = 1,2$) then the vector sum $K_1 + K_2$ carries $\mu_1 \star \mu_2$; and that the Laplace—Borel transform of the convolution $\mu_1 \star \mu_2$ is the product $\hat{\mu}_1 \hat{\mu}_2$ of the Laplace—Borel transforms.

Next let us look at $\mathbb{R}^m$ as a subset of $\mathbb{C}^m$; and let $\mathcal{O}'(\mathbb{R}^m)$ be the space of analytic functionals in $\mathbb{C}^m$ carried by $\mathbb{R}^m$. By the Weierstrass approximation theorem the restriction map from $\mathbb{C}^m$ to $\mathbb{R}^m$ maps the space of holomorphic polynomials, hence the space $\mathcal{O}(\mathbb{C}^m)$ of entire functions, in $\mathbb{C}^m$, onto a dense subspace of $C^\infty(\mathbb{R}^m)$. Its transpose defines an *injection* $\mathcal{E}'(\mathbb{R}^m) \to \mathcal{O}'(\mathbb{R}^m)$. This is not a surjection; for instance, there exist analytic functionals carried by $\{0\}$ that are not distributions:

EXAMPLE I.2.1.— Consider the analytic functional in $\mathbb{C}^m$,

$$h \to \sum_{\alpha \in \mathbb{Z}_+^m} c_\alpha h^{(\alpha)}(0)/\alpha! \tag{I.2.9}$$

where the sequence $\{c_\alpha\}_{\alpha\in\mathbb{Z}^m_+}$ has the following property: to each $\epsilon > 0$ there is $C_\epsilon > 0$ such that

(I.2.10) $$|c_\alpha| \le C_\epsilon \epsilon^{|\alpha|}, \forall \alpha \in \mathbb{Z}^m_+.$$

Let $r > 0$ be arbitrary and select $\epsilon < r$; the Cauchy inequalities entail

(I.2.11) $$\left| \sum_{\alpha\in\mathbb{Z}^m_+} c_\alpha h^{(\alpha)}(0)/\alpha! \right| \le C_\epsilon (1-\epsilon/r)^{-m} \operatorname*{Max}_{|z_j|=r,\ j=1,\dots,m} |h|.$$

This shows that the analytic functional (I.2.9) is carried by $\{0\}$. On the other hand, it defines a distribution if and only if $c_\alpha \ne 0 \Rightarrow |\alpha| \le k_0 < +\infty$. □

If a distribution $u \in \mathcal{E}'(\mathbb{R}^m)$ is identified to an analytic functional μ_u in $\mathbb{C}^m$ then it is easy to see, by the Cauchy inequalities, that the compact set supp u carries μ_u.

Analytic functionals carried by a Runge compact subset of the complex plane. The Cauchy transform

In the remainder of this section we assume m = 1. Let K be a compact subset of the complex plane and let $\mathcal{O}'(K)$ denote the space of analytic functionals in $\mathbb{C}$ carried by K, equipped with the Fréchet space structure defined after the statement of Corollary I.1.1. Then, given any w

$\in \mathbb{C}\backslash K$, the function $z \to (z-w)^{-1}$ is holomorphic in an open neighborhood of K.

We shall hypothesize that $\mathbb{C}\backslash K$ is connected. This is equivalent to the Runge property understood as follows: *every holomorphic function in an open neighborhood* U *of* K *is the uniform limit on* K *of entire functions* (see e. g., Theorem 1.3.1, [HÖRMANDER, 1967]). If K is a *Runge compact subset* of the plane, as thus defined, and if the analytic functional μ in $\mathbb{C}$ is carried by K, ie., $\mu \in \mathcal{O}'(K)$, and if $w \in \mathbb{C}\backslash K$, we have the right to consider

$$\Gamma\mu(w) = (2\imath\pi)^{-1} \langle \mu_z, (w-z)^{-1} \rangle. \tag{I.2.12}$$

The right hand side is the limit of $\langle \mu_z, P_\nu(w;z) \rangle$ for any sequence of polynomials with respect to z (depending on w) $\{P_\nu(w;z)\}_{\nu=1,2...}$ that converge uniformly to $(2\imath\pi)^{-1}(w-z)^{-1}$ in some compact neighborhood of K on $\mathbb{C}\backslash\{w\}$.

The function $\Gamma\mu$ (defined in $\mathbb{C}\backslash K$) is called the *Cauchy transform* of μ.

Let $\chi \in C_c^\infty(\mathbb{R}^2)$, $\chi \equiv 1$ in some neighborhood of K. Then

$$\langle \mu, h \rangle = \int h(w)\bar{\partial}\chi(w)\Gamma\mu(w)\mathrm{d}w\wedge\mathrm{d}\bar{w}, \quad \forall\, h \in \mathcal{O}(\mathbb{C}). \tag{I.2.13}$$

Proof of (I.2.13): Since supp $\bar{\partial}\chi$ stays away from K there is an open neighborhood Ω of K in which we have, thanks to the inhomogeneous Cauchy formula,

$$h(z) = (2\imath\pi)^{-1} \int h(w)\bar{\partial}\chi(w)(w-z)^{-1}\mathrm{d}w\wedge\mathrm{d}\bar{w}.$$

This entails

$$<\mu,h> = (2i\pi)^{-1}<\mu_z, \int h(w)\bar{\partial}\chi(w)(w-z)^{-1}dw\wedge d\bar{w}>,$$

where the duality bracket is the one between $\mathcal{O}(\Omega)$ and $\mathcal{O}'(\Omega)$ (thus z stands for the variable point in Ω). But we can interchange the integral and the duality bracket, whence (I.2.13). □

We shall denote by $\mathcal{O}_0(\mathbb{C}\backslash K)$ the space of holomorphic functions in $\mathbb{C}\backslash K$ that tend to zero as $|z| \to +\infty$. The space $\mathcal{O}_0(\mathbb{C}\backslash K)$ will be equipped with the topology of uniform convergence on the closed subsets whose complements are bounded and contain K. The space $\mathcal{O}_0(\mathbb{C}\backslash K)$ is a Fréchet space; its topology is strictly finer than the one inherited from $\mathcal{O}(\mathbb{C}\backslash K)$. It is convenient to identify $\mathcal{O}_0(\mathbb{C}\backslash K)$ to the space of holomorphic functions in the complement of K in the Riemann sphere **S** that vanish at infinity. Then the topology of $\mathcal{O}_0(\mathbb{C}\backslash K)$ is simply the one inherited from the topology of uniform convergence on compact subsets on $\mathcal{O}(S\backslash K)$; evidently $\mathcal{O}_0(\mathbb{C}\backslash K)$ is a closed vector subspace of $\mathcal{O}(S\backslash K)$.

THEOREM I.2.3.— *Let* K *be a Runge compact subset of* $\mathbb{C}$. *The Cauchy transform defines an isomorphism of* $\mathcal{O}'(K)$ *onto* $\mathcal{O}_0(\mathbb{C}\backslash K)$.

Proof: By letting $\bar{\partial}_w$ act under the duality bracket at the right in (I.2.12) we see right—away that $\Gamma\mu \in \mathcal{O}(\mathbb{C}\backslash K)$. Set $K_\epsilon = \{ z \in \mathbb{C};\ \mathrm{dist}(z,K) \leq \epsilon \}$ ($\epsilon > 0$). According to (I.2.12) there is a constant $C_\epsilon > 0$ such that

$$\text{(I.2.14)} \qquad |\Gamma\mu(w)| \leq C_\epsilon \operatorname*{Max}_{z\in K_\epsilon} |w-z|^{-1},\ \forall\, w \in \mathbb{C}\backslash K_\epsilon.$$

It follows at once from this that $|\Gamma\mu(w)| \to 0$ as $|w| \to +\infty$; and that the map $\mathcal{O}'(K) \ni \mu \to \Gamma\mu \in \mathcal{O}_0(\mathbb{C}\backslash K)$ is continuous. That this map is injective follows at once from (I.2.13).

Now consider $g \in \mathcal{O}_0(\mathbb{C}\backslash K)$ and let χ be as in (I.2.13). For any $h \in \mathcal{O}(\mathbb{C})$ define

$$(I.2.15) \qquad <\mu_g, h> = -\int g(z)\bar{\partial}\chi(z)h(z)dz\wedge d\bar{z}.$$

Suppose we replace χ by a different function $\chi_1 \in C_c^\infty(\mathbb{R}^2)$ with $\chi_1 \equiv 1$ in a neighborhood of K. Integration by parts shows that

$$\int g(z)\bar{\partial}(\chi-\chi_1)(z)h(z)dz\wedge d\bar{z} = \int (\bar{\partial}g)(z)(\chi-\chi_1)(z)h(z)dz\wedge d\bar{z} = 0$$

and thus that the definition of μ_g is independent of the choice of the cutoff χ.

Now select $h(z) = (2i\pi)^{-1}(w-z)^{-1}$ for a fixed $w \in \mathbb{C}\backslash K$. Then select χ so that $\chi(w) = 0$. We have, by (I.2.15),

$$<\mu_g, (2i\pi)^{-1}(w-z)^{-1}> = (2i\pi)^{-1}\int g(z)\bar{\partial}(1-\chi)(z)(w-z)^{-1}dz\wedge d\bar{z} = g(w),$$

again thanks to the inhomogeneous Cauchy formula and to the fact that g vanishes at infinity. We have thus proved that $\mu \to \Gamma\mu$ is a continuous linear bijection of $\mathcal{O}'(K)$ onto $\mathcal{O}_0(\mathbb{C}\backslash K)$. Since both these spaces are Fréchet spaces, the Cauchy transform defines a homeomorphism. □

COROLLARY I.2.1.— *Let* K_1 *and* K_2 *be two Runge compact subsets of* $\mathbb{C}$ *such that* $K_1 \cup K_2$ *is Runge, and let* μ *be an analytic functional in* $\mathbb{C}$. *The following is valid:*

If μ *is carried by both* K_1 *and* K_2 *then* μ *is carried by the intersection* $K_1 \cap K_2$.

If μ *is carried by* $K_1 \cup K_2$ *then there are analytic functionals* μ_1 *and* μ_2 *in* $\mathbb{C}$, *carried by* K_1 *and* K_2 *respectively, such that* $\mu = \mu_1 + \mu_2$.

The intersection of two Runge compact subsets K_1 and K_2 of $\mathbb{C}$ is Runge. Indeed, the union $(\mathbb{C}\backslash K_1)\cup(\mathbb{C}\backslash K_2) = \mathbb{C}\backslash(K_1\cap K_2)$ is a connected neighborhood of ∞ since each set $\mathbb{C}\backslash K_i$ $(i = 1,2)$ is one.

Proof: First suppose μ is carried by K_1 and by K_2. The Cauchy transform $\Gamma\mu$ defines a holomorphic function g_j in $\mathcal{O}_0(\mathbb{C}\backslash K_j)$ for each $j = 1, 2$. We have $g_1 = g_2$ in $(\mathbb{C}\backslash K_1)\cap(\mathbb{C}\backslash K_2) = \mathbb{C}\backslash(K_1\cup K_2)$, since $\mathbb{C}\backslash(K_1\cup K_2)$ is a connected neighborhood of ∞, according to the hypothesis. Therefore there exists a holomorphic function $g \in \mathcal{O}_0(\mathbb{C}\backslash(K_1\cap K_2))$ that extends both g_1 and g_2. Theorem I.2.3 implies that it is the Cauchy transform of an analytic functional carried by $K_1\cap K_2$, necessarily equal to μ.

Now suppose μ is carried by $K_1\cup K_2$. Let Ω_1 and Ω_2 be two arbitrary open subsets of $\mathbb{C}$ and h a holomorphic function in $\Omega_1\cap\Omega_2$. Let $\chi \in C^\infty(\Omega_1\cup\Omega_2)$ be such that supp $\chi \subset \Omega_1$ and supp$(1-\chi) \subset \Omega_2$; $(1-\chi)h$ can be extended as a C^∞ function in Ω_1, χh as a C^∞ function in Ω_2 and $h(\partial\chi/\partial\bar{z})$ as a C^∞ function in $\Omega_1\cup\Omega_2$, by setting each one of these functions to be zero off $\Omega_1\cap\Omega_2$. Select $u \in C^\infty(\Omega_1\cup\Omega_2)$ such that $\partial u/\partial\bar{z} = h(\partial\chi/\partial\bar{z})$ and define $h_1 = (1-\chi)h + u$ in Ω_1, $h_2 = \chi h - u$ in Ω_2. We have $h = h_1 + h_2$ in $\Omega_1\cap\Omega_2$, $h_j \in \mathcal{O}(\Omega_j)$, $j = 1,2$. Apply this to $\Omega_j = \mathbb{C}\backslash K_j$ and $h = \Gamma\mu$; we get $h_j = \mathcal{O}(\mathbb{C}\backslash K_j)$. Since $|h(z)| \to 0$ as $|z| \to +\infty$ the nonnegative power series in the Laurent expansions of h_1 and h_2 cancel out, and we can eliminate them, ie. we may assume $h_j \in \mathcal{O}_0(\mathbb{C}\backslash K_j)$. Then, for $j = 1, 2$, h_j is the Cauchy transform of an analytic functional $\mu_j \in$

$\mathcal{O}'(K_j)$; and $\mu = \mu_1 + \mu_2$. □

If the requirement that $K_1 \cup K_2$ be Runge is lifted the first assertion in Corollary I.2.1 is not any more valid, as shown by

EXAMPLE I.2.2.— Consider the analytic functional μ in $\mathbb{C}$ defined by

$$<\mu,h> = \int_{\gamma_{01}} h(z)dz, \quad h \in \mathcal{O}(\mathbb{C}), \tag{I.2.16}$$

where γ_{01} is any smooth curve joining 0 to 1. It is clear that μ is carried by all curves γ_{01} but that it is not carried by their intersection, the set $\{0\} \cup \{1\}$: μ defines a compactly supported distribution on $\mathbb{R}$, the one defined by the characteristic function of the unit interval [0,1].

What fails in the proof of Corollary I.2.1 when we lift the condition that $K_1 \cup K_2$ be Runge, is the extension to $\mathbb{C} \backslash (K_1 \cap K_2)$ of the Cauchy transforms g_1 and g_2. Make the following two choices for the path γ_{01} in (I.2.16): first take $\gamma_{01} = \gamma^+$, the arc of the circle $|z-\frac{1}{2}| = \frac{1}{2}$ joining 0 to 1 in the upper half–plane; second, take $\gamma_{01} = \gamma^-$, the arc of the same circle joing 0 to 1 in the lower half–plane. But $g_1 - g_2 = 0$ in the region $|z-\frac{1}{2}| > \frac{1}{2}$ whereas $g_1 - g_2 = 1$ in the disk $|z-\frac{1}{2}| < \frac{1}{2}$. This confirms that there is no common extension of g_1 and g_2 to $\mathbb{C} \backslash (\gamma^+ \cap \gamma^-)$. □

Hyperfunctions on the real line

Let U be a bounded and open interval in $\mathbb{R}$, $\mathscr{C}\ell$U its closure, ∂U its boundary in $\mathbb{R}$. Obviously $\mathscr{C}\ell$U is a Runge compact set since its complement is connected. By definition a *hyperfunction u in* U will be an element of the quotient space

$$\mathcal{B}(\mathrm{U}) = \mathcal{O}'(\mathscr{C}\ell\mathrm{U})/\mathcal{O}'(\partial\mathrm{U}).$$

According to Theorem I.2.3 the Cauchy transform defines an isomorphism

$$\mathcal{B}(\mathrm{U}) \cong \mathcal{O}_0(\mathbb{C}\backslash(\mathscr{C}\ell\mathrm{U}))/\mathcal{O}_0(\mathbb{C}\backslash\partial\mathrm{U}). \tag{I.2.17}$$

The inclusion $\mathcal{O}_0(\mathbb{C}\backslash(\mathscr{C}\ell\mathrm{U})) \subset \mathcal{O}(\mathbb{C}\backslash(\mathscr{C}\ell\mathrm{U}))$ induces an isomorphism

$$\mathcal{O}_0(\mathbb{C}\backslash(\mathscr{C}\ell\mathrm{U}))/\mathcal{O}_0(\mathbb{C}\backslash\partial\mathrm{U}) \cong \mathcal{O}(\mathbb{C}\backslash(\mathscr{C}\ell\mathrm{U}))/\mathcal{O}(\mathbb{C}\backslash\partial\mathrm{U}), \tag{I.2.18}$$

as one sees by taking Laurent expansions. Now let $\tilde{\mathrm{U}}$ be a connected open subset of $\mathbb{C}$ such that $\mathrm{U} = \tilde{\mathrm{U}}\cap\mathbb{R}$. By the reasoning in the proof of Corollary I.2.1, given any $g \in \mathcal{O}(\tilde{\mathrm{U}}\backslash\mathrm{U})$ there are $G \in \mathcal{O}(\mathbb{C}\backslash(\mathscr{C}\ell\mathrm{U}))$ and $F \in \mathcal{O}(\tilde{\mathrm{U}})$ such that $g = F - G$ in $\tilde{\mathrm{U}}\backslash\mathrm{U}$. It follows from this that the restriction mapping $\mathcal{O}(\mathbb{C}\backslash(\mathscr{C}\ell\mathrm{U})) \to \mathcal{O}(\tilde{\mathrm{U}}\backslash\mathrm{U})$ induces a surjection

$$\mathcal{O}(\mathbb{C}\backslash(\mathscr{C}\ell\mathrm{U})) \to \mathcal{O}(\tilde{\mathrm{U}}\backslash\mathrm{U})/\mathcal{O}(\tilde{\mathrm{U}}) \tag{I.2.19}$$

and consequently an isomorphism

$$\mathcal{O}(\mathbb{C}\backslash(\mathscr{C}\ell\mathrm{U}))/\mathcal{O}(\mathbb{C}\backslash\partial\mathrm{U}) \cong \mathcal{O}(\tilde{\mathrm{U}}\backslash\mathrm{U})/\mathcal{O}(\tilde{\mathrm{U}}). \tag{I.2.20}$$

By combining the isomorphisms (I.2.17), (I.2.18) and (I.2.20) we obtain

$$\mathcal{B}(\mathrm{U}) \cong \mathcal{O}(\tilde{\mathrm{U}}\backslash\mathrm{U})/\mathcal{O}(\tilde{\mathrm{U}}). \tag{I.2.21}$$

Call Π^+ (resp., Π^-) the open upper (resp. lower) half plane. Notice that, since U is an interval, $\tilde{U}\backslash U$ has two connected components, $\tilde{U}^+ = \tilde{U}\cap\Pi^+$ and $\tilde{U}^- = \tilde{U}\cap\Pi^-$. In other words, we may identify $\mathcal{O}(\tilde{U}\backslash U)$ to $\mathcal{O}(\tilde{U}^+)\times\mathcal{O}(\tilde{U}^-)$. With this identification the restrictions $\mathcal{O}(\mathbb{C}\backslash(\mathcal{C}\ell\, U)) \to \mathcal{O}(\Pi^{\pm}) \to \mathcal{O}(\tilde{U}^{\pm})$ yield the factorization

$$\mathcal{O}(\mathbb{C}\backslash(\mathcal{C}\ell U)) \to \mathcal{O}(\Pi^+)\times\mathcal{O}(\Pi^-) \to \mathcal{O}(\tilde{U}^+)\times\mathcal{O}(\tilde{U}^-)/\Delta(\tilde{U}),$$

of the map (I.2.19), where $\Delta(\tilde{U})$ is the linear subspace of $\mathcal{O}(\Pi^+)\times\mathcal{O}(\Pi^-)$ consisting of the pairs (g^+,g^-) such that $g^{\pm}$ can be extended holomorphically to $\Pi^{\pm}\cup\tilde{U}$ and that their extensions are equal in $\tilde{U}$. Finally we obtain a "natural" isomorphism

$$\mathcal{B}(U) \cong \mathcal{O}(\Pi^+)\times\mathcal{O}(\Pi^-)/\Delta(\tilde{U}). \tag{I.2.22}$$

The assertion of the isomorphisms (I.2.21) or (I.2.22) is sometimes referred to as the "jump theorem" for hyperfunctions, by which it is meant that $u \in \mathcal{B}(U)$ corresponding to a pair $(g^+,g^-) \in \mathcal{O}(\Pi^+)\times\mathcal{O}(\Pi^-)$ can be equated to the difference between the "boundary value" of g^+ and that of g^-. Indeed, by retracing the argument leading to (I.2.22) via the Cauchy transform, one can prove that u has a representative $\mu \in \mathcal{O}'(\mathcal{C}\ell U)$ given by

$$\langle\mu,h\rangle = \lim_{t\to+0} \int_{U+\imath t} hg^+dz - \lim_{t\to+0} \int_{U-\imath t} hg^-dz.$$

Notice also that, according to (I.2.22), u extends as a hyperfunction to the whole real line: it suffices to keep the pair $(g^+,g^-) \in \mathcal{O}(\Pi^+)\times\mathcal{O}(\Pi^-)$ fixed, and to let U exhaust $\mathbb{R}$.

We propose to generalize (when possible) the preceding definitions and results to higher dimensions, with $\mathbb{R}$ replaced by an m–dimensional totally real submanifold $\mathcal{X}$ of complex space $\mathbb{C}^m$ ($m \geq 2$). What we must do before anything else is to generalize, somehow, the Cauchy transform, and to prove the analogue of Corollary I.2.1. In the following sections we are going to see that that it is feasible, provided we limit our attention to analytic functionals in $\mathbb{C}^m$ carried by sufficiently small compact subsets of the submanifold $\mathcal{X}$.

I.3 INTEGRAL REPRESENTATIONS OF ANALYTIC FUNCTIONALS CARRIED BY SMALL COMPACT SUBSETS OF A MAXIMALLY REAL SUBMANIFOLD

In the present text, following the terminology of [TREVES, 1992] (Definition I.3.3) we shall always refer to a totally real submanifold of $\mathbb{C}^m$ whose real dimension is equal to m, as a *maximally real submanifold of* $\mathbb{C}^m$.

Our purpose in this section is to generalize the Cauchy transform to higher dimensions — more precisely to obtain convenient integral representations of analytic functionals carried by a sufficiently small compact subset of a maximally real submanifold $\mathcal{X}$ of $\mathbb{C}^m$. We are going to construct such representations by modifying an argument in [HARVEY—WELLS, 1972].

We must begin by stating precisely what we mean by a "small" compact subset of $\mathcal{X}$. We shall reason in the neighborhood of a point of $\mathcal{X}$ which we take to be the origin in $\mathbb{C}^m$. We may select the complex coordinates $z_j = x_j + \imath y_j$ ($j = 1,\dots,m$) so that the tangent space to $\mathcal{X}$ at 0 be the real space $\mathbb{R}^m$ (defined by $y_1 = \cdots = y_m = 0$). In other words, we shall postulate that there are two open balls B_0, $B_1 \subset \mathbb{R}^m$, both centered at the origin, and a $\mathcal{C}^\infty$ map $\Phi = (\Phi_1,\dots,\Phi_m)$: $B_0 \to B_1$, such that

$$\mathcal{X} \cap (B_0 + \imath B_1) = \{\, z = x + \imath y \in \mathbb{C}^m;\ x \in B_0,\ y = \Phi(x) \,\}, \tag{I.3.1}$$

and such that, furthermore,

$$\Phi_j(0) = 0,\ d\Phi_j(0) = 0,\ j = 1,\dots,m. \tag{I.3.2}$$

Possibly after decreasing the radius of B_0 we may, and shall, also assume that

(I.3.3) $$\forall\ x,\ x' \in B_0,\ |\Phi(x)-\Phi(x')| \leq \tfrac{1}{2}|x-x'|.$$

We shall systematically denote by Ω the image of B_0 under the map $x \to Z(x) = x + i\Phi(x)$. Calling B_{01} the "biball" B_0+iB_1 we note that $\Omega = Z(B_0) = \mathcal{X}\cap B_{01}$.

LEMMA I.3.1.— *Every compact subset of Ω has the Runge property.*

Proof: We use the notation of Chapter II, [TREVES, 1992]: for any $u \in C_c^0(\Omega)$ we write

(I.3.4) $$\mathcal{E}_\nu u(z) = (\nu/\pi)^{m/2}\int_\Omega e^{-\nu\langle z-z'\rangle^2} u(z')dz',$$

where $z \in \mathbb{C}^m$ and $\langle\zeta\rangle^2 = \zeta_1^2+\cdots+\zeta_m^2$.

Let then K be an arbitrary compact subset of Ω and $h \in \mathcal{O}(\mathcal{U})$ with $\mathcal{U}$ an open neighborhood of K in $\mathbb{C}^m$. Select a function $\chi \in C_c^\infty(\mathcal{U})$, $\chi \equiv 1$ in a open neighborhod $\mathcal{U}' \subset \mathcal{U}$ of K in $\mathbb{C}^m$. We form (I.3.4) with $u = \chi h$. Let then $\psi \in C_c^\infty(\mathcal{U}')$, $0 \leq \psi \leq 1$ everywhere, $\psi = 1$ in a neighborhood $\mathcal{U}''$ $\subset \mathcal{U}'$ of K. Let B denote the unit ball in $\mathbb{R}^m$. If $\epsilon > 0$ is sufficiently small, $\operatorname{supp}\psi + i\epsilon' B \subset \mathcal{U}'$ for all ϵ', $0 \leq \epsilon' \leq \epsilon$. We can deform the domain of integration in the integral (I.3.4) from Ω to the image of Ω under the map $z \to H(z,t) = z + i\psi(z)t$, with $t \in \mathbb{R}^m$, $|t| \leq \epsilon$. We obtain

$$\mathcal{E}_\nu u(H(z,t)) = (\nu/\pi)^{m/2}\int_\Omega e^{-\nu\langle H(z,t)-H(z',t)\rangle^2} u(H(z',t))dH(z',t).$$

Condition (I.3.3) allows us to apply Lemma II.2.1 in [TREVES, 1992] with $Z(x,t)$ replaced by $H(x+\imath\Phi(x),t)$ and to show that $\mathcal{E}_\nu u(H(z,t))$ converges to $u(H(z,t))$ as $\nu \to +\infty$, uniformly with respect to z in $\Omega \cap \mathcal{U}'$ and $t \in \mathbb{R}^m$, $|t| \leq \epsilon$ ($\epsilon << 1$). The latter property implies that $\mathcal{E}_\nu u \in \mathcal{O}(\mathbb{C}^m)$ converges to h uniformly on the image of the map $(\Omega \cap \mathcal{U}'') \times \epsilon B \ni (z,t) \to z+\imath t$, which is a neighborhood of K in $\mathbb{C}^m$. □

In general, large compact subsets of a maximally real submanifold of $\mathbb{C}^m$ will not have the Runge property, as shown by

EXAMPLE I.3.1.— The unit circle in $\mathbb{C}$ is maximally real but not Runge. In higher dimensions (m > 1) an example is provided by $K = \mathbf{S}^1 \times \{0\} \subset \mathcal{X} = \mathbf{S}^1 \times \mathbb{R}^{m-1} \subset \mathbb{C} \times \mathbb{C}^{m-1}$. □

However there are maximally real submanifolds $\mathcal{X}$ of $\mathbb{C}^m$ in which every compact subset has the Runge property:

EXAMPLE I.3.2.— Let $\mathcal{X} = \mathbb{R}^m$ and take $Z(x) \equiv x$, ie., $\Phi \equiv 0$, and thus $\Omega = B_0$. The reader will easily check that the proof of Lemma I.3.1 extends without any limitation on the radius of the ball B_0. We conclude that *every compact subset of* $\mathbb{R}^m$ *has the Runge property.* □

Next we look at certain integral kernels which define linear operators acting on differential forms and currents of type (p,q). We shall use the terminology and the notation of [TREVES, 1992] (see particularly section VI.1): Let $\mathcal{A}$ be an open subset of $\mathbb{C}^m$. An element of $\mathcal{C}^k(\mathcal{A};\Lambda^{p,q})$ ($0 \leq k \leq +\infty$) is a differential form

(I.3.5)
$$f = \sum_{|I|=p} \sum_{|J|=q} f_{I,J} dz_I \wedge d\bar{z}_J$$

with $f_{I,J} \in \mathcal{C}^k(\mathcal{A})$. We are using standard multi–index notation: I is a p–tuple of integers i_α: $1 \le i_1 < \cdots < i_p \le m$; $dz_I = dz_{i_1} \wedge \cdots \wedge dz_{i_p}$; like–wise with J and $d\bar{z}_J$. We shall also deal with currents $f \in \mathcal{D}'(\mathcal{A};\Lambda^{p,q})$: this simply means that the coefficients $f_{I,J}$ are distributions in $\mathcal{A}$. When all the coefficients have compact support we write $f \in \mathcal{C}^k_c(\mathcal{A};\Lambda^{p,q})$ or, in the case of distributions, $f \in \mathcal{E}'(\mathcal{A};\Lambda^{p,q})$. The action of the Cauchy–Riemann operator on a form or a current (I.3.5) is given by

(I.3.6)
$$\bar{\partial} f = \sum_{|I|=p} \sum_{|J|=q} \sum_{\ell=1}^{m} (\partial f_{I,J}/\partial \bar{z}_\ell) d\bar{z}_\ell \wedge dz_I \wedge d\bar{z}_J.$$

Of course $\bar{\partial}^2 = 0$ and $\mathcal{D}'(\mathcal{A};\Lambda^{p,q}) = 0$ if either $p > m$ or $q > m$. This gives rise to the various Dolbeault (or Cauchy–Riemann) complexes (for each $p = 0,1,..,m$):

(I.3.7)
$$\bar{\partial}: \mathcal{C}^\infty(\mathcal{A};\Lambda^{p,q}) \to \mathcal{C}^\infty(\mathcal{A};\Lambda^{p,q+1}),\ q = 0,1,\ldots,m;$$

(I.3.8)
$$\bar{\partial}: \mathcal{D}'(\mathcal{A};\Lambda^{p,q}) \to \mathcal{D}'(\mathcal{A};\Lambda^{p,q+1}),\ q = 0,1,\ldots,m.$$

The cohomology spaces of the differential complex (I.3.7) will be called $H^{p,q}(\mathcal{A})$. It does not matter whether one uses forms with $\mathcal{C}^\infty$ or with dis–tribution coefficients in the Dolbeault resolution of the sheaf ${}_m\mathcal{O}^{(p)}$ of germs of holomorphic p–forms [ie., $\bar{\partial}$–closed (0,p)–forms (I.3.5)]. There–fore the space $H^{p,q}(\mathcal{A})$ is naturally isomorphic to the qth cohomology space of the differential complex (I.3.8). We might also consider the sub–

complexes of (I.3.7) and (I.3.8) defined by requiring that the forms or currents have compact supports. Their cohomology spaces will be denoted by $H_c^{p,q}(\mathcal{A})$; $H_c^{p,q}(\mathcal{A})$ is often interpreted as a "*homology*" space, of bidegree (m—p,m—q).

Consider m functions $k_j \in \mathcal{C}^\infty(\mathcal{A}\times\mathcal{A})$ (j = 1,...,m) and define

$$k(w,z) = \sum_{j=1}^{m} k_j(w,z)(w_j - z_j).$$

In the open set $\mathcal{R}_k = \{\ (w,z) \in \mathcal{A}\times\mathcal{A};\ k(w,z) \neq 0\ \}$ we form the kernel

(I.3.9)
$$N_k(w,z) = \frac{(m-1)!}{(2\imath\pi k)^m} \sum_{j=1}^{m} (-1)^{j-1} k_j dk_1 \wedge \cdots \wedge \widehat{dk_j} \wedge \cdots \wedge dk_m \wedge d(w_1 - z_1) \wedge \cdots \wedge d(w_m - z_m),$$

where d stands for the "total" differential, ie., the differential in (w,z)—space. The kernel N_k is a smooth differential form of degree 2m—1 in the region $\mathcal{R}_k$. According to Lemma VI.2.2 [TREVES, 1992], we also have (always in $\mathcal{R}_k$)

(I.3.10)
$$N_k(w,z) = \frac{(m-1)!}{(2\imath\pi k)^m} \sum_{j=1}^{m} (-1)^{j-1} u_j du_1 \wedge \cdots \wedge \widehat{du_j} \wedge \cdots \wedge du_m \wedge d(w_1 - z_1) \wedge \cdots \wedge d(w_m - z_m),$$

where $u_j = k_j/k$.

LEMMA I.3.2.— *In the region* $\mathcal{R}_k$, $dN_k \equiv 0$.

Proof: We limit the variation of (w,z) to $\mathcal{R}_k$. Then, by (I.3.10),

$$dN_k = m!(2\imath\pi)^{-m}du_1\wedge\cdots\wedge du_m\wedge d(w_1-z_1)\wedge\cdots\wedge d(w_m-z_m).$$

But $\sum_{j=1}^{m} u_j(w_j-z_j) \equiv 1 \Rightarrow \sum_{j=1}^{m} u_j d(w_j-z_j) = -\sum_{j=1}^{m}(w_j-z_j)du_j$. Multiplying both sides of this equation by

$$(-1)^{i-1}u_i du_1\wedge\cdots\wedge\widehat{du_i}\wedge\cdots\wedge du_m\wedge d(w_1-z_1)\wedge\cdots\wedge d(w_m-z_m)$$

and summing with respect to $i = 1,\ldots,m$, shows that

$$du_1\wedge\cdots\wedge du_m\wedge d(w_1-z_1)\wedge\cdots\wedge d(w_m-z_m) \equiv 0.$$

□

We are going to apply Lemma I.3.2 with a special choice of the functions k_j. As said the analysis will be carried out in the neighborhood of an arbitrary point of $\mathcal{X}$ taken to be the origin. We select the complex coordinates and the defining equations of $\mathcal{X}$ near 0 as described at the beginning of this section. In particular, (I.3.1), (I.3.2) and (I.3.3) will hold. Recall that $\Omega = \mathcal{Z}(B_0) = \mathcal{X}\cap B_{01}$.

Now set $\rho(w) = |v-\Phi(u)|^2$ $(w = u+\imath v)$ and, for each $w \in B_{01}$, consider the quadratic form in z–space,

$$\mathcal{Q}(w,z) = \sum_{j,k=1}^{m} \frac{\partial^2\rho}{\partial w_j\partial\bar{w}_k}(w)z_j\bar{z}_k.$$

Taylor's expansion yields

$$\text{(I.3.11)}\quad \rho(w) = \rho(z) + 2\,\mathcal{R}e\,\lambda(w,w-z) - \mathcal{Q}(w,w-z) + O(|w-z|^3),$$

where

$$\lambda(w,z) = \sum_{j=1}^{m} \lambda_j(w,z)z_j,$$

$$\lambda_j(w,z) = \frac{\partial \rho}{\partial w_j}(w) - \tfrac{1}{2}\sum_{k=1}^{m} \frac{\partial^2 \rho}{\partial w_j \partial w_k}(w) z_k \quad (j = 1,\ldots,m).$$

In the sequel we must keep in mind that $\lambda(w,z)$ is a holomorphic polynomial of degree two with respect to z whose coefficients are C^∞ functions of w.

We have $\rho(w) = |v|^2 + O(|u|^3+|v|^3)$. It follows that the Hermitian quadratic form $z \to \mathcal{Q}(0,z)$ is definite positive. After contracting B_0 and B_1 about 0 we may assume that, for some $c > 0$,

$$\mathcal{Q}(w,z) \geq c|z|^2, \ \forall\, z \in \mathbb{C}^m,\ w \in B_{01}. \tag{I.3.12}$$

After further contracting B_0 and B_1 one derives from (I.3.11) and (I.3.12):

$$2\,\mathcal{R}e\lambda(w,w-z) \geq \rho(w) - \rho(z) + \tfrac{1}{2}c|w-z|^2,\ \forall\, w,\ z \in B_{01}.$$

In turn this entails

$$2\,\mathcal{R}e\lambda(w,w-z) \geq \tfrac{1}{2}c|w-z|^2,\ \forall\, w \in B_{01},\ z \in \Omega. \tag{I.3.13}$$

The following kernel in $B_{01}\times\mathbb{C}^m$ will play an important role in connection with the maximally real submanifold $\mathcal{X}$,

$$G_\lambda(w,z) = \tag{I.3.14}$$

$$(m-1)!(2\imath\pi\lambda)^{-m}\sum_{j=1}^{m}(-1)^{j-1}\lambda_j\bar\partial_w\lambda_1\wedge\cdots\wedge\widehat{\bar\partial_w\lambda_j}\wedge\cdots\wedge\bar\partial_w\lambda_m\wedge dw_1\wedge\cdots\wedge dw_m.$$

In (I.3.14) $\lambda = \lambda(w,z)$, $\lambda_j = \lambda_j(w,z)$. We are going to make use of the kernel $G_\lambda(w, w-z)$ obtained by substituting $w-z$ for z in the coefficients of $G_\lambda(w,z)$. Since the λ_j are affine functions of z, $G_\lambda(w,w-z)$ is the pullback of $G_\lambda(w,z)$ under the map $(w,z) \to (w,w-z)$ (of $B_{01}\times\mathbb{C}^m$ into itself). Note that $G_\lambda(w,w-z)$ is a current of bidegree (m, m—1) in the biball B_{01} in w—space, depending on $z \in \mathbb{C}^m$. By (I.3.13) the coefficients of $G_\lambda(w,w-z)$ are well defined when $(w,z) \in B_{01}\times\Omega$ and $w \neq z$.

REMARK I.3.1.— When m = 1,

$$G_\lambda(w,w-z) = (2i\pi)^{-1}\lambda(w,w-z)^{-1}\lambda_1(w,w-z)dw.$$

Since $\lambda(w,w-z) = \lambda_1(w,w-z)(w-z)$, $G_\lambda(w,w-z)$ is the Cauchy kernel—form $(2i\pi)^{-1}(w-z)^{-1}dw$. □

Finally we define the kernel N_k that is really going to be used as a substitute for the Cauchy kernel. It will be associated to a given compact subset K of Ω and also to a pair of open subsets $\mathcal{U}$ and $\mathcal{U}'$ of B_{01} such that $K \subset \mathcal{U}' \subset\subset \mathcal{U} \subset\subset B_{01}$. Select a cutoff function $\psi \in C^\infty(B_{01})$, $0 \leq \psi \leq 1$, $\psi = 0$ in $\mathcal{U}'$, $\psi = 1$ in $B_{01}\backslash\mathcal{U}$. We define

$$k_j(w,z) = \psi(w)\lambda_j(w,w-z) + [1-\psi(w)](\bar{w}_j-\bar{z}_j),\ j = 1,\ldots,m.$$

Then

$$k(w,z) = \sum_{j=1}^{m} k_j(w,z)(w_j-z_j) = \psi(w)\lambda(w,w-z) + [1-\psi(w)]\,|w-z|^2.$$

We define the kernel N_k in accordance with (I.3.9) or (I.3.10). At this stage (I.3.13) allows merely to assert that its coefficients are well defined when $w \in B_{01}$, $z \in \Omega$ and $w \neq z$.

PROPOSITION I.3.1.— *Let* $K \subset \Omega$ *be compact, and let* $\mathcal{U}$ *and* $\mathcal{U}'$ *be open sets in* $\mathbb{C}^m$ *such that* $K \subset \mathcal{U}' \subset\subset \mathcal{U} \subset\subset B_{01}$. *There is an open neighborhood* $\mathcal{U}''$ $\subset\subset \mathcal{U}'$ *of* K *such that the kernel* N_k *satisfies the following conditions:*

$$\text{(I.3.15)} \quad |N_k(w,z)| \leq \textit{const.}\,|w-z|^{1-2m}, \ \forall\, w \in B_{01},\ z \in \mathcal{U}'',\ w \neq z;$$

$$\text{(I.3.16)} \quad \mathrm{d}N_k(w,z) = (2\imath)^{-m}\delta(w-z)\mathrm{d}(\bar{w}-\bar{z})\wedge\mathrm{d}(w-z) \ \textit{ in } B_{01}\times\mathcal{U}'';$$

$$\text{(I.3.17)} \quad N_k(w,z)\wedge\mathrm{d}z = G_\lambda(w,w-z)\wedge\mathrm{d}z \ \textit{ if } w \in B_{01}\backslash\mathcal{U},\ z \in \mathcal{U}'';$$

and such, furthermore, that the coefficients of $G_\lambda(w,w-z)$ *are holomorphic functions of* $z \in \mathcal{U}''$ *when* $w \in B_{01}\backslash\mathcal{U}$.

In (I.3.15) $|N_k(w,z)|$ stands for the Euclidean norm of the element $N_k(w,z) \in \Lambda^{2m-1}\mathbb{C}T^*_{(w,z)}(\mathbb{R}^{2m}\times\mathbb{R}^{2m})$ with respect to the basis $\mathrm{d}w_i$, $\mathrm{d}\bar{w}_j$, $\mathrm{d}z_k$, $\mathrm{d}\bar{z}_{k'}$ $(1 \leq i, j, k, k' \leq m)$.

Proof: The inequality (3.13) enables us to select the open set $\mathcal{U}''$ in $\mathbb{C}^m$, $K \subset \mathcal{U}'' \subset\subset \mathcal{U}'$, and a number $c_0 > 0$ so as to have

$$\mathcal{R}e\,\lambda(w,w-z) \geq c_0,\ \forall\, w \in B_{01}\backslash\mathcal{U},\ z \in \mathcal{U}'',$$

which implies at once the last assertion in the statement.

Another consequence of (I.3.13) is that there is a constant $c_1 > 0$ such that

$$\mathcal{R}e\,k(w,z) \geq c_1|w-z|^2,\ \forall\, w \in B_{01},\ z \in K,$$

which in turn allows us to select $\mathcal{U}''$ and $c_2 > 0$ so as to have

(I.3.18) $$\mathcal{R}e\, k(w,z) \geq c_2 \text{ if } w \in B_{01}\backslash\mathcal{U}',\ z \in \mathcal{U}''.$$

This implies the estimate in (I.3.15) when $w \notin \mathcal{U}'$, $z \in \mathcal{U}''$, $w \neq z$. On the other hand, when $w \in \mathcal{U}'$ then $\psi(w) \equiv 0$, $k(w,z) = |w-z|^2$. In other words,

(I.3.19) $$N_k(w,z) = \mathcal{E}(w-z),\ \forall\ w \in \mathcal{U}',\ z \in \mathbb{C}^m,$$

with $\mathcal{E}$ the Bochner—Martinelli kernel in $\mathbb{C}^m$. The well known expression of the latter proves the estimate in (I.3.15) also when $w \in \mathcal{U}'$, and therefore for all $w \in B_{01}$, $z \in \mathcal{U}''$, $w \neq z$.

Lemma I.3.2 and (I.3.18) entail

(I.3.20) $$dN_k(w,z) \equiv 0 \quad \text{if } w \in B_{01}\backslash\mathcal{U}',\ z \in \mathcal{U}''.$$

Combining (I.3.19) with (I.3.20), and taking into account the properties of $\mathcal{E}$ (see, e.g., Corollary VI.1.1 in [TREVES, 1992]) yields (I.3.16).

Let us use the notation $\ell_j(w,z) = \lambda_j(w,w-z)$ and $\ell(w,z) = \lambda(w,w-z)$. Since $\psi = 1$ in $B_{01}\backslash\mathcal{U}$ we have $k = \ell$ in $(B_{01}\backslash\mathcal{U})\times\mathbb{C}^m$, hence $N_k(w,z) = N_\ell(w,z)$ if $w \in B_{01}\backslash\mathcal{U}$, $z \in \mathcal{U}''$. According to (I.3.9) we have

$$N_\ell(w,z)\wedge dz = (m-1)!(2\iota\pi\ell)^{-m}\sum_{j=1}^{m}(-1)^{j-1}\ell_j\, d\ell_1\wedge\cdots\wedge\widehat{d\ell_j}\wedge\cdots\wedge d\ell_m\wedge dw\wedge dz,$$

where $dz = dz_1\wedge\cdots\wedge dz_m$ and likewise for dw. But each $\ell_j(w,z)$ is holo—morphic with respect to z, and therefore $d\ell_j\wedge dw\wedge dz = (\bar\partial_w\ell_j)\wedge dw\wedge dz$. We get

(I.3.21) $$N_\ell(w,z)\wedge dz = G_\lambda(w,w-z)\wedge dz.$$

If we return to the definition of k and take $w \in B_{01}\backslash\mathcal{U}$ we obtain $k_j(w,z) = \lambda_j(w,w-z) = \ell_j(w,z)$; thus (I.3.21) $\Rightarrow$ (I.3.17). □

Let $K \subset\subset \mathcal{U}'' \subset\subset B_{01}$ and the functions $k_j \in C^\infty(B_{01}\times\mathbb{C}^m)$ be as in Proposition I.3.1. For $1 \leq q \leq m$ we define a bounded linear operator

$$\mathcal{T}^q\colon C_c^\infty(\mathcal{U}'';\Lambda^{m,q}) \to C^\infty(B_{01};\Lambda^{m,q-1})$$

by the formula

(I.3.22) $$(\mathcal{T}^q\varphi)(w) = (-1)^{m-q}\int \varphi(z)\wedge N_k(w,z),$$

with the integration carried out in z—space (only the component of degree 2m with respect to z of the current $\varphi(z)\wedge N_k(w,z)$ contributes to the inte— gral). Thanks to (I.3.15) we know that $\varphi(z)\wedge N_k(w,z)$ is of class L^1 with respect to z in $\mathcal{U}''$. As in the case of the Bochner—Martinelli kernel differ— entiation under the integral sign followed by integration by parts shows easily that $\mathcal{T}^q$ transforms smooth forms [with compact support, of type (m,q)] into smooth forms [of type (m,q—1)].

PROPOSITION I.3.2.— *Assume* $1 \leq q \leq m$ *and* $\varphi \in C_c^\infty(\mathcal{U}'';\Lambda^{m,q})$; *then*

(I.3.23) $$\bar\partial\varphi = 0 \Rightarrow \bar\partial\mathcal{T}^q\varphi = \varphi \text{ in } B_{01}.$$

Moreover,

(I.3.24) $$\mathcal{T}^q\varphi = 0 \quad \text{in } B_{01}\backslash\mathcal{U}$$

if either $q \leq m-1$ *or, when* $q = m$, *if*

(I.3.25) $$\int h\varphi = 0, \quad \forall\, h \in \mathcal{O}(\mathcal{U}'').$$

Proof: We have, by (I.3.16),

$$\bar\partial_w(\mathcal{T}^q\varphi)(w) = d_w(\mathcal{T}^q\varphi)(w) =$$

$$\int \varphi(z)\wedge(d_w + d_z)N_k(w,z) - \int \varphi(z)\wedge d_z N_k(w,z).$$

But, still with integration restricted to z–space,

$$\int \varphi(z)\wedge(d_w + d_z)N_k(w,z) = (2\imath)^{-m}\int \delta(w-z)\varphi(z)\wedge d(\bar w - \bar z)\wedge dw.$$

If $\varphi(z) = \sum_{|J|=q} \varphi_J(z)d\bar z_J\wedge dz$ $(dz = dz_1\wedge\cdots\wedge dz_m)$ a straightforward computation shows that

$$\int \delta(w-z)\varphi(z)\wedge d(\bar w - \bar z)\wedge dw =$$

$$\sum_{|J|=q}\int \delta(w-z)\varphi_J(z)d\bar z\wedge dz\wedge d\bar w_J\wedge dw = (2\imath)^m\varphi(w).$$

We obtain $\bar\partial_w(\mathcal{T}^q\varphi)(w) = \varphi(w) - \int \varphi(z)\wedge d_z N_k(w,z)$. Integration by parts shows that

$$\bar\partial\varphi \equiv 0 \Rightarrow \int \varphi(z)\wedge d_z N_k(w,z) \equiv 0$$

whence (I.3.23).

To prove (I.3.24) we apply (I.3.17):

(I.3.26) $$N_k(w,z)\wedge\varphi(z) = G_\lambda(w,w-z)\wedge\varphi(z), \quad w \in B_{01}\backslash\mathcal{U}.$$

In view of (I.3.14) we note that the right–hand side in (I.3.26) is a current of bidegree (m,q) in z–space (like φ) and therefore its integral with respect to z must vanish unless q = m. When q = m the result follows from the last assertion in Proposition I.3.1 combined with (I.3.25) and (I.3.26). □

By K we continue to denote a compact subset of $\Omega = Z(B_0)$. We define the *Cauchy–Fantappie transform relative to* $\mathcal{X}$ of an analytic functional μ carried by K:

$$\Gamma_{\mathcal{X}}\mu(w) = <\mu_z, G_\lambda(w,w-z)>. \tag{I.3.27}$$

The function $\Gamma_{\mathcal{X}}\mu$ is well defined, and belongs to $\in \mathcal{C}^\infty(B_{01}\backslash K;\Lambda^{m,m-1})$, due to the fact that $\mu \in \mathcal{O}'(K)$ and that, for fixed $w \in B_{01}\backslash K$, $z \to G_\lambda(w,w-z)$ is a holomorphic function in some open neighborhood of K (Proposition I.3.1). Note also that $\bar\partial\Gamma_{\mathcal{X}}\mu \equiv 0$ since $\bar\partial_w[G_\lambda(w,w-z)] \equiv 0$ in its domain of definition.

Remark I.3.1 shows that if m = 1, the Cauchy–Fantappiè transform (relative to any smooth curve through the origin) coincides with $\Gamma\mu(w)dw$, where $\Gamma\mu$ is the Cauchy transform (see (I.2.12)).

THEOREM I.3.1.– *Let* K *be a compact subset of* $\Omega = Z(B_0)$ *and let* μ *be an analytic functional in* $\mathbb{C}^m$ *carried by* K. *If* $\chi \in \mathcal{C}_c^\infty(B_{01})$, $\chi \equiv 1$ *in an open neighborhood* $\mathcal{U}$ *of* K, *then, for all entire holomorphic functions h,*

$$<\mu,h> = -\int h(w)\bar\partial\chi(w)\wedge\Gamma_{\mathcal{X}}\mu(w). \tag{I.3.28}$$

Proof: The assertion will follow if we prove that there is an open neigh— borhood $\mathcal{U}'' \subset \mathcal{U}$ of K such that, for all $h \in \mathcal{O}(\mathbb{C}^m)$,

$$h(z) = -\int h(w)\bar{\partial}\chi(w)\wedge G_\lambda(w,w-z), \ \forall\ z \in \mathcal{U}''. \tag{I.3.29}$$

In (I.3.29) the integration is carried out in w—space. We introduce the functions k_j (j = 1,...,m) and the associated kernel N_k of Proposition I.3.2. It follows at once from (I.3.17) that (I.3.29) is equivalent to the validity of

$$h(z)dz = -\int h(w)\bar{\partial}\chi(w)\wedge N_k(w,z)\wedge dz, \ \forall\ z \in \mathcal{U}''. \tag{I.3.30}$$

In turn (I.3.30) is equivalent to the validity of

$$\int h\varphi = -\int\int h(w)\bar{\partial}\chi(w)\wedge N_k(w,z)\wedge\varphi(z),$$

for all $\varphi \in C_c^\infty(\mathcal{U}'';\Lambda^{m,m})$ [with the integration at the right carried out in (w,z) space]; or, equivalently, in the notation of (I.3.22),

$$\int h\varphi = -\int \bar{\partial}(\chi h)\wedge \mathcal{T}^m\varphi. \tag{I.3.31}$$

To get (I.3.31) it suffices to integrate by parts at the right, and to apply (I.3.23) — recalling that $\bar{\partial}C^\infty(\mathcal{U}'';\Lambda^{p,m}) = 0$. □

REMARK I.3.2.— Inspection of the reasoning throughout section I.3 shows that it applies to $\mathcal{X} = \mathbb{R}^m$ with no limitation on the size of Ω. □

I.4 SUPPORT OF AN ANALYTIC FUNCTIONAL CARRIED BY SMALL COMPACT SUBSETS OF A MAXIMALLY REAL SUBMANIFOLD

We continue to reason in the framework of section I.3. As a general rule, and unless otherwise specified, we shall assume $m \geq 2$. We have seen that, if $\mu \in \mathcal{O}'(\mathbb{C}^m)$ is carried by a compact set $K \subset \Omega = \mathcal{Z}(B_0) = \mathcal{X} \cap B_{01}$, the Cauchy—Fantappiè transform $\Gamma_{\mathcal{X}}\mu \in C^\infty(B_{01}\backslash K; \Lambda^{m,m-1})$ is $\bar\partial$–closed and therefore defines a cohomology class $[\Gamma_{\mathcal{X}}\mu] \in H^{m,m-1}(B_{01}\backslash K) = \{ f \in C^\infty(B_{01}\backslash K; \Lambda^{m,m-1});\ \bar\partial f = 0 \}/\bar\partial C^\infty(B_{01}\backslash K; \Lambda^{m,m-2})$.

Conversely there is a natural map $H^{m,m-1}(B_{01}\backslash K) \ni [g] \to \mu_g \in \mathcal{O}'(K)$: consider a representative of the class $[g]$, $g \in C^\infty(B_{01}\backslash K; \Lambda^{m,m-1})$ such that $\bar\partial g = 0$. Select $\chi \in C_c^\infty(B_{01})$, $\chi \equiv 1$ in an open neighborhood of K. The linear functional

$$\mathcal{O}(\mathbb{C}^m) \ni h \to \int h \bar\partial\chi \wedge g \in \mathbb{C} \tag{I.4.1}$$

does not depend on the choice of the representative g nor of the cutoff function χ. Indeed, suppose we replace them by g_1 and χ_1 respectively. Then $g_1 = g + \bar\partial v$ with $v \in C^\infty(B_{01}\backslash K; \Lambda^{m,m-2})$. Consequently,

$$\int h\bar\partial\chi_1\wedge g_1 - \int h\bar\partial\chi\wedge g = \int h\bar\partial(\chi_1-\chi)\wedge g + \int h\bar\partial\chi_1\wedge(g_1-g) =$$
$$\int h\bar\partial[(\chi_1-\chi)g] - \int h\bar\partial(\bar\partial\chi_1\wedge v) = 0$$

since both $(\chi_1-\chi)g$ and $\bar\partial\chi_1\wedge v$ have compact support in $B_{01}\backslash K$. This allows us to select supp χ as close to K as we wish. But then we see that (I.4.1) defines an analytic functional μ_g carried by K. Since K is Runge

(Lemma I.3.1) we have the right to identify μ_g to an element of of the strong dual of $\mathcal{O}(K)$. And as seen in the proof of Theorem I.2.3 the above argument, with minor modifications, also works in the case $m = 1$.

THEOREM I.4.1.— *Suppose* $m \geq 2$ *and let* $g \in C^\infty(B_{01}\backslash K;\Lambda^{m,m-1})$ *be such that* $\bar\partial g = 0$. *If* $\mu_g = 0$ *there is* $u \in C^\infty(B_{01}\backslash K,\Lambda^{m,m-2})$ *satisfying* $\bar\partial u = g$ *in* $B_{01}\backslash K$.

Proof: We construct a basis of open neighborhoods of K in $\mathbb{C}^m$, $\{\mathcal{U}_\nu\}_{\nu=1,2,...}$, with $\mathcal{U}_{\nu+1} \subset\subset \mathcal{U}_\nu \subset\subset B_{01}$, as follows. For each $\nu = 1,2,...$, we select an open neighborhood $\mathcal{U}''_\nu \subset\subset \mathcal{U}_\nu$ of K such that the conclusions in Proposition I.3.2 are valid if $\mathcal{U} = \mathcal{U}_\nu$, $\mathcal{U}'' = \mathcal{U}''_\nu$ and with C^∞ functions $k_{\nu j}$ in $B_{01}\times\mathbb{C}^m$. By reasoning inductively on ν we select $\mathcal{U}_{\nu+1} \subset \mathcal{U}''_\nu$. We use the notation

$$k_\nu(w,z) = \sum_{j=1}^{m} k_{\nu j}(w,z)(w_j - z_j)$$

and we denote by $\mathcal{T}^q_\nu$ the restriction to $C^\infty_c(\mathcal{U}_{\nu+1};\Lambda^{m,q})$ of the operator $\mathcal{T}^q$ defined in (I.3.22) with $k = k_\nu$.

We shall now avail ourselves of the following consequence of Lemma I.3.1: *Given any open neighborhood* $\mathcal{U}$ *of* K *in* $\mathbb{C}^m$ *there is another open subset of* $\mathbb{C}^m$, *such that* $K \subset \mathcal{V} \subset \mathcal{U}$ *and that every holomorphic func— tion in* $\mathcal{U}$ *is the uniform limit in* $\mathcal{V}$ *of entire functions.* Select $\psi_\nu \in C^\infty(B_{01})$ with $\psi_\nu \equiv 1$ in a neighborhood of $B_{01}\backslash\mathcal{U}_{\nu+1}$, $\psi_\nu \equiv 0$ in a neighborhood of K, and such that every holomorphic function in $\mathcal{U}_{\nu+1}$ is a uniform limit, on supp $d\psi_\nu$ ($\subset\subset \mathcal{U}_{\nu+1}$), of a sequence of entire functions. If $\chi_\nu = 1-\psi_\nu$, it follows from (I.3.17) that

$$\text{(I.4.2)}\quad \mathcal{T}^{m}_{\nu}(\bar\partial\psi_\nu\wedge g)(w) = -\int \bar\partial\chi_\nu(z)\wedge g(z)\wedge G_\lambda(w,w-z) \quad \text{if } w \in \mathrm{B}_{01}\backslash\mathcal{U}_\nu.$$

If $w \notin \mathcal{U}_\nu$, the function $\mathcal{U}_{\nu+1} \ni z \to G_\lambda(w,w-z)$ is holomorphic. For each fixed $w \in \mathrm{B}_{01}\backslash\mathcal{U}_\nu$ the function $z \to G_\lambda(w,w-z)$ is the uniform limit on supp $\bar\partial\chi_\nu$, of a sequence of entire functions $f_{\nu k}$ (k = 1,2,...). Since $\mu_g = 0$ we have

$$\int \bar\partial\chi_\nu\wedge g\wedge f_{\nu k} = 0,$$

and thus we reach the conclusion that $\mathcal{T}^{m}_{\nu}(\bar\partial\psi_\nu\wedge g) \equiv 0$ in $\mathrm{B}_{01}\backslash\mathcal{U}_\nu$. We apply (I.3.23) (with q = m); we obtain $\bar\partial[\mathcal{T}^{m}_{\nu}(\bar\partial\psi_\nu\wedge g)] = \bar\partial\psi_\nu\wedge g = \bar\partial(\psi_\nu g)$. As a consequence we can find $v_\nu \in \mathcal{C}^\infty(\mathrm{B}_{01};\Lambda^{m,m-2})$ such that $\bar\partial v_\nu = \psi_\nu g - \mathcal{T}^{m}_{\nu}(\bar\partial\psi_\nu\wedge g)$ in B_{01}. This reduces to $\bar\partial v_\nu = g$ in $\mathrm{B}_{01}\backslash\mathcal{U}_\nu$.

Select $\phi_\nu \in \mathcal{C}^\infty(\mathrm{B}_{01})$, $\phi_\nu \equiv 1$ in $\mathrm{B}_{01}\backslash\mathcal{U}_{\nu+1}$, $\phi_\nu \equiv 0$ in a neighborhood of $\mathcal{C}\ell\,\mathcal{U}_{\nu+2}$, with supp $\bar\partial\phi_\nu \subset \mathcal{U}_{\nu+1}$. We now make use of the following remark: if $v \in \mathcal{C}^\infty(\mathrm{B}_{01};\Lambda^{m,m-2})$ satisfies $\bar\partial v \equiv 0$ in $\mathrm{B}_{01}\backslash\mathcal{U}_{\nu+2}$ and if $f = \bar\partial(\phi_\nu v) \in \mathcal{C}^\infty_c(\mathrm{B}_{01};\Lambda^{m,m-1})$ then supp $f \subset \mathcal{U}_{\nu+1}$ and by Proposition I.3.2, $\bar\partial(\mathcal{T}^{m-1}_{\nu} f_\nu) = f_\nu$ in B_{01} and $\mathcal{T}^{m-1}_{\nu} f_\nu \equiv 0$ in $\mathrm{B}_{01}\backslash\mathcal{U}_\nu$.

We define $u_1 = v_3$ and

$$u_2 = v_4 - \phi_1(v_4-u_1) + \mathcal{T}^{m-1}_{1}\bar\partial[\phi_1(v_4-u_1)].$$

We have $u_2 = u_1$ in $\mathrm{B}_{01}\backslash\mathcal{U}_1$ and $\bar\partial u_2 = g$ in $\mathrm{B}_{01}\backslash\mathcal{U}_4$. We reason by induction on $\nu \geq 2$, by defining

$$u_{\nu+1} = v_{\nu+3} - \phi_\nu(v_{\nu+3}-u_\nu) + \mathcal{T}^{m-1}_{\nu}\bar\partial[\phi_\nu(v_{\nu+3}-u_\nu)].$$

We have $u_{\nu+1} = u_\nu$ in $B_{01}\backslash\mathcal{U}_\nu$ and $\bar\partial u_{\nu+1} = g$ in $B_{01}\backslash\mathcal{U}_{\nu+3}$. As $\nu \to +\infty$ the forms $u_\nu \in \mathcal{C}^\infty(B_{01};\Lambda^{m,m-1})$ converge in $B_{01}\backslash K$ to the sought solution u. □

We are now in a position to state the sought generalization of Theorem I.2.3:

THEOREM I.4.2.— *If* m $\geq$ 2 *the map* $[g] \to \mu_g$ *is an isomorphism of* $H^{m,m-1}(B_{01}\backslash K)$ *onto* $\mathcal{O}'(K)$.

Proof: We have already proved the surjectivity of the map under con— sideration since, according to Theorem I.3.1, $\mu_g = \mu$ if $g = \Gamma_\chi\mu$. The in— jectivity follows at once from Theorem I.4.1. □

REMARK I.4.1.— When m = 1 the map $[g] \to \mu_g$ is not an isomorphism of $H^{1,0}(\Delta\backslash K)$ onto $\mathcal{O}'(K)$ [here it is convenient to replace the rectangle B_{01} by an open disk $\Delta \supset K$]. It is surjective since it remains true that $\mu = \mu_g$ if $g = (\Gamma\mu)dz$. But it is not injective. Indeed, suppose that $\mu_g = 0$ for some $g = g_0 dz$, $g \in \mathcal{O}(\Delta\backslash K)$. This means that, for any $\chi \in \mathcal{C}_c^\infty(\Delta)$, $\chi \equiv 1$ in a neighborhood of K, and any $h \in \mathcal{O}(\mathbb{C})$, we have

$$\int (\partial\chi/\partial\bar z) h g_0 d\bar z \wedge dz = 0.$$

We may let χ converge to the characteristic function of an arbitrary open subset $\mathcal{U}$ of Δ, with $K \subset \mathcal{U}$ and $\mathfrak{c} = \partial\mathcal{U}$ a smooth curve. We conclude that

$$\int_{\mathfrak{c}} h g_0 dz = 0, \forall\, h \in \mathcal{O}(\mathbb{C}).$$

This means that g_0 extends holomorphically across $\mathfrak{c}$ to $\mathcal{U}$ (as one see, for instance, by using a Laurent expansion of g) and therefore $g_0 \in \mathcal{O}(\Delta)$. We conclude that $g \to \mu_g$ defines an isomorphism of $\mathcal{O}(\Delta\backslash K)/\mathcal{O}(\Delta)$ onto $\mathcal{O}'(K)$. The reader will easily ascertain that restriction from $\mathbb{C}\backslash K$ to $\Delta\backslash K$ induces an isomorphism of $\mathcal{O}_0(\mathbb{C}\backslash K)$ onto $\mathcal{O}(\Delta\backslash K)/\mathcal{O}(\Delta)$. We reach the desired conclusion, namely that $g \to \mu_g$ defines an isomorphism of $\mathcal{O}_0(\mathbb{C}\backslash K)$ onto $\mathcal{O}'(K)$ — whose inverse is the Cauchy transform. □

We are now in a position to generalize the first part of Corollary I.2.1:

PROPOSITION I.4.1.— *Let* K_1 *and* K_2 *be two compact subsets of* $\Omega = Z(B_0)$. *If* $\mu \in \mathcal{O}'(\mathbb{C}^m)$ *is carried by* K_1 *and by* K_2 *then* μ *is carried by* $K_1 \cap K_2$.

Proof: The case m = 1 has been settled by Corollary I.2.1. We shall therefore assume m ≥ 2. We avail ourselves of Theorem I.4.2. For each j = 1,2, let $[g_j] \in H^{m,m-1}(B_{01}\backslash K_j)$ be such that $\mu = \mu_{g_j}$. Representatives $g_j \in \mathcal{C}^\infty(B_{01}\backslash K_j;\Lambda^{m,m-1})$ define, by restriction, $\bar\partial$—closed (m,m—1)—forms in $B_{01}\backslash(K_1 \cup K_2)$. In turn the latter define analytic functionals carried by $K_1 \cup K_2$, both perforce equal to μ, as one sees by evaluating them on any entire function. We apply Theorem I.4.1 with $K = K_1 \cup K_2$ and conclude that, in $B_{01}\backslash(K_1 \cup K_2)$, $g_1 = g_2 + du$ with $u \in \mathcal{C}^\infty(B_{01}\backslash(K_1 \cup K_2);\Lambda^{m,m-2})$. Let $\Omega_j = B_{01}\backslash K_j$, j = 1,2. Select $\phi_j \in \mathcal{C}^\infty(\Omega_1 \cup \Omega_2)$, supp $\phi_j \subset \Omega_j$ (j = 1,2), $\phi_1 + \phi_2 \equiv 1$ in $\Omega_1 \cup \Omega_2$. If we define

$$u_1 = \phi_2 u \text{ in } \Omega_1 \cap \Omega_2,\ u_1 = 0 \text{ in } \Omega_1 \backslash (\Omega_2 \cap \Omega_2),$$

$$u_2 = \phi_1 u \text{ in } \Omega_1 \cap \Omega_2,\ u_2 = 0 \text{ in } \Omega_2 \backslash (\Omega_2 \cap \Omega_2),$$

then $u_j \in \mathcal{C}^\infty(B_{01} \backslash K_j;\ \Lambda^{m,m-2})$, $j = 1,2$, and $u = u_1 + u_2$ in $\Omega_1 \cap \Omega_2$. Finally define $\tilde{g}_1 = g_1 - du_1$ in $B_{01} \backslash K_1$; $\tilde{g}_2 = g_2 + du_2$ in $B_{01} \backslash K_2$. Since $\tilde{g}_1 = \tilde{g}_2$ in $B_{01} \backslash (K_1 \cup K_2)$ there is $g \in \mathcal{C}^\infty(B_{01} \backslash (K_1 \cap K_2); \Lambda^{m,m-1})$ whose restriction to $B_{01} \backslash K_j$ is equal to $\tilde{g}_j$ for each $j = 1,2$. Moreover $\bar{\partial} g \equiv 0$ in $B_{01} \backslash (K_1 \cap K_2)$. Thus g defines a class $[g] \in H^{m,m-1}(B_{01} \backslash (K_1 \cap K_2))$ whose restriction map to $B_{01} \backslash K_j$ is equal to $[g_j]$ $(j = 1,2)$. If $\chi \in \mathcal{C}_c^\infty(B_{01} \backslash (K_1 \cup K_2))$, $\chi \equiv 1$ in a neighborhood of $K_1 \cup K_2$, we have $<\mu,h> = \int h(\bar{\partial}\chi) \wedge g$. This means that $\mu_g = \mu$, thereby proving that μ is carried by $K_1 \cap K_2$. □

COROLLARY I.4.1.— *Let $\mu \in \mathcal{O}'(\mathbb{C}^m)$ be carried by Ω. The intersection of all the compact subsets of Ω that carry μ is a compact subset K_μ of Ω that also carries μ.*

In the sequel we shall refer to K_μ as the *support* of μ in Ω. Instead of K_μ we shall write $\operatorname{supp}_\Omega \mu$, or simply $\operatorname{supp} \mu$ if there is no danger of confusion.

REMARK I.4.2.— When $\mathcal{X} = \mathbb{R}^m$ all the preceding results are valid without limitation of the size of Ω (cf. Remarks I.3.1, I.3.2). Thus the support in $\mathbb{R}^m$ of an analytic functional carried by real space is well defined. □

What we must prove next is the decomposition of an analytic functional μ carried by the union of two compact subsets K_1 and K_2 of Ω, namely the fact that $\mu = \mu_1+\mu_2$, with μ_j carried by K_j (j = 1,2). In order to prove this we need the following result from [MALGRANGE, 1955]:

THEOREM I.4.3.— *Whatever the open subset* $\mathcal{A}$ *of* $\mathbb{C}^m$ *and the integer* p, $0 \leq p \leq m$,

$$\bar{\partial}\mathcal{C}^\infty(\mathcal{A};\Lambda^{p,m-1}) = \mathcal{C}^\infty(\mathcal{A};\Lambda^{p,m}).$$

Proof: Let $f = \sum_{|I|=p} f_I(z)dz_I\wedge d\bar{z}_1\wedge\cdots\wedge d\bar{z}_m \in \mathcal{C}^\infty(\mathcal{A};\Lambda^{p,m})$. We make use of Lemma I.4.1 below: for each I, $|I| = p$, we can solve the equation $\Delta u_I = f_I$ with $u_I \in \mathcal{C}^\infty(\mathcal{A})$. Then the equation $\bar{\partial}v = f$ in $\mathcal{A}$ has the solution

$$v = 4 \sum_{|I|=p} \sum_{j=1}^{m} (-1)^{p+j-1}(\partial u_I/\partial z_j)dz_I\wedge d\bar{z}_1\wedge\cdots\wedge d\bar{z}_{j-1}\wedge\bar{z}_{j+1}\wedge\cdots\wedge d\bar{z}_m.$$

□

LEMMA I.4.1.— *Whatever the open set* $\mathcal{A} \subset \mathbb{R}^N$, *the Laplace operator* Δ *maps* $\mathcal{C}^\infty(\mathcal{A})$ *onto itself.*

Proof: We first consider the case of a bounded open set $\mathcal{A}$. Let $K \subset \mathcal{A}$ be compact and such that no connected component of $\mathcal{A}\backslash K$ is relatively compact in $\mathcal{A}$. We claim that each harmonic function in some open neighborhood of K is the uniform limit on K of a sequence of harmonic functions in $\mathcal{A}$. By the Hahn—Banach theorem we must show that every Radon measure μ supported by K and orthogonal to all harmonic functions in $\mathcal{A}$ is also orthogonal to every function harmonic in a neighborhood of K.

Let $E(x)$ be a fundamental solution of Δ; the potential

$$U(x) = \int E(x-y)d\mu(y)$$

is harmonic in $\mathbb{R}^N\backslash K$. Furthermore, for any $x \in \mathbb{R}^N\backslash K$ and any $\alpha \in \mathbb{Z}^N_+$,

$$U^{(\alpha)}(x) = \int E^{(\alpha)}(x-y)d\mu(y).$$

If $x \notin \mathcal{A}$, $E^{(\alpha)}(x-y)$ is a harmonic function of y in $\mathcal{A}$, and therefore, by the defining property of μ, $U^{(\alpha)}(x) = 0$. We conclude that $U(x)$ vanishes to infinite order in $\mathbb{R}^N \backslash \mathcal{A}$. But since $\mathcal{A}$ is bounded, the closure of each connected component of $\mathcal{A} \backslash K$ intersects $\mathbb{R}^N \backslash \mathcal{A}$. As a consequence of this and of the unique continuation property of harmonic functions we conclude that $U \equiv 0$ in $\mathbb{R}^N \backslash K$. But then $\mu = \Delta U$ is orthogonal to every function that is harmonic in some open neighborhood of K.

Next consider a possibly unbounded open set $\mathcal{A}$. Select a sequence of compact subsets $\{K_\nu\}_{\nu=1}^{+\infty}$ of $\mathcal{A}$ having the following properties: $\mathcal{A} = \bigcup_{\nu=1}^{+\infty} K_\nu$; for each $\nu = 1,2,\ldots$, $K_\nu \subset \mathcal{Int} K_{\nu+1}$ and no connected component of $\mathcal{A} \backslash K_\nu$ has a compact closure contained in $\mathcal{A}$. In order to complete the proof it suffices to show that to any $f \in C^\infty(\mathcal{A})$ there is a sequence $\{u_\nu\}_{\nu=1,2,\ldots}$ in $C^\infty(\mathcal{A})$ such that: (a) $\Delta u_\nu = f$ in a neighborhood of $K_{\nu+1}$; (b) $|u_\nu - u_{\nu-1}| \leq 2^{-\nu}$ in K_ν. For then the series $u_1 + \sum_{\nu=1}^{\infty} (u_{\nu+1} - u_\nu)$ converges uniformly on K_ν for every ν to a distribution solution $u \in C^0(\mathcal{A})$ of $\Delta u = f$ in $\mathcal{A}$. The hypo–ellipticity of the Laplacian (see e.g. Theorem 2.1 in [TREVES, 1975]) entails $u \in C^\infty(\mathcal{A})$.

Let $\chi_\nu \in C_c^\infty(\mathcal{Int} K_{\nu+2})$, $\chi_\nu \equiv 1$ in a neighborhood of $K_{\nu+1}$, and set $v_\nu = E \star (\chi_\nu f)$, $\nu = 1,2,\ldots$. Define $u_1 = v_1$ and if $\nu_0 > 1$ suppose there are $u_\nu \in C^\infty(\mathcal{A})$, $\nu < \nu_0$, with Properties (a) and (b) above. Then $\Delta(v_{\nu_0} - u_{\nu_0 - 1}) = 0$ in a neighborhood of K_{ν_0}. The first part of the proof tells us that there is a harmonic function h_{ν_0} in $\mathcal{Int} K_{\nu_0+2}$ satisfying $|v_{\nu_0} - u_{\nu_0-1} - h_{\nu_0}| < 2^{-\nu_0}$ in K_{ν_0}. Then $u_{\nu_0} = v_{\nu_0} - \chi_{\nu_0} h_{\nu_0}$ satisfies

$\Delta u_{\nu_0} = f$ in a neighborhood of K_{ν_0+1}; and in K_{ν_0}, $|u_{\nu_0} - u_{\nu_0-1}| < 2^{-\nu_0}$.

□

We are now in a position to generalize the second part of Corollary I.2.1:

PROPOSITION I.4.2.— *Let* K_1 *and* K_2 *be two compact subsets of* $\Omega = Z(B_0)$. *If* $\mu \in \mathcal{O}'(\mathbb{C}^m)$ *is carried by* $K_1 \cup K_2$ *then there are two analytic functionals* $\mu_j \in \mathcal{O}'(\mathbb{C}^m)$ *carried by* K_j $(j = 1,2)$ *such that* $\mu = \mu_1 + \mu_2$.

Proof: Select $[g] \in H^{m,m-1}(B_{01} \backslash (K_1 \cup K_2))$ such that $\mu = \mu_g$ (see Theorems I.2.3 and I.4.2, also Remark I.4.1). We decompose g in exactly the same fashion as we have decomposed u in the proof of Proposition I.4.1:

$$g = g_1 + g_2, \text{ with } g_1 = \phi_2 g,\ g_2 = \phi_1 g \text{ in } \Omega_1 \cap \Omega_2,\ g_j = 0 \text{ in } \Omega_j \backslash (\Omega_1 \cap \Omega_2);$$

$$g_j \in C^\infty(B_{01} \backslash K_j; \Lambda^{m,m-1})\ (j = 1,2).$$

Thanks to the fact that $\bar\partial g = 0$ we have $\bar\partial g_1 = -\bar\partial g_2$ in $B_{01} \backslash (K_1 \cup K_2)$. This allows us to define a differential form $f \in C^\infty(B_{01} \backslash (K_1 \cap K_2); \Lambda^{m,m})$ by the equations $f = -\bar\partial g_1$ in $B_{01} \backslash K_1$, $f = \bar\partial g_2$ in $B_{01} \backslash K_2$. By Theorem I.4.3 there is $v \in C^\infty(B_{01} \backslash (K_1 \cap K_2); \Lambda^{m,m-1})$ such that $\bar\partial v = f$. Let v_j denote the restriction of $(-1)^j v$ to $B_{01} \backslash K_j$ $(j = 1,2)$. Notice that $v_1 + v_2 = 0$ in $B_{01} \backslash (K_1 \cup K_2)$. We have $\bar\partial (g_j - v_j) = 0$ in $B_{01} \backslash K_j$: the class $[g_j - v_j]$ defines an element of $H^{m,m-1}(B_{01} \backslash K_j)$ and therefore, by Theorems I.2.3 and I.4.1 (see also Remark I.4.1), an analytic functional μ_j carried by K_j. Let $\chi \in C_c^\infty(B_{01})$, $\chi \equiv 1$ in some neighborhood of $K_1 \cup K_2$. If $h \in \mathcal{O}(\mathbb{C}^m)$, we have

$$<\mu,h> = \int h(\bar{\partial}\chi)\wedge g = \int h(\bar{\partial}\chi)\wedge(g_1-v_1) + \int h(\bar{\partial}\chi)\wedge(g_2-v_2) =$$

$$<\mu_1,h> + <\mu_2,h>,$$

ie., $\mu = \mu_1 + \mu_2$. □

Let $\mathcal{E}_\nu$ denote the operator in (1.3.4). If $u \in \mathcal{C}^\infty(\Omega)$ and $\chi \in \mathcal{C}^\infty_c(\Omega)$ then $\mathcal{E}_\nu(\chi u)\circ Z \to \chi u$ in $\mathcal{C}^\infty(\Omega)$ as $\nu \to +\infty$ (Section II.2, [TREVES, 1992]). This proves that the restrictions to Ω of the entire functions are dense in $\mathcal{C}^\infty(\Omega)$. Transposing the restriction map to Ω yields an injection of the space of compactly supported distributions in Ω, $\mathcal{E}'(\Omega)$, into the space $\mathcal{O}'(\Omega)$ of analytic functionals in $\mathbb{C}^m$ carried by Ω.

THEOREM I.4.4.— *Let $\Omega = Z(B_0) \subset \mathcal{X}$. The support of any compactly supported distribution u in Ω is equal to its support as an analytic functional carried by Ω.*

Proof: If $u \in \mathcal{E}'(\Omega)$ defines an analytic functional μ_u carried by Ω it is obvious that μ_u is carried by supp u; thus, by Corollary I.4.1, we have K = supp $\mu_u \subset$ supp u. Let now $\phi \in \mathcal{C}^\infty_c(\Omega\backslash K)$ be arbitrary; we know that, as $\nu \to +\infty$, $(\mathcal{E}_\nu\phi)\circ Z \to \phi$ in $\mathcal{C}^\infty(\Omega)$ whence

$$<u,\phi> = \lim_\nu <u,(\mathcal{E}_\nu\phi)\circ Z> = \lim_\nu <\mu_u,\mathcal{E}_\nu\phi>.$$

By virtue of (I.3.3), if $z = Z(x) \in K$ and $z' = Z(x') \in$ supp ϕ, then

$$\mathcal{R}e<z-z'>^2 = |x-x'|^2 - |\Phi(x)-\Phi(x')|^2 \geq \tfrac{3}{4}[\mathrm{dist}(K,\mathrm{supp}\ \phi)]^2.$$

It follows that there is an open neighborhood U of K in $\mathbb{C}^m$ such that

$$\mathcal{R}e\langle z-z'\rangle^2 \geq c > 0, \ \forall\, z \in \mathrm{U},\ z' \in \operatorname{supp} \phi.$$

The definition (I.3.4) of $\mathcal{E}_\nu$ shows that $\mathcal{E}_\nu\phi$ converges to zero uniformly in U, and therefore $\langle\mu_u, \mathcal{E}_\nu\phi\rangle \to 0$, which implies $\langle u,\phi\rangle = 0$. □

I.5 HYPERFUNCTIONS ON A MAXIMALLY REAL SUBMANIFOLD OF COMPLEX SPACE. HYPERFUNCTIONS ON A MANIFOLD EQUIPPED WITH A MAXIMAL HYPO—ANALYTIC STRUCTURE

We shall say that an open subset U of the maximally real submanifold $\mathcal{X}$ of $\mathbb{C}^m$ is *holomorphically small* if the closure of U, $\mathcal{C}\ell$U, is compact and contained in an open subset Ω of $\mathcal{X}$ of the type considered in the previous sections: possibly after an affine change of the complex coordinates in $\mathbb{C}^m$, $\Omega = \mathcal{X} \cap (B_{01}) = Z(B_0)$. As before $Z(x) = x + i\Phi(x)$; the map Φ: $B_0 \to B_1$ is C^∞; $\Phi(0) = 0$, $d\Phi(0) = 0$; (I.3.3) hold.

Note that every bounded open subset Ω of $\mathbb{R}^m$ is holomorphically small.

DEFINITION I.5.1.— *Let* U *be a holomorphically small open subset of* $\mathcal{X}$. *We denote by* $\mathcal{B}$(U) *the quotient space* $\mathcal{O}'(\mathcal{C}\ell\mathrm{U})/\mathcal{O}'(\partial\mathrm{U})$. *An element of* $\mathcal{B}$(U) *is called a* ***hyperfunction*** *in* U.

We have denoted by ∂U the boundary of U relative to $\mathcal{X}$.

REMARK I.5.1.— Both spaces $\mathcal{O}'(\mathcal{C}\ell\mathrm{U})$ and $\mathcal{O}'(\partial\mathrm{U})$ have natural Fréchet space structures. But $\mathcal{B}$(U) cannot be equipped, in any natural manner with a Hausdorff locally convex topology. For $\mathcal{O}'(\partial\mathrm{U})$ is not closed in $\mathcal{O}'(\mathcal{C}\ell\mathrm{U})$. As a matter of fact, $\mathcal{O}'(\partial\mathrm{U})$ is dense in $\mathcal{O}'(\mathcal{C}\ell\mathrm{U})$, as implied by the following

LEMMA I.5.1.— *Let* V *be an open subset of* U *such that no connected component of* U\V *is compact. Then the natural injection* $\mathcal{O}'(\partial\mathrm{U}) \to$

$\mathcal{O}'((\mathcal{C}\ell U)\backslash V)$ *has a dense image.*

Proof: By the Hahn–Banach theorem (and the reflexivity of the spaces $\mathcal{O}(K)$) the claim is equivalent to the injectivity of the restriction map $\mathcal{O}((\mathcal{C}\ell U)\backslash V) \to \mathcal{O}(\partial U)$; we must show that if a holomorphic function h in an open neighborhood of $(\mathcal{C}\ell U)\backslash V$ vanishes identically in some neighbor–hood of ∂U then it vanishes identically in a neighborhood of $(\mathcal{C}\ell U)\backslash V$. But this is evident by the unique continuation of holomorphic functions and since each connected component of the latter intersects ∂U. □

If U is holomorphically small and if V is an open subset of U we can define the restriction map ρ^U_V: $\mathcal{B}(U) \to \mathcal{B}(V)$ as follows. Let $\mu \in \mathcal{O}'(\mathcal{C}\ell U)$ represent $u \in \mathcal{B}(U)$. The sets $K_1 = \mathcal{C}\ell V$ and $K_2 = (\mathcal{C}\ell U)\backslash V$ are compact. Proposition I.4.2 allows us to decompose $\mu = \mu_1+\mu_2$ with $\mu_j \in \mathcal{O}'(K_j)$, $j = 1,2$. Suppose we changed representative and decomposition, to $\mu' = \mu'_1+\mu'_2$; then $\mu-\mu' \in \mathcal{O}'(\partial U)$ and $\mu_1- \mu'_1 = \mu - \mu' - (\mu_2-\mu'_2) \in \mathcal{O}'(K_2)$. Proposition I.4.1 entails $\mu_1- \mu'_1 \in \mathcal{O}'(K_1\cap K_2)$. Since $K_1\cap K_2 = \partial V$ μ_1 defines an element of $\mathcal{B}(V)$ which depends solely on u, and which we set to be $\rho^U_V u$. The restriction map ρ^U_V: $\mathcal{B}(U) \to \mathcal{B}(V)$ is linear.

We leave as an exercise to the reader the proof of the fact that, if W is a another open subset of $\mathcal{X}$, such that $W \subset V \subset U$, then $\rho^U_W = \rho^U_V \circ \rho^V_W$.

Because U is holomorphically small we can talk of the support of an analytic functional carried by $\mathcal{C}\ell U$ (Corollary I.4.1). With u, μ and μ' as above we note that $\text{supp}(\mu-\mu') \subset \partial U$. As a consequence, the set $U\cap(\text{supp}\ \mu)$ is independent of the choice of the representative μ. We call it the *support* of $u \in \mathcal{B}(U)$ and denote it by supp u.

By the same token one sees that

(I.5.1) *if* $V \subset U$, $\rho_V^U u = 0 \Leftrightarrow (\text{supp } u) \cap V = \emptyset$.

If K is a compact subset of U we call $\mathcal{B}(K)$ the linear subspace of $\mathcal{B}(U)$ consisting of the hyperfunctions u in U such that supp $u \subset K$. The quotient map $\mathcal{O}'(\mathscr{C}\ell U) \to \mathcal{B}(U)$ induces a linear map $\mathcal{O}'(K) \to \mathcal{B}(K)$. The latter map is a bijection. Indeed, let $u \in \mathcal{B}(K)$ and let $\mu \in \mathcal{O}'(\mathscr{C}\ell U)$ be one of its representatives; write $\mu = \mu_1 + \mu_2$ with supp $\mu_1 \subset K$, supp $\mu_2 \subset \partial U$. If there is another representative $\mu' = \mu_1' + \mu_2'$ of u, with supp $\mu_1' \subset K$, supp $\mu_2' \subset \partial U$, then supp $(\mu_1 - \mu_1') \subset \partial U$. Since also supp $(\mu_1 - \mu_1') \subset K$ it follows from Proposition I.4.1 that $\mu_1 = \mu_1'$. The map $u \to \mu_1$ is the inverse of the natural map $\mathcal{O}'(K) \to \mathcal{B}(K)$. In the sequel, we shall always identify $\mathcal{B}(K)$ to $\mathcal{O}'(K)$.

The next result represents the first step in proving that the presheaf $\{\mathcal{B}(U), \rho_V^U\}$ is in fact a sheaf. Throughout this we shall always assume that U is holomorphically small. The result involves a special kind of exhausting sequences $\{U_j\}_{j=1,2,\ldots}$ of open subsets of U. As usual we assume that $U = \bigcup_{j=1}^{+\infty} U_j$ and that $U_j \subset U_{j+1}$ for each j = 1,2,.... But now we shall also put the following requirement:

(I.5.2) $\forall$ j = 1,2,..., *no connected component of* $U \backslash U_j$ *is compact.*

LEMMA I.5.2.— *Let be given, for each* j = 1,2,..., *a hyperfunction* $f_j \in \mathcal{B}(U_j)$ *whose restriction to* U_k *is equal to* f_k, *for every* $k \leq j$. *Then there is a hyperfunction* f *in* U *whose restriction to* U_j *is equal to* f_j *for every* j = 1,2,....

Proof: According to Lemma I.5.1, whatever $j = 1,2,...$, the natural injections

$$\mathcal{O}'(\partial U) \subset \mathcal{O}'((\mathcal{C}\ell U)\backslash U_j) \subset \mathcal{O}'((\mathcal{C}\ell U)\backslash U_{j-1}) \subset \cdots \subset \mathcal{O}'((\mathcal{C}\ell U)\backslash U_1)$$

have dense images. Given any $\mu \in \mathcal{O}'((\mathcal{C}\ell U)\backslash U_j)$, any $k \leq j$ and any open neighborhood $\mathcal{N}_k$ of μ in $\mathcal{O}'((\mathcal{C}\ell U)\backslash U_k)$, there is $\omega \in \mathcal{O}'(\partial U)\cap\mathcal{N}_k$ with ω independent of $k \leq j$. As a matter of fact, we may assume $\mathcal{N}_k \supset \mathcal{N}_l$ if $1 \leq k \leq l \leq j$: it suffices to replace $\mathcal{N}_2$ by $\mathcal{N}_2\cap\mathcal{N}_1$, $\mathcal{N}_3$ by $\mathcal{N}_3\cap\mathcal{U}_2\cap\mathcal{N}_1$, etc. [This is permitted because the inclusion maps are continuous.] In such a set up we select $\omega \in \mathcal{O}'(\partial U)\cap\mathcal{N}_j$.

It is convenient to define the topology of the Fréchet space $\mathcal{O}'((\mathcal{C}\ell U)\backslash U_j)$ by means of a translation invariant metric d_j. For each $j = 1,2,...$, select a representative $\mu_j \in \mathcal{O}'(\mathcal{C}\ell U_j)$ of $f_j \in \mathcal{B}(U_j)$. Since the restrictions of f_j and of f_{j+1} to U_j coincide, we get

$$\mu_{j+1}-\mu_j \in \mathcal{O}'((\mathcal{C}\ell U_{j+1})\backslash U_j) \subset \mathcal{O}'((\mathcal{C}\ell U)\backslash U_j).$$

By the discussion that precedes we can select $\omega_j \in \mathcal{O}'(\partial U)$ such that

$$d_k(\mu_{j+1}-\mu_j,\omega_j) < 2^{-j},\ k = 1,...,j. \tag{I.5.3}$$

We may then define the following analytic functionals supported in $\mathcal{C}\ell U$: $\mu'_1 = \mu_1$ and if $j \geq 2$, $\mu'_j = \mu_j - \sum_{k=1}^{j-1} \omega_k$. By the triangle inequality and (I.5.3) we have $d_1(\mu'_{j+p}-\mu'_j,0) < 2^{-j+1}$ whatever p, j. The fact that the inclusion map $\mathcal{O}'((\mathcal{C}\ell U)\backslash U_1) \to \mathcal{O}'(\mathcal{C}\ell U)$ is continuous entails that $\{\mu'_j\}_{j=1,2,...}$ is a Cauchy sequence and therefore converges to an element $\mu \in \mathcal{O}'(\mathcal{C}\ell U)$. Denote by f the hyperfunction in U defined by μ. It remains to show that the restriction of f to U_j is equal to f_j.

We see that $\mu - \mu_j$ is the limit, as $j \le p \to +\infty$, of the analytic functionals

$$\mu_{p+1} - \mu_j - \sum_{k=1}^{p} \omega_k = \sum_{q=j}^{p} (\mu_{q+1} - \mu_q - \omega_q) - \sum_{k=1}^{j-1} \omega_k .$$

By (I.5.3) we see that $\sum_{\ell=j}^{p} (\mu_{\ell+1} - \mu_\ell - \omega_\ell)$ converges in $\mathcal{O}'((\mathcal{Cl}\,U)\backslash U_j)$, which proves that $\mu - \mu_j \in \mathcal{O}'((\mathcal{Cl}\,U)\backslash U_j)$. □

THEOREM I.5.1.— *Let* U *be a holomorphically small open subset of* $\mathcal{X}$ *and let* $\{U_\iota\}_{\iota \in I}$ *be an open covering of* U*. Suppose given, for each* $\iota \in I$*, a hyperfunction* $f_\iota \in \mathcal{B}(U_\iota)$ *such that, for each pair of indices* $\iota, \kappa \in I$*, the restrictions to* $U_\iota \cap U_\kappa$ *of* f_ι *and* f_κ *are equal. Then there exists a unique hyperfunction* $f \in \mathcal{B}(U)$ *whose restriction to* U_ι *is equal to* f_ι*, for every* $\iota \in I$.

Proof: The uniqueness is evident: if g is a hyperfunction in U whose restriction to each U_ι is equal to zero, the support of any representative $\nu \in \mathcal{O}'(\mathcal{Cl}\,U)$ does not intersect any U_ι, hence $\operatorname{supp} \nu \subset \partial U$, ie., $g = 0$.

To prove the existence of f we begin by looking at the case where the index set I consists of two elements only: $I = \{1,2\}$. Call $v \in \mathcal{B}(U_1 \cap U_2)$ the restriction of f_j to $U_1 \cap U_2$; let $\mu_j \in \mathcal{O}'(\mathcal{Cl}\,U_j)$, $\nu \in \mathcal{O}'(\mathcal{Cl}(U_1 \cap U_2))$ be representatives of f_j ($j = 1,2$) and v respectively. We have

$$\operatorname{supp}(\mu_j - \nu) \subset (\mathcal{Cl}\,U_j)\backslash(U_1 \cap U_2) = \mathcal{Cl}(U_j \backslash (U_1 \cap U_2)) \cup \partial U_j .$$

As a consequence we may write

$$\mu_j - \nu = \alpha_j + \beta_j$$

with

$$\alpha_j \in \mathcal{O}'(\mathcal{C}\ell(U_j\backslash(U_1\cap U_2))),\ \beta_j \in \mathcal{O}'(\partial U_j).$$

The hyperfunction f in the statement will be the one defined by

$$\mu = \nu + \alpha_1 + \alpha_2 \in \mathcal{O}'(\mathcal{C}\ell(U_1\cup U_2)).$$

We must show that

$$\mathrm{supp}(\mu-\mu_j) \subset [\mathcal{C}\ell(U_1\cup U_2)]\backslash U_j.$$

We have

$$\mu-\mu_1 = \alpha_2-\beta_1,$$

$$\mathrm{supp}\ \alpha_2 \subset \mathcal{C}\ell(U_2\backslash(U_1\cap U_2)) \subset (\mathcal{C}\ell(U_1\cup U_2))\backslash U_1,$$

$$\mathrm{supp}\ \beta_1 \subset \partial U_1 \subset (\mathcal{C}\ell(U_1\cup U_2))\backslash U_1.$$

The analogous property holds for $\mu-\mu_2 = \alpha_1-\beta_2$,. This proves the result when $I = \{1,2\}$, and then also, of course, for any finite index set I.

Back to the general case, first we replace $\{U_\iota\}_{\iota\in I}$ by a countable subcovering, ie., we assume $I = \mathbb{Z}_+$. For any integer $j_0 \geq 1$ set

$$V_{j_0} = U_1\cup\ldots\cup U_{j_0}\cup K_{j_0}$$

where K_{j_0} is the union of all the compact connected components of $U\backslash(U_1\cup\cdots\cup U_{j_0})$; V_{j_0} is an open subset of U and we may extract from the sequence of sets $\{U_j\cap V_{j_0}\}_{j=1,2,\ldots}$ a finite open covering of V_{j_0}. Applying what precedes to V_{j_0} we see that there is a hyperfunction f_{j_0} in V_{j_0} whose restriction to every set $U_j\cap V_{j_0}$ is equal to the restriction of f_j (regardless of the value of j thanks to uniqueness). Again thanks to uniqueness we can claim that $f_{j_0+1} \equiv f_{j_0}$ in V_{j_0} for all j_0. Replacing U_j by V_j brings us to the case in which no set $U\backslash U_j$ has any compact component. Theorem I.5.1 follows then at once from Lemma I.5.2. □

LEMMA I.5.3.— *Let* U *be holomorphically small and* V ⊂ U *be an open subset of* $\mathcal{X}$. *The restriction map* ρ_V^U *is a surjection of* $\mathcal{B}$(U) *onto* $\mathcal{B}$(V).

Proof: If we regard a representative $\nu \in \mathcal{O}'(\mathscr{C}\ell V)$ of $g \in \mathcal{B}(V)$ as an element of $\mathcal{O}'(\mathscr{C}\ell U)$ it defines a hyperfunction in U whose restriction to V is equal to g. □

The time has now come to define the *sheaf of hyperfunctions* over $\mathcal{X}$, $\mathscr{B}_{\mathcal{X}}$ (called $\mathscr{B}$ when there is no risk of confusion), ie., the *sheaf of germs of hyperfunctions* in $\mathcal{X}$: the stalk $\mathscr{B}_z$ of $\mathscr{B}$ at an arbitrary point $z \in \mathcal{X}$, will be the inductive (direct) limit of the pairs $\{\mathcal{B}(U), \rho_V^U\}$ with U and V ⊂ U holomorphically small open neighborhoods of z in $\mathcal{X}$. A germ $\dot{f} \in \mathscr{B}_z$ is an equivalence class of elements $f_U \in \mathcal{B}(U)$, where U ranges over the set of all holomorphically small open neighborhoods of z in $\mathcal{X}$; if U′ is another such neighborhood, and $f_{U'}$ the corresponding representative of $\dot{f}$, there is an open neighborhood of z in $\mathcal{X}$, V ⊂ U∩U′, such that $\rho_V^U f_U = \rho_V^{U'} f_{U'}$. The definition of $\mathscr{B}_z$ does not depend on the choice of the local representation $B_0 \ni x \to Z(x) \in B_0 + iB_1$ since $\mathcal{O}'(\mathscr{C}\ell U)$ and $\mathcal{O}'(\partial U)$ are well defined vector subspaces of $\mathcal{O}'(\mathbb{C}^m)$.

For U holomorphically small we know, by Theorem I.5.1 and by general sheaf theory, that there is a "canonical" isomorphism

$$\Gamma(U, \mathscr{B}) \cong \mathcal{B}(U) = \mathcal{O}'(U)/\mathcal{O}'(\partial U).$$

It is therefore natural to define the hyperfunctions in an arbitrary open subset of $\mathcal{X}$ as the continuous sections of the sheaf $\mathscr{B}$ over that open subset:

DEFINITION I.5.2.— *Let* U *be an open subset of the maximally real submanifold* $\mathcal{X}$ *of* $\mathbb{C}^m$. *A continuous section of the sheaf* $\mathcal{B}$ *over* U *is called a* ***hyperfunction in*** U. *The space of hyperfunctions in* U [that is, $\Gamma(U,\mathcal{B})$] *is* (also) *denoted by* $\mathcal{B}(U)$.

A hyperfunction in U can be thought of as a collection $\{f_\iota, U_\iota\}_{\iota\in I}$: for each index $\iota \in I$ we are given the element U_ι of a covering of U by holomorphically small open subsets of $\mathcal{X}$, and an element f_ι of $\mathcal{B}(U_\iota)$ such that, for any $\kappa \in I$, the restrictions to $U_\iota \cap U_\kappa$ of f_ι and f_κ are equal. A different collection $\{f'_{\iota'}, U'_{\iota'}\}_{\iota'\in I'}$ defines the same hyperfunction in U if the joint collection $\{f_\iota, f'_{\iota'}, U_\iota, U'_{\iota'}\}_{\iota\in I, \iota'\in I'}$ defines a hyperfunction in U (ie., if the restrictions of the hyperfunctions in the joint collection agree on overlaps).

THEOREM I.5.2.— *The sheaf* $\mathcal{B}$ *is flabby.*

Proof: Given an arbitrary open subset U of $\mathcal{X}$ we must show that the restriction map $\mathcal{B}(\mathcal{X}) \to \mathcal{B}(U)$ is surjective. Let $f \in \mathcal{B}(U)$ and call $\mathcal{F}$ the family of all pairs (V,g) with $U \subset V$ and $g \in \mathcal{B}(V)$ whose restriction to U is equal to f . We define on $\mathcal{F}$ the obvious (partial) order: $(V_1,g_1) \preceq (V_2,g_2)$ means that $V_1 \subset V_2$ and that the restriction of g_2 is equal to g_1 in V_1. Every chain (ie., totally ordered subfamily) $\mathcal{F}_1$ of $\mathcal{F}$ has an upper bound: the pair (V_0,g_0) such that V_0 is the union of all the open sets V in the pairs $(V,g) \in \mathcal{F}_1$ and g_0 is the hyperfunction in V_0 whose restriction to each such set V is equal to the corresponding hyperfunction g in V. By Zorn's lemma there is a maximal element $(\tilde{V},\tilde{f})$ in $\mathcal{F}$.

Suppose $\tilde{V} \neq \mathcal{X}$ and let z_0 be a point in $\mathcal{X}\backslash\tilde{V}$. Select a holomorphically small W open neighborhood of z_0 in $\mathcal{X}$. By Lemma I.5.3 there is $g \in \mathcal{B}(W)$ whose restriction to $\tilde{V}\cap W$ is equal to $\tilde{f}$. But then there is $\tilde{\tilde{f}} \in \mathcal{B}(\tilde{V}\cup W)$ whose restriction to $\tilde{V}$ is equal to $\tilde{f}$, contradicting the maximality of $(\tilde{V},\tilde{f})$. □

When U is an arbitrary open subset of $\mathcal{X}$ we define the support (in U) of a hyperfunction $f \in \mathcal{B}(U)$ as the support of the section f of $\mathscr{B}$. The reader will check easily that this notion coincides with the notion of support as defined earlier, when U is holomorphically small.

Let now $\mathscr{D}'$ denote the sheaf of germs of distributions in $\mathcal{X}$, and $\mathcal{D}'(\mathcal{X})$ the space of distributions in $\mathcal{X}$ (ie., the space of sections of $\mathscr{D}'$ over $\mathcal{X}$).

THEOREM I.5.3.— *There is a natural sheaf homomorphism $\mathscr{D}' \to \mathscr{B}$; it is injective.*

Proof: Let $z_0 \in \mathcal{X}$ and let $\mathscr{D}'_{z_0}$ denote the space of the germs at z_0 of distributions on $\mathcal{X}$. Let U be a holomorphically small open neighborhood of z_0 in $\mathcal{X}$ in which a germ $\dot{u} \in \mathscr{D}'_{z_0}$ has a representative u. Select a cutoff function $\chi \in C_c^\infty(U)$, $\chi \equiv 1$ in a neighborhood of z_0 in U. Then $\chi u \in \mathcal{E}'(U)$ may be regarded as an element of $\mathcal{O}'(\mathscr{C}\ell U)$ and therefore defines an element of $\mathcal{B}(U)$. In turn this defines a germ $\tilde{u} \in \mathscr{B}_{z_0}$. Suppose we change the pair (U,u) to (U_1,u_1) with u_1 representing $\dot{u}$ in U_1; let $\chi_1 \in C_c^\infty(U_1)$, $\chi \equiv 1$ in a neighborhood of z_0 in U_1. Then $\chi u - \chi_1 u_1 \in \mathcal{E}'(U_1 \cup U)$ and $z_0 \notin$

$\mathrm{supp}(\chi u - \chi_1 u_1)$. By Theorem I.4.4 z_0 does not belong to the support of the analytic functional $\chi u - \chi_1 u_1$, which entails that the germ of hyperfunction at z_0 defined by $\chi u - \chi_1 u_1$ vanishes. This proves that the map $\mathscr{D}'_{z_0} \ni \dot{u} \to \tilde{u} \in \mathscr{B}_{z^0}$ is well defined. It is injective, again by Theorem I.4.4. □

COROLLARY I.5.1.— *There is a natural linear injection* $\mathcal{D}'(\mathcal{X}) \to \mathcal{B}(\mathcal{X})$; *it preserves the supports.*

The fact that the sheaf of hyperfunctions $\mathscr{B}$ is flabby stands in sharp contrast with the properties of the subsheaf $\mathscr{D}'$. To see this it suffices to look at $\mathcal{X} = \mathbb{R}^m$:

EXAMPLE I.5.1.— Let Ω be any open subset of $\mathbb{R}$, $\Omega \neq \mathbb{R}$. If $x_0 \in \partial\Omega$ and if x_ν $(\nu = 1, 2, \ldots)$ is a sequence of points in Ω converging to x_0 the series $\sum_{\nu=1}^{+\infty} \delta^{(\nu)}(x - x_\nu)$ ($\delta(x)$: Dirac's distribution) converges in $\mathcal{D}'(\Omega)$ but admits no distribution extension to $\mathbb{R}$, since such an extension would be of finite order in a neighborhood of x_0. □

Example I.5.1 was that of a distribution that becomes ever more singular as the boundary is approached, but it is easy to construct $\mathcal{C}^\infty$, or even real–analytic functions in an open set that admit no distribution extensions to the whole space, as shown in

EXAMPLE I.5.2.— Let u_0 denote the distribution in $\mathbb{R}^m \backslash \{0\}$ defined by the $\mathcal{C}^\infty$ function $\exp(1/|x|)$. Suppose there were a distribution u in $\mathbb{R}^m$ that extends u_0. Since locally every distribution has finite order there would then be a constant $C > 0$ and an integer $r \geq 0$ such that, for every $\phi \in C_c^\infty(\mathbb{R}^m)$ such that $\phi(x) = 0$ if $|x| > 1$,

$$\text{(I.5.4)} \qquad |\langle u, \phi\rangle| \leq C \operatorname{Max} \sum_{|\alpha| \leq r} |\partial^\alpha \phi|.$$

In particular, (I.5.4) should be valid if supp $\phi \subset \{ x \in \mathbb{R}^m;\ 0 < |x| \leq 1 \}$, in which case it reads

$$\text{(I.5.5)} \qquad \left|\int \phi(x) e^{1/|x|} dx\right| \leq C \operatorname{Max} \sum_{|\alpha| \leq r} |\partial^\alpha \phi|.$$

Consider $\chi \in C^\infty(\mathbb{R})$, $\chi(s) = 0$ unless $1 < s < 4$, $\chi(s) = 1$ if $2 < s < 3$, $\chi \geq 0$ everywhere. Then, in (I.5.5) put $\phi(x) = \chi(\rho|x|)$ with $\rho > 4$; we get, for some constant $C' > 0$ and all $\rho > 4$,

$$\int \phi(x) e^{1/|x|} dx = |S^{m-1}| \rho^{-m} \int \chi(s) e^{\rho/s} s^{m-1} ds \leq C' \rho^r.$$

But

$$\int \chi(s) e^{\rho/s} s^{m-1} ds \geq \int_2^3 e^{\rho/s} s^{m-1} ds \geq 2^{m-1} e^{\rho/3},$$

whence a contradiction as $\rho \to +\infty$. □

Lastly we discuss the action of the holomorphic differential operators on the hyperfunctions in $\mathcal{X}$. As before let U be a holomorphically small open subset of $\mathcal{X}$. Consider a differential operator

$$P(z, \partial_z) = \sum_{|\alpha| \leq r} c_\alpha(z)(\partial/\partial z)^\alpha$$

with coefficients $c_\alpha \in \mathcal{O}(\mathscr{C}\ell \mathrm{U})$. If a hyperfunction f in U is represented by the analytic functional $\mu \in \mathcal{O}'(\mathscr{C}\ell \mathrm{U})$ we define $P(z, \partial_z)f$ as the hyper—

function defined by the analytic functional $P(z,\partial_z)\mu$. This is a meaningful definition since $\nu \in \mathcal{O}'(\partial U) \Rightarrow P(z,\partial_z)\nu \in \mathcal{O}'(\partial U)$. A similar argument shows that the action of $P(z,\partial_z)$ commutes with restrictions to open sub-sets. As a consequence, if now we assume that all the coefficients c_α are holomorphic in an open neighborhood of $\mathcal{X}$ we obtain, for each $z_0 \in \mathcal{X}$, an endomorphism $P(z,\partial_z)\big|_{z_0}$ of the vector space $\mathcal{B}_{z_0}$. As z_0 ranges over $\mathcal{X}$, these linear maps make up a sheaf homomorphism of $\mathcal{B}$ into itself. The latter defines an endomorphism of $\mathcal{B}(\mathcal{X})$; of course,

(I.5.6) $$\operatorname{supp} P(z,\partial_z)f \subset \operatorname{supp} f, \ \forall f \in \mathcal{B}(\mathcal{X}).$$

Hyperfunctions on a manifold equipped with a maximal hypo–analytic structure

We recall the definition (Definition III.1.4, [TREVES, 1992]) of a maximal hypo–analytic structure on a C^∞ manifold $\mathcal{M}$. We begin by recalling the notion of *a hypo–analytic structure* on $\mathcal{M}$ of rank $m \in \mathbb{Z}_+$ (see section III.1, *loc. cit.*). We are given an open covering $\{U^\alpha\}_{\alpha\in A}$ of $\mathcal{M}$, and for each $\alpha \in A$, a C^∞ map $Z^\alpha : U^\alpha \to \mathbb{C}^m$ such that

(I.5.7) $$dZ_1^\alpha \wedge \cdots \wedge dZ_m^\alpha \neq 0 \textit{ at any point of } U^\alpha;$$

(I.5.8) *For any pair of indices $\alpha, \beta \in A$ such that $U^\alpha \cap U^\beta \neq \emptyset$ there is a holomorphic mapping F_β^α of an open neighbor-hood of $Z^\beta(U^\alpha \cap U^\beta)$ in $\mathbb{C}^m$ onto one of $Z^\alpha(U^\alpha \cap U^\beta)$ such*

$$\textit{that } Z^\alpha = F^\alpha_\beta \circ Z^\beta \textit{ in } U^\alpha \cap U^\beta.$$

A function h in an open set $\Omega \subset \mathcal{M}$ is said to be *hypo–analytic* in Ω if, whatever $\alpha \in A$, each point $\wp \in \Omega \cap U^\alpha$ has a neighborhood $V^\alpha_\wp \subset \Omega \cap U^\alpha$ in which $h = \tilde{h} \circ Z^\alpha$ for some holomorphic function $\tilde{h}$ in an open neighborhood of $Z^\alpha(V^\alpha_\wp)$ in $\mathbb{C}^m$. We can talk of a general *hypo–analytic chart* in $\mathcal{M}$: it is a pair (U,Z) consisting of an open subset U of $\mathcal{M}$ and of a C^∞ map $Z = (Z_1,...,Z_m)$: $U \to \mathbb{C}^m$ which is hypo–analytic, ie., whose components are all hypo–analytic, and satisfy $dZ_1 \wedge \cdots \wedge dZ_m \neq 0$ at every point of U.

The hypo–analytic structure of $\mathcal{M}$ is said to be *maximal* if $m = \dim_{\mathbb{R}} \mathcal{M}$. For the remainder of this section we shall hypothesize that this is the case. Let (U,Z) be a hypo–analytic chart in $\mathcal{M}$. The differential DZ is injective and therefore Z realizes an *immersion* of U into $\mathbb{C}^m$. After contracting U about any one of its points we may assume that Z is a *diffeomorphism* of U onto a C^∞ submanifold of $\mathbb{C}^m$; $Z(U)$ is maximally real, since the pullback to $Z(U)$ of $dz_1 \wedge \cdots \wedge dz_m$ does not vanish at any point. Henceforth this property will be part of our definition of a hypo–analytic chart (U,Z) in $\mathcal{M}$ equipped with a maximal hypo–analytic structure: Z is a hypo–analytic isomorphism of the open set U onto $Z(U)$ when $Z(U)$ is equipped with the (maximal) hypo–analytic structure inherited from $\mathbb{C}^m$. [In general, this cannot be achieved in a hypo–analytic structure of rank $m < \dim_{\mathbb{R}} \mathcal{M}$, as shown by the Mizohata structure on $\mathbb{R}^2$, defined by the single "first integral" $Z = x + iy^2$.]

If (U,Z) and (U,Z') are two hypo–analytic charts in $\mathcal{M}$ with the same domain U, the compose $Z' \circ Z^{-1}$ extends as a biholomorphism of an open neighborhood of $Z(U)$ in $\mathbb{C}^m$ onto one of $Z'(U)$ and yields an iso–

open neighborhood of $Z(\mathrm{U})$ in $\mathbb{C}^{\mathrm{m}}$ onto one of $Z'(\mathrm{U})$ and yields an iso—morphism $\varphi_{Z'}^{Z}$ of the space $\mathcal{B}(Z(\mathrm{U}))$ of hyperfunctions in $Z(\mathrm{U})$ onto $\mathcal{B}(Z'(\mathrm{U}))$. There is a natural equivalence relation on the disjoint union of the spaces $\mathcal{B}(Z(\mathrm{U}))$ as Z ranges over the set of all hypo—analytic isomor—phisms of U onto a maximally real submanifold of $\mathbb{C}^{\mathrm{m}}$: two elements, $u \in \mathcal{B}(Z(\mathrm{U}))$ and $u' \in \mathcal{B}(Z'(\mathrm{U}))$, are equivalent if $u' = \varphi_{Z'}^{Z}(u)$. The quotient space $\mathcal{B}(\mathrm{U})$ is the space of hyperfunctions on U. As U varies these spaces and the obvious restriction mappings define a presheaf, which in turn defines the *sheaf of* (germs of) *hyperfunctions on* $\mathcal{M}$, $\mathcal{B}_{\mathcal{M}}$. The sheaf $\mathcal{B}_{\mathcal{M}}$ is flabby. By definition a hyperfunction in $\mathcal{M}$ will be a continuous section over $\mathcal{M}$ of the sheaf $\mathcal{B}_{\mathcal{M}}$. The *space* of hyperfunctions in $\mathcal{M}$ will be denoted by $\mathcal{B}(\mathcal{M})$.

A special case of a maximal hypo—analytic structure on $\mathcal{M}$ is that of a *real analytic structure*. This means that every point of $\mathcal{M}$ lies in the domain U of a hypo—analytic chart (U,Z) such that $Z(\mathrm{U}) \subset \mathbb{R}^{\mathrm{m}}$ (now with $\mathrm{m} = \dim \mathcal{M}$). In this case our construction recovers the sheaf of hyper—functions on a $\mathcal{C}^{\omega}$ manifold as originally introduced in [SATO, 1960].

I.6 HYPERFUNCTION THEORY, POLYNOMIAL CONVEXITY AND SERRE DUALITY

The purpose of this section and of the next one is to establish a link between the definition of hyperfunctions in section I.5 and some of the more general aspects of the theory of several complex variables. At the end of section I.2 we have raised the question of how to generalize the Cauchy transform to higher dimensions. To this end we have replaced (in section I.3) the concept of a Runge compact subset of the complex plane by that of a compact subset of a holomorphically small open subset of the maximally real submanifold $\mathcal{X}$. But what really underpins the success of our approach is the fact that in the complex plane, the Runge property is equivalent to polynomial convexity; and that every compact subset of a holomorphically small open subset of $\mathcal{X}$ is polynomially convex. We shall now repeat the approach to hyperfunctions from this more general stand–point.

Let us denote by $\mathbb{C}[z]$ the set of polynomials with complex coeffi–cients in the variables $z_1,\dots,z_m$. Throughout this section K will denote a compact subset of $\mathbb{C}^m$. The set

$$\hat{K} = \{\ z \in \mathbb{C}^m;\ \forall\ P \in \mathbb{C}[z],\ |P(z)| \leq \max_K |P|\ \}$$

is called the *polynomial convex hull* of K. Note that it makes no difference whether in the definition of $\hat{K}$ we use the space of polynomials $\mathbb{C}[z]$ or that of all entire functions, $\mathcal{O}(\mathbb{C}^m)$ – since the former is dense in the latter. The set $\hat{K}$ is obviously closed; $\hat{K}$ is contained in the convex hull of K, since $\exp(\mathcal{R}e\, z\cdot\zeta) \leq \max_{w\in K} \exp(\mathcal{R}e\, w\cdot\zeta)$ for all $z \in \hat{K}$ and all $\zeta \in \mathbb{C}^m$. Thus $\hat{K}$ is compact.

EXAMPLE I.6.1.– Let $\mathcal{A}$ be an open and bounded subset of the complex plane, $K = (\mathscr{C}\ell\mathcal{A}) \cap (\mathbb{C}\backslash\mathcal{A})$ its boundary. The maximum principle entails that $\mathcal{A} \subset \hat{K}$. In order that $\hat{K} = \mathcal{A} \cup K$ it is necessary and sufficient that the complement of $\mathscr{C}\ell\mathcal{A}$ not have any bounded component. □

A compact subset K of $\mathbb{C}^m$ is said to be *polynomially convex* if

$$K = \hat{K}.$$

When K is polynomially convex it is possible to select each open set U in a basis of neighborhoods of K, to be a *polynomial polyhedron*, ie., to be defined by finitely many polynomial inequalities: there are polyno— mials $P_1,\ldots,P_\nu \in \mathbb{C}[z]$ such that

$$U = \{ z \in \mathbb{C}^m;\ |P_k(z)| < 1,\ k = 1,\ldots,\nu \}. \tag{I.6.1}$$

Indeed, given any point $z_0 \notin K$, there is a $P \in \mathbb{C}[z]$ such that $|P(z_0)| > 1$ and $|P(z)| \leq 1$ for all $z \in K$. It follows at once that K is equal to the intersection of the closure of all polynomial polyhedrons that contain K.

We recall that $K \subset\subset \mathbb{C}^m$ is said to have the *Cousin property* if, given any integers p, q, $0 \leq p \leq m$, $1 \leq q \leq m$, to every differential form $f \in C^\infty(U;\Lambda^{p,q})$ with U open, $K \subset U$, and $\bar{\partial} f = 0$ in U, there is an open set $V \subset U$, $V \supset K$, and a form $u \in C^\infty(V;\Lambda^{p,q-1})$ such that $\bar{\partial} u = f$ in V [see (I.3.5) and (I.3.6)].

Of great importance in the sequel will be the following classical result of several complex variable theory:

THEOREM I.6.1.— *Any polynomially convex compact subset of $\mathbb{C}^m$ has both the Runge and the Cousin properties.*

Proof: It suffices to show that any polynomial polyhedron of the kind

$$V = \{ z \in \mathbb{C}^m;\ \epsilon_j |z_j| < 1,\ j = 1,...,m,\ |P_k(z)| < 1,\ k = 1,...,\nu \}$$

($\epsilon_j > 0$, j = 1,...,m, $P_k \in \mathbb{C}[z]$, k = 1,...,ν) has the Runge property, as well as the following property (for any pair of integers p, q $\geq$ 0):

(I.6.2) *Given any* $f \in C^\infty(V;\Lambda^{p,q+1})$ *such that* $\bar\partial f = 0$ *in* V *there is* $u \in C^\infty(V;\Lambda^{p,q})$ *such that* $\bar\partial u = f$ *in* V.

Indeed, the sets V form a basis of neighborhoods of K: if U is the open neighborhood of K defined in (I.6.1) then V $\subset$ U; on the other hand, if the numbers ϵ_j are sufficiently small, K $\subset$ V.

We add one complex dimension to the picture and consider the polynomial polyhedron in $\mathbb{C}^{m+1}$,

$$\tilde V = \{ z \in \mathbb{C}^{m+1};\ \epsilon_j |z_j| < 1,\ j = 1,...,m+1,\ |P_k(z')| < 1,\ k = 1,...,\nu-1 \},$$

where $\epsilon_{m+1} = 1$ and $z' = (z_1,...,z_m)$. Notice that the coordinate projection π: $(z',z_{m+1}) \to z'$ induces a biholomorphism of the complex hypersurface $\Sigma = \{ z \in \tilde V;\ z_{m+1} = P_\nu(z') \}$ onto V.

We reason by induction on ν (for any dimension m). When $\nu = 0$, ie. when V is an open polydisk, the claims are classical. We begin by proving (I.6.2); for a proof of (I.6.2) when $\nu = 0$ see, e.g., Theorem 5, Section D, Chapter I in [GUNNING–ROSSI, 1965]. Suppose the result true up to $\nu-1$. Let f^* denote the pullback of f to $\pi^{-1}(V)$ under the projection π. If $\psi \in C^\infty(\tilde V)$, supp $\psi \subset \tilde V \cap \pi^{-1}(V)$ and $\psi \equiv 1$ in a neighborhood of Σ, then $\bar\partial(\psi f^*) = \bar\partial\psi \wedge f^* \in C^\infty(\tilde V;\Lambda^{p,q+2})$ vanishes in a full neighborhood of Σ. The form $[z_{m+1}-P_\nu(z')]^{-1}\bar\partial(\psi f^*)$ is $\bar\partial$–closed in $\tilde V$. By the induction hypothesis we can find $g \in C^\infty(\tilde V;\Lambda^{p,q+1})$ such that

$$F = \psi f^* + [z_{m+1}-P_\nu(z')]g$$

is $\bar\partial$–closed. Using again the induction hypothesis we can solve $\bar\partial U = F$ in

$\tilde{V}$. Pushing down to V the pullback of U to Σ yields the sought form u.

For the Runge property take $f \in \mathcal{O}(V)$; F defined as before (with q = —1) is now a holomorphic function in $\tilde{V}$. By the induction hypthesis F is the limit in $\mathcal{O}(\tilde{V})$ of a sequence of polynomials in $\mathbb{C}^{m+1}$. Pushing down to V the restrictions to Σ of those polynomials yields a sequence of polynomials in $\mathbb{C}^m$ that converges to f in $\mathcal{O}(V)$. □

We derive at once from Theorem I.6.1 that, when K is polynomi— ally convex, the strong dual of $\mathcal{O}(K)$ can be identified to the space $\mathcal{O}'(K)$ of analytic functionals in $\mathbb{C}^m$ carried by K (see end of section I.1).

The following criterion of polynomial convexity will be useful:

PROPOSITION I.6.1.— *If the restriction to* K *maps* $\mathcal{O}(\mathbb{C}^m)$ *onto a dense subspace of* $\mathcal{C}(K)$ *then* K *is polynomially convex.*

Here $\mathcal{C}(K)$ denotes the Banach space of all continuous functions in K, equipped with the maximum norm.

Proof: Let $\mathcal{O}(\mathbb{C}^m)\big|_K$ denote the subspace of $\mathcal{C}(K)$ made up of the restric— tions to K of functions $h \in \mathcal{O}(\mathbb{C}^m)$. Notice that if $h_1, h_2 \in \mathcal{O}(\mathbb{C}^m)$ and if $h_1 \equiv h_2$ in K, then $h_1 \equiv h_2$ in $\hat{K}$ and thus, if $z^* \in \hat{K}$ the formula $\lambda_{z^*}(h\big|_K) = h(z^*)$ defines a linear functional on $\mathcal{O}(\mathbb{C}^m)\big|_K$. Since

$$|\lambda_{z^*}(h\big|_K)| \leq \operatorname*{Max}_K |h|,$$

λ_{z^*} extends as a linear functional to the whole of $\mathcal{C}(K)$. Assume $\mathcal{O}(\mathbb{C}^m)\big|_K$ dense in $\mathcal{C}(K)$; then the extension of λ_{z^*} is unique (we denote it also by λ_{z^*}) and it must be multiplicative:

$$\lambda_{z^*}(f\,g) = \lambda_{z^*}(f)\lambda_{z^*}(g).$$

At this juncture we recall a standard argument about the Banach algebra $\mathcal{C}(K)$. The null space $\mathcal{I}_{z^*}$ of λ_{z^*} is a closed hyperplane, as well as an ideal (perforce maximal). Suppose that to every point $z \in K$ there were an element $g \in \mathcal{I}_{z^*}$ such that $g(z) \neq 0$. We would have $g(w) \neq 0$ for all w in a neighborhood U of z in K. A finite number of such neighborhoods, $U_1,\ldots,U_\nu$, would form a covering of K; for each $j = 1,\ldots,\nu$, let $g_j \in \mathcal{I}_{z^*}$ be such that $g_j(w) \neq 0$ for all $w \in U_j$. Then $f = \sum_{j=1}^{\nu} \bar{g}_j g_j \in \mathcal{I}_{z^*}$; $f(z) \neq 0$ for all $z \in K$; and as a consequence $1/f \in \mathcal{C}(K)$, hence $1 \in \mathcal{I}_{z^*}$, an absurdity. Thus there is a point $z' \in K$ such that $f(z') = 0$ for all $f \in \mathcal{I}_{z^*}$. If $z_j^* \neq z_j'$ for some j, $1 \leq j \leq n$, the function $z_j - z_j^*$ would belong to $\mathcal{I}_{z^*}$ but would not vanish at z'. We conclude that $z^* = z' \in K$. □

COROLLARY I.6.1.— *If* $K \subset\subset \mathbb{R}^m$ *then* K *is polynomially convex.*

Follows at once from the Weierstrass approximation theorem. Corollary I.6.1 and the results that follow, about the maximally real submanifolds of $\mathbb{C}^m$, justify the introduction of

DEFINITION I.6.1.– *We shall say that a subset* S *of* $\mathbb{C}^m$ *is* ***strongly polynomially convex*** *if every compact subset of* S *is polynomially convex.*

By Corollary I.6.1 the real space $\mathbb{R}^m$ is strongly polynomially convex.

Let us provisionally go back to the open subset $\Omega = \mathcal{Z}(B_0)$ of the maximally real manifold $\mathcal{X}$ (see beginning of section I.3).

LEMMA I.6.1.— *The open subset Ω of $\mathcal{X}$ is strongly polynomially convex.*

Proof: For a compactly supported $u \in C(\Omega)$ let $\mathcal{E}_\nu u$ be defined as in (I.3.4). It is a particular case (when there are no variables t) of Lemma II.2.1 in [TREVES, 1992] that $\mathcal{E}_\nu u$ converges uniformly to u in Ω as $\nu \to +\infty$. Since $\mathcal{E}_\nu u$ is an entire function, Lemma I.6.1 is a direct consequence of Proposition I.6.1. □

Lemma I.6.1 entails at once

THEOREM I.6.2.— *An arbitrary maximally real submanifold $\mathcal{X}$ of $\mathbb{C}^m$ admits a covering by strongly polynomially convex open subsets of $\mathcal{X}$.*

Next, we introduce the following sheaves over $\mathbb{C}^m$:

$\mathscr{C}^\infty\Lambda^{p,q}$: the sheaf of germs of C^∞ sections of $\Lambda^{p,q}$;

$\mathscr{D}'\Lambda^{p,q}$: the sheaf of germs of distribution sections of $\Lambda^{p,q}$;

$\mathcal{O}^{(p)}$: the sheaf of germs of C^∞ sections of $\Lambda^{p,0}$ that are $\bar\partial$–closed.

Let us point out that any $\bar\partial$–closed distribution section of $\Lambda^{p,0}$ over an open set $U \subset \mathbb{C}^m$ is a differential form of the type

$$h = \sum_{|J|=p} h_J(z)dz_J$$

whose coefficients h_J are holomorphic functions in U. Of course $\mathcal{O}^{(0)}$ is the sheaf of germs of holomorphic functions in $\mathbb{C}^m$, also called $\mathcal{O}$.

The validity of the Poincaré lemma for the Cauchy–Riemann complex has the following immediate consequence:

PROPOSITION I.6.2.— *The sequences of sheaf homomorphisms*

$$(I.6.3) \qquad 0 \to \mathcal{O}^{(p)} \xrightarrow{\imath} \mathscr{C}^{\infty}\Lambda^{p,0} \xrightarrow{\bar\partial} \mathscr{C}^{\infty}\Lambda^{p,1} \xrightarrow{\bar\partial} \cdots \xrightarrow{\bar\partial} \mathscr{C}^{\infty}\Lambda^{p,m} \to 0,$$

$$(I.6.4) \qquad 0 \to \mathcal{O}^{(p)} \xrightarrow{\imath} \mathscr{D}'\Lambda^{p,0} \xrightarrow{\bar\partial} \mathscr{D}'\Lambda^{p,1} \xrightarrow{\bar\partial} \cdots \xrightarrow{\bar\partial} \mathscr{D}'\Lambda^{p,m} \to 0,$$

are exact.

REMARK I.6.1.— We have tacitly availed ourselves of the fact that, if u is a scalar distribution in an open set U, $\bar\partial u = 0$ in U entails that u is a holomorphic function in U. □

Since both $\mathscr{C}^{\infty}\Lambda^{p,q}$ and $\mathscr{D}'\Lambda^{p,q}$ are fine sheaves the exact sequences (I.6.3) and (I.6.4) provide cohomological resolutions of $\mathcal{O}^{(p)}$. Given any open set $\mathcal{A} \subset \mathbb{C}^m$, consider the Dolbeault complexes (I.3.7) and (I.3.8). If $q \in \mathbb{Z}_+$ the cohomology space $H^q(\mathcal{A};\mathcal{O}^{(p)})$ is naturally isomorphic to the qth cohomology space of either one of the complexes (I.3.7) or (I.3.8), $H^{p,q}(\mathcal{A})$. The analogous statement is valid for the cohomology space $H^q_c(\mathcal{A};\mathcal{O}^{(p)})$: it is naturally isomorphic to the qth cohomology space of either one of the following differential complexes

$$(I.6.5) \qquad \bar\partial\colon C^{\infty}_c(\mathcal{A};\Lambda^{p,q}) \to C^{\infty}_c(\mathcal{A};\Lambda^{p,q+1}),\ q = 0,1,\ldots,m;$$

$$(I.6.6) \qquad \bar\partial\colon \mathcal{E}'(\mathcal{A};\Lambda^{p,q}) \to \mathcal{E}'(\mathcal{A};\Lambda^{p,q+1}),\ q = 0,1,\ldots,m.$$

The $(p,q)^{th}$ cohomology space of these complexes, $H^{p,q}_c(\mathcal{A})$, is often interpreted as a *homology* space of the complementary bidegree, $H_{m-p,m-q}(\mathcal{A})$. We have (cf. p. 370, [TREVES, 1992]):

(I.6.7) $$H^q(\mathcal{A};\mathcal{O}^{(p)}) \cong H^{p,q}(\mathcal{A});$$

(I.6.8) $$H_c^{m-q}(\mathcal{A};\mathcal{O}^{m-p}) \cong H_{p,q}(\mathcal{A}).$$

We recall the essential content of Proposition VIII.1.2, [TREVES, 1992]. The natural "duality bracket"

$$(u,f) \to \int u\wedge f$$

from $\mathcal{E}'(\mathcal{A};\Lambda^{m-p,m-q})\times\mathcal{C}^{\infty}(\mathcal{A};\Lambda^{p,q})$ to $\mathbb{C}$, identifies the dual of each one of the spaces $\mathcal{C}^{\infty}(\mathcal{A};\Lambda^{p,q})$ and $\mathcal{E}'(\mathcal{A};\Lambda^{m-p,m-q})$ to the other one. [By "dual" we always mean the topological dual, ie., the space of *continuous* linear functionals.] Likewise, the same duality bracket but now on the product space $\mathcal{D}'(\mathcal{A};\Lambda^{m-p,m-q})\times\mathcal{C}_c^{\infty}(\mathcal{A};\Lambda^{p,q})$ identifies the dual of each one of the spaces $\mathcal{C}_c^{\infty}(\mathcal{A};\Lambda^{p,q})$ and $\mathcal{D}'(\mathcal{A};\Lambda^{m-p,m-q})$ to the other one. Moreover, under this identification and if $0 \leq p \leq m$, $0 \leq q \leq m-1$, the transpose of the differential operator $\bar\partial$: $\mathcal{C}_c^{\infty}(\mathcal{A};\Lambda^{p,q}) \to \mathcal{C}_c^{\infty}(\mathcal{A};\Lambda^{p,q+1})$ is the operator

$$(-1)^{p+q}\bar\partial\colon \mathcal{D}'(\mathcal{A};\Lambda^{m-p,m-q-1}) \to \mathcal{D}'(\mathcal{A};\Lambda^{m-p,m-q}).$$

Indeed, if $u \in \mathcal{D}'(\mathcal{A};\Lambda^{m-p,m-q-1})$ and $f \in \mathcal{C}^{\infty}(\mathcal{A};\Lambda^{p,q})$ and if at least one of these two currents, u or f, has a compact support, then by Stokes' formula,

$$\int \bar\partial u\wedge f + (-1)^{p+q-1}\int u\wedge\bar\partial f = \int d(u\wedge f) = 0.$$

We come now to the Serre duality theorem. For the terminology and results about Fréchet spaces which are now going to be used, we refer the reader to the Appendix to the present section. Serre's lemma (Theorem I.6A.2) allows us to state

THEOREM I.6.3.— *Let* p, q *be two integers,* $0 \leq p, q \leq m$. *Suppose the following hypothesis holds:*

(I.6.9) *Each one of the two maps in the sequence*

$$\mathcal{C}^{\infty}(\mathcal{A};\Lambda^{p,q-1}) \xrightarrow{\bar{\partial}} \mathcal{C}^{\infty}(\mathcal{A};\Lambda^{p,q}) \xrightarrow{\bar{\partial}} \mathcal{C}^{\infty}(\mathcal{A};\Lambda^{p,q+1})$$

has a closed range.

Then $H^{p,q}(\mathcal{A})$ *is a Fréchet space and the natural isomorphism of* $\mathcal{E}'(\mathcal{A};\Lambda^{m-p,m-q})$ *onto the dual of* $\mathcal{C}^{\infty}(\mathcal{A};\Lambda^{p,q})$ *induces an isomorphism of* $H_{p,q}(\mathcal{A})$ *onto the dual of* $H^{p,q}(\mathcal{A})$.

When $q = 0$, $\mathcal{C}^{\infty}(\mathcal{A};\Lambda^{p,q-1}) = \{0\}$; when $q = m$, $\mathcal{C}^{\infty}(\mathcal{A};\Lambda^{p,q+1}) = \{0\}$.

EXAMPLE I.6.2.— Consider an open subset $\mathcal{A}$ of $\mathbb{C}^m$ in which the sequence in (I.6.9) is exact for all $q \geq 1$. For instance $\mathcal{A}$ could be a polydisk or a ball. The result in the case of a polydisk is classical (as already indicated a proof can be found in Section D, Chapter I of [GUNNING–ROSSI, 1965]). The argument that works for a polydisk can be adapted to a ball: it suffices to use the Bochner—Martinelli formula (see Section VI.1, [TREVES, 1992]) in place of the Cauchy—Riemann formula.

It follows that, for $q \geq 1$, $H^{p,q}(\mathcal{A}) = \{0\}$. It is then a consequence of Theorem I.6.1 that $H_{p,q}(\mathcal{A}) = \{0\}$. On the other hand, $H^{p,0}(\mathcal{A}) = \mathcal{O}^p(\mathcal{A})$, the space of continuous sections of the sheaf $\mathcal{O}^{(p)}$ over $\mathcal{A}$; and $H_{p,0}(\mathcal{A}) \cong \mathcal{O}^p(\mathcal{A})'$, the dual of $\mathcal{O}^p(\mathcal{A})$. The latter can be interpreted as the space $\mathcal{O}'(\mathcal{A};\Lambda^{m-p,m})$ of "currents"

$$\mu = \sum_{|J|=m-p} \mu_J(z)dz_J \wedge d\bar{z}_1 \wedge \cdots \wedge d\bar{z}_m$$

with coefficients that are analytic functionals in $\mathcal{A}$.

In particular $H_{0,0}(\mathcal{A}) \cong \mathcal{O}'(\mathcal{A})$. The meaning of this, if we make use of the differential complex (I.6.6), is that each coset modulo

$\bar{\partial}\mathcal{E}'(\mathcal{A};\Lambda^{m,m-1})$ of currents $u = u_0 dz_1 \wedge \cdots \wedge dz_m \wedge d\bar{z}_1 \wedge \cdots \wedge d\bar{z}_m$, with $u_0 \in \mathcal{E}'(\mathcal{A})$, can be identified to a unique analytic functional in $\mathcal{A}$. The latter implies that $\langle u,h\rangle = \int h u_0 dz_1 \wedge \cdots \wedge dz_m \wedge d\bar{z}_1 \wedge \cdots \wedge d\bar{z}_m = 0$ for all $h \in \mathcal{O}(\mathcal{A})$ if and only $u = \bar{\partial}v$ for some $v \in \mathcal{E}'(\mathcal{A};\Lambda^{m,m-1})$. Same conclusion if we make use of the differential complex (I.6.5) and replace $\mathcal{E}'$ by $\mathcal{C}^\infty_c$. □

In dealing with "top—degree" forms we avail ourselves of Theorem I.4.3. In connection with the latter we call the attention of the reader to the following

EXAMPLE I.6.3.— Suppose m = 1 and let $\mathcal{A}$ be an arbitrary open subset of $\mathbb{C}$. Then (Theorem I.4.3) the map $\mathcal{C}^\infty(\mathcal{A};\Lambda^{p,0}) \xrightarrow{\bar{\partial}} \mathcal{C}^\infty(\mathcal{A};\Lambda^{p,1})$ is surjective (p = 0,1). By Theorem I.6.3 we conclude that, whatever the open set $\mathcal{A} \subset \mathbb{C}$, $H_{0,0}(\mathcal{A})$ is naturally isomorphic to the dual of $H^{0,0}(\mathcal{A}) \cong \mathcal{O}(\mathcal{A})$ and $H_{1,0}(\mathcal{A})$ is naturally isomorphic to the dual of $H^{1,0}(\mathcal{A}) \cong \mathcal{O}^1(\mathcal{A})$, the space of (1,0)—forms $h(z)dz$, with $h \in \mathcal{O}(\mathcal{A})$. □

To deal with less—than—top degree forms we shall take special open sets $\mathcal{A}$: let K be a polynomially convex compact set in $\mathbb{C}^m$ and B any open ball containing K; we are going to apply Theorem I.6.3 to $\mathcal{A} = B\backslash K$. This is possible thanks to

THEOREM I.6.4.— *If* K *is a polynomially convex compact set in* $\mathbb{C}^m$ *and* B *an open ball that contains* K *the differential operator*

$$\bar{\partial}\colon \mathcal{C}^\infty(B\backslash K;\Lambda^{m,m-2}) \to \mathcal{C}^\infty(B\backslash K;\Lambda^{m,m-1}) \tag{I.6.10}$$

has a closed range.

Proof: The assertion is trivial when m = 1. We shall therefore assume m $\geq$ 2. Up to sign the transpose of the map (I.6.10) is the map

$$\bar{\partial}\colon \mathcal{E}'(B\backslash K;\Lambda^{0,1}) \to \mathcal{E}'(B\backslash K;\Lambda^{0,2}). \tag{I.6.11}$$

By Theorem I.6A.1 it suffices to show that the map (I.6.11) has a closed range. This claim will be proved if we show that the range is equal, when m $\geq$ 3, to the set of currents $f \in \mathcal{E}'(B\backslash K;\Lambda^{0,2})$ such that $\bar{\partial}f = 0$; and when m = 2, to the set of currents $f \in \mathcal{E}'(B\backslash K;\Lambda^{0,2})$ such that

$$\int hf \wedge dz_1 \wedge dz_2 = 0,\ \forall\ h \in \mathcal{O}(B). \tag{I.6.12}$$

Now, the second isomorphism (I.6.8) shows that there exists $\phi \in C_c^\infty(B\backslash K;\Lambda^{0,2})$ such that $f - \phi = \bar{\partial}v$ with $v \in \mathcal{E}'(B\backslash K;\Lambda^{0,1})$. Observe that $\bar{\partial}\phi = 0$ and that, when m = 2 and (I.6.12) holds, we have

$$\int h\phi \wedge dz_1 \wedge dz_2 = 0,\ \forall\ h \in \mathcal{O}(B). \tag{I.6.13}$$

It will therefore suffice to prove that, under the above hypotheses, there is $\lambda \in C_c^\infty(B\backslash K;\Lambda^{0,1})$ such that $\bar{\partial}\lambda = \phi$. For then $f = \bar{\partial}(\lambda+v)$ and $\lambda+v \in \mathcal{E}'(B\backslash K;\Lambda^{0,1})$.

We refer the reader to the result in Example I.6.2, according to which $H_{m,m-2}(B) = 0$ if m $\geq$ 3. By (I.6.8) this is equivalent to $H_c^2(B;\mathcal{O}) = 0$; $H_c^q(B;\mathcal{O})$ is the qth cohomology space of the differential complex (I.6.5) when $\mathcal{A} = B$. According to this, there is $\psi \in C_c^\infty(B;\Lambda^{0,1})$ such that $\bar{\partial}\psi = \phi$.

When m = 2 we use the property mentioned at the end of Example I.6.2, according to which $\phi = \bar{\partial}\psi$ for some $\psi \in C_c^\infty(B;\Lambda^{0,1})$, if and only if (I.6.12) holds.

Now let $\psi \in \mathcal{C}^{\infty}_{c}(B;\Lambda^{0,1})$ be such that $\bar{\partial}\psi = \phi$; then $\bar{\partial}\psi = 0$ in an open neighborhood $U \subset B$ of K. To complete the proof of Theorem I.6.3 we apply Theorem I.6.1: there is an open neighborhood $V \subset U$ of K and an element $\chi \in \mathcal{C}^{\infty}(V)$ such that $\psi = \bar{\partial}\chi$ in V. Let $g \in \mathcal{C}^{\infty}_{c}(V)$, $g \equiv 1$ in some neighborhood of K. This means that $\psi - \bar{\partial}(g\chi) \in \mathcal{C}^{\infty}_{c}(B\backslash K;\Lambda^{0,1})$, and $\phi = \bar{\partial}[\psi - \bar{\partial}(g\chi)]$. □

Combining Theorems I.4.3 and I.6.4 allows us to state

THEOREM I.6.5.— *If* K *is polynomially convex and if* $B \supset K$ *is an open ball, then* $H^{m,m-1}(B\backslash K)$ *is a Frechet space. The map*

$$[g] \to ([f] \to \int f\wedge g) \tag{I.6.14}$$

is an isomorphism of $H_{m,m-1}(B\backslash K)$ *onto the dual of* $H^{m,m-1}(B\backslash K)$.

We continue to assume that the compact set K is polynomially convex and contained in the open ball B. Consider a cohomology class $[g] \in H^{m,m-1}(B\backslash K)$. It is represented by a smooth form $g \in \mathcal{C}^{\infty}(B\backslash K;\Lambda^{m,m-1})$ such that $\bar{\partial}g = 0$. Select $\chi \in \mathcal{C}^{\infty}_{c}(B)$, $\chi \equiv 1$ in an open neighborhood of K. The linear functional

$$\mathcal{O}(\mathbb{C}^m) \ni h \to \int h\bar{\partial}\chi\wedge g \in \mathbb{C} \tag{I.6.15}$$

defines an analytic functional μ_g carried by K which depends solely on the cohomology class $[g]$ of g (see beginning of section I.4).

We shall now describe the *transpose* of the map $[g] \to \mu_g$. According to Theorem I.6.3 and to the fact that $\mathcal{O}(K)$ is reflexive, this will be a map $\mathcal{O}(K) \to H_{m,m-1}(B\backslash K)$.

Let $\dot{h} \in \mathcal{O}(K)$ be the germ (at K) of a holomorphic function h in an open neighborhood U of K in B. Select $\chi \in C_c^\infty(U)$, $\chi = 1$ in some neighborhood of K. Then $\bar\partial(\chi h) \in C_c^\infty(B\backslash K;\Lambda^{0,1})$ is clearly $\bar\partial$–closed in $B\backslash K$. Suppose we change our choices of h and χ to $h_1 \in \mathcal{O}(U_1)$ and $\chi_1 \in C_c^\infty(U_1)$, with $h = h_1$ in some neighborhood of K contained in $U \cap U_1$, and $\chi_1 \equiv 1$ in a neighborhood of K; then $\bar\partial(\chi_1 h_1) = \bar\partial(\chi h) + \bar\partial(\chi_1 h_1 - \chi h)$ is homologous to $\bar\partial(\chi h)$ in $B\backslash K$. Thus the class $[\bar\partial(\chi h)] \in H_{m,m-1}(B\backslash K)$ is defined independently of the choice of the representative h of $\dot{h}$ and of the cutoff function χ. We shall denote it by $[\mathfrak{D}\dot{h}]$.

If $g \in C^\infty(B\backslash K;\Lambda^{m,m-1})$ and $h \in \mathcal{O}(\mathbb{C}^m)$ then (calling $\dot{h}$ the image of h in $\mathcal{O}(K)$)

$$<\mu_g,\dot{h}> = \int h\bar\partial\chi\wedge g = \int \bar\partial(\chi h)\wedge g = <[\mathfrak{D}\dot{h}],[g]>.$$

Since, by our hypothesis that K is polynomially convex, the map $h \to \dot{h}$ maps $\mathcal{O}(\mathbb{C}^m)$ onto a dense subspace of $\mathcal{O}(K)$ this proves that $[g] \to \mu_g$ is the transpose of the map $\dot{h} \to [\mathfrak{D}\dot{h}]$.

REMARK I.6.2.— The map $\mathcal{O}(K) \ni \dot{h} \to [\mathfrak{D}\dot{h}] \in H_c^{0,1}(B\backslash K) \cong H_{m,m-1}(B\backslash K)$ is well defined even when K is not polynomially convex. However the poly–nomial convexity of K ensures that it be the transpose of the map $H^{m,m-1}(B\backslash K) \ni [g] \to \mu_g$. □

THEOREM I.6.6.— *Suppose* $m \geq 2$. *Let* K *be a polynomially convex compact subset of* $\mathbb{C}^m$ *and let* B *be an open ball that contains* K. *The linear map* $[g]$

$\to \mu_g$ *is a Fréchet space isomorphism of* $H^{m,m-1}(B\backslash K)$ *onto* $\mathcal{O}'(K)$.

Thanks to the above remarks Theorem I.6.6 follows from

PROPOSITION I.6.3.— *Let* B *be an open ball in* $\mathbb{C}^m$ $(m \geq 2)$ *and* K *be any compact subset of* B. *The map* $\dot{h} \to [\partial\dot{h}]$ *is an isomorphism of* $\mathcal{O}(K)$ *onto* $H_{m,m-1}(B\backslash K)$.

Proof: First we prove the injectivity of the map $\dot{h} \to [\partial\dot{h}]$. Let h be holomorphic in the open set $U \supset K$ and represent the germ $\dot{h}$; and let $\chi \in C_c^\infty(U)$, $\chi \equiv 1$ in a neighborhood of K. If $[\bar{\partial}(\chi h)] = 0$ then $\bar{\partial}(\chi h) = \bar{\partial}v$ for some $v \in C_c^\infty(B\backslash K)$ and $\chi h - v$ is both holomorphic and compactly supported in B, hence $\chi h \equiv v$. But this means that $h \equiv 0$ in a neighborhood of K, ie., $\dot{h} = 0$.

Next we show the surjectivity of the map $\dot{h} \to [\partial\dot{h}]$. Consider a homology class $[g\,] \in H_{m,m-1}(B\backslash K)$ represented by $g = \sum_{j=1}^{m} g_j d\bar{z}_j$ with $g_j \in C_c^\infty(B\backslash K)$ and $\bar{\partial}g = 0$. There is (see Example I.6.2) a solution $u \in C_c^\infty(B)$ of the equation $\bar{\partial}u = g$; u is holomorphic in $B\backslash(\text{supp } g)$, and defines a germ $\dot{u}$ at K. We have $[\partial\dot{u}] = [g]$. □

REMARK I.6.3.— When $m = 1$ we cannot expect the range of the map $\dot{h} \to [\partial\dot{h}]$ to be the whole of $H_{1,0}(B\backslash K)$: every (0,1)—form $\bar{\partial}(\chi h)$ is orthogonal to all (1,0)—forms $f(z)dz$, $f \in \mathcal{O}(B)$. The class $[\partial\dot{h}]$ lies in the "reduced homology" space $\tilde{H}_{1,0}(B\backslash K)$ (cf. Definition VIII.4.2, [TREVES, 1992]), ie., the subspace of $H_{1,0}(B\backslash K)$ consisting of the homology classes $[u d\bar{z}]$ such that $u \in C_c^\infty(B\backslash K)$ and $\int u d\bar{z} \wedge f\, dz = 0$ for all $f \in \mathcal{O}(B)$. According to Example I.6.2 $H_{1,0}(B\backslash K) \cong \mathcal{O}^1(B\backslash K)'$ and therefore $\tilde{H}_{1,0}(B\backslash K)$ is natu—

rally isomorphic to the dual of the Frechet space $\mathcal{O}^1(B\backslash K)/\mathcal{O}^1(B)$ ($\cong \mathcal{O}(B\backslash K)/\mathcal{O}(B)$ if we make use of the isomorphism $f \to f\,dz$). Here we may state:

PROPOSITION I.6.4.— *When* m = 1 *the map* $\dot{h} \to [\partial\dot{h}]$ *is an isomorphism of* $\mathcal{O}(K)$ *onto* $\tilde{H}_{1,0}(B\backslash K)$.

Proof: The proof of the injectivity is the same as in the case m $\geq$ 2. To prove the surjectivity when m = 1 we also use the Cauchy kernel. That $u \in C^\infty_c(B)$ follows from the additional hypothesis that g is orthogonal to all one–forms $h dz$, $h \in \mathcal{O}(B)$ (see Example I.6.2). From there one concludes as in the proof of Proposition I.6.3. □

Assume now that K is polynomially convex. In the remarks pre–ceding Theorem I.6.6 the transpose of the map $\dot{h} \to [\partial\dot{h}]$ is interpreted as a linear map $H^{1,0}(B\backslash K) \cong \mathcal{O}^1(B\backslash K) \to \mathcal{O}'(K)$. But if $g_0 \in \mathcal{O}(B)$ then

$$\int h\bar{\partial}\chi\wedge g_0 dz = \int \bar{\partial}(\chi h)\wedge g_0 dz = 0, \quad \forall\, h \in \mathcal{O}(\mathbb{C}).$$

As a consequence the map $g_0 dz = g \to \mu_g$ induces a linear map

$$\mathcal{O}(B\backslash K)/\mathcal{O}(B) \to \mathcal{O}'(K).$$

This is the transpose of the map $\mathcal{O}(K) \ni \dot{h} \to [\partial\dot{h}] \in \tilde{H}_{1,0}(B\backslash K) \cong (\mathcal{O}(B\backslash K)/\mathcal{O}(B))'$. But if K is polynomially convex, ie., $\mathbb{C}\backslash K$ is connected, Laurent expansion shows at once that $\mathcal{O}(B\backslash K)/\mathcal{O}(B) \cong \mathcal{O}_0(\mathbb{C}\backslash K)$ (Definition I.2.1). The reader will check easily that the map $\mathcal{O}_0(\mathbb{C}\backslash K) \to \mathcal{O}'(K)$ in–duced by the map $g \to \mu_g$ is the inverse of the Cauchy transform. In sum–mary, Theorem I.2.3 is the "correct" version of Theorem I.6.6 in the case m = 1. □

Theorem I.6.6 allows us to prove the analogues of Propositions I.4.1 and I.4.2:

PROPOSITION I.6.5.— *Let* K_1 *and* K_2 *be two polynomially convex compact subsets of* $\mathbb{C}^m$ *such that* $K_1 \cup K_2$ *is polynomially convex. Then the following two properties holds*:

(I.6.16) *if* $\mu \in \mathcal{O}'(\mathbb{C}^m)$ *is carried by* K_1 *and by* K_2 *then* μ *is carried by* $K_1 \cap K_2$;

(I.6.17) *if* $\mu \in \mathcal{O}'(\mathbb{C}^m)$ *is carried by* $K_1 \cup K_2$ *then there are two analytic functionals* $\mu_j \in \mathcal{O}'(\mathbb{C}^m)$ *carried by* K_j $(j = 1,2)$ *such that* $\mu = \mu_1 + \mu_2$.

When $m = 1$, (I.6.16) has been already proved (Corollary I.2.1). When $m \geq 2$ the proof of (I.6.16) is identical to that of Proposition I.4.1. Whatever $m \geq 1$ the proof of (I.6.17) is identical to that of Proposition I.4.2. The key to those proofs is the validity of Theorem I.6.6 for $K = K_j$ $(j = 1,2)$ as well as for $K = K_1 \cup K_2$.

Returning to the strongly polynomially convex open subset $\Omega = \mathcal{Z}(B_0)$ of $\mathcal{X}$ (Lemma I.6.1) we can apply Proposition I.6.4 to any pair of compact subsets K_1 and K_2 of Ω. This is the only property needed to develop hyperfunction theory on Ω, and from there onto $\mathcal{X}$, along the lines described in section I.5.

APPENDIX TO SECTION I.6:

HOMOMORPHISMS OF FRÉCHET SPACES

Let E, F be Fréchet spaces. A continuous linear map u: E $\to$ F is said to be a *homomorphism* if the map $\tilde{u}$: E/Ker $u \to u$(E) defined by u is an isomorphism, ie., a bijection and a homeomorphism. Here E/Ker u is equipped with the quotient topology; it is a Fréchet space; $u(\mathrm{E}) \subset \mathrm{F}$ is equipped with the topology induced by F. In general, u(E) is merely metrizable. But if u is a homomorphism, u(E) is complete (as a copy of E/Ker u) hence closed in F. Conversely, if u(E) is a closed subspace of F it is also a Fréchet space and $\tilde{u}$ must perforce be open by the open mapping theorem, and therefore must be a homeomorphism.

The following result is due to S. Banach:

THEOREM I.6A.1.— *Let* E, F *be Fréchet spaces,* E′, F′ *their duals. For a continuous linear map* u: E $\to$ F *to be a homomorphism it is necessary and sufficient that its transpose* ${}^t u$: F′ $\to$ E′ *have a weakly closed image or equivalently, that the following hold:*

$$(\mathrm{Ker}\ u)^\perp = {}^t u(\mathrm{F}'). \tag{I.6A.1}$$

Proof: The Hahn—Banach theorem implies that Ker $u = {}^t u(\mathrm{F}')^\perp$; for if $x \in \mathrm{E}$, $\langle {}^t u(y'), x\rangle = 0$ for all $y' \in \mathrm{F}'$ implies $x \in$ Ker u. By taking the orthogonals of both sides we obtain that $(\mathrm{Ker}\ u)^\perp$ is equal to the weak closure of ${}^t u(\mathrm{F}')$.

First assume the map $\tilde{u}$: E/Ker $u \to u$(E) defined by u is an isomorphism. The transpose of the quotient map E $\to$ E/Ker u defines an isomorphism $(\mathrm{Ker}\ u)^\perp \cong \mathrm{E}/\mathrm{Ker}\ u)'$ ($\cong (u(\mathrm{E}))'$ via ${}^t\tilde{u}$). To $\lambda \in (\mathrm{Ker}\ u)^\perp$ these isomorphisms assign a unique $\tilde{\lambda} \in (u(\mathrm{E}))'$. By the Hahn—Banach theorem there is $\hat{\lambda} \in \mathrm{F}'$ that extends $\tilde{\lambda}$; it is checked at once that ${}^t u(\hat{\lambda}) =$

λ. This proves (I.6A.1).

Now assume (I.6A.1) holds. Then

$$(E/\mathrm{Ker}\ u)' \cong (\mathrm{Ker}\ u)^{\perp} = {}^t u(F').$$

A consequence of this is that the inverse map $\tilde{u}^{-1}$: $u(E) \to E/\mathrm{Ker}\ u$ is weakly continuous. Indeed, we derive from (I.6A.1) that the sets

$$\mathcal{N}(\lambda_1,\dots,\lambda_N;\epsilon) = \{\ \dot{z} \in E/\mathrm{Ker}\ u;\ |\lambda_j(\dot{z})| < \epsilon,\ j = 1,\dots,N\ \},$$

with $\epsilon > 0$ and $\lambda_j = {}^t u(y_j')$ for some $y_j' \in F'$, make up a basis of neigh– borhoods of 0 in the weak topology on $E/\mathrm{Ker}\ u$. Then if we set

$$\mathcal{N}'(y_1',\dots y_N';\epsilon) = \{\ y \in F;\ |\langle y_j',y\rangle| < \epsilon,\ j = 1,\dots,N\ \}$$

we see that $u(E) \cap \mathcal{N}'(y_1',\dots,y_N';\epsilon) \subset \tilde{u}(\mathcal{N}(\lambda_1,\dots,\lambda_N;\epsilon))$.

Now suppose $\tilde{u}^{-1}$ were not "strongly" continuous. Select a coun– table basis of neighborhoods of 0 in $u(E)$, $\{V_n\}_{n=1,2,\dots}$. There would be a neighborhood W of 0 in $E/\mathrm{Ker}\ u$ and, for each $n = 1,2,\dots$, a point $x_n \in n^{-1}V_n$ such that $\tilde{u}^{-1}(x_n) \notin W$. We may as well assume $cW \subset W$ for all c, $0 \leq c \leq 1$. The sequence $\{nx_n\}_{n=1,2,\dots}$ converges to 0 in $u(E)$, hence $n\tilde{u}^{-1}(x_n)$ converges weakly to 0 in $E/\mathrm{Ker}\ u$. As a consequence of this and of Mackey's theorem (Theorem 36.2 in [TREVES, 1967]) the sequence $\{n\tilde{u}^{-1}(x_n)\}$ must be bounded in $E/\mathrm{Ker}\ u$, which implies that there is a number $\lambda > 0$ such that $n\tilde{u}^{-1}(x_n) \in \lambda W$ for all n. We reach a contradic– tion as soon as $n \geq \lambda$. □

Now consider a sequence of homomorphisms

$$E \xrightarrow{u} F \xrightarrow{v} G \tag{I.6A.2}$$

such that $v \circ u = 0$. Since $u(E)$ is closed, $\mathrm{Ker}\ v/u(E)$ is a Fréchet space. We can define a natural map

$$\text{(I.6A.3)} \qquad \lambda\colon \operatorname{Ker} {}^t u / {}^t v(G') \to (\operatorname{Ker} v / u(E))'$$

as follows: $y' \in \operatorname{Ker} {}^t u$ ($\subset F'$) defines, by restriction, a continuous linear functional on Ker v, clearly orthogonal to $u(E)$, and thus an element $\hat{y}'$ of $(\operatorname{Ker} v/u(E))'$. Obviously $\hat{y}' = 0$ when $y' \in {}^t v(G')$, which means that the map $y' \to \hat{y}'$ induces a map (I.6A.3).

The following result is known as *Serre's lemma* (see [SERRE, 1955]):

THEOREM I.6A.2.— *Suppose the linear maps u and v in* (I.6A.2) *are homomorphisms and $v \circ u = 0$. Then the map* (I.6A.3) *is an isomorphism.*

Proof: Denote by $[y']$ the equivalence class of $y' \in \operatorname{Ker} {}^t u$ modulo ${}^t v(G')$. The map λ is injective: if $\lambda([y']) = 0$ then the restriction of y' to Ker v vanishes identically and therefore, by (I.6A.1), $y' \in {}^t v(G')$, ie., $[y'] = 0$. The map λ is surjective: denote by y'_0 the pullback to Ker v, under the quotient map, of $\hat{y}' \in (\operatorname{Ker} v/u(E))'$; by the Hahn—Banach theorem there is $y' \in F'$ whose restriction to Ker v is equal to y'_0; $y'_0 = 0$ on $u(E)$ $\Rightarrow y' \in \operatorname{Ker} {}^t u$, and it is checked at once that $\lambda([y']) = \hat{y}'$. □

CHAPTER II

MICROLOCAL THEORY OF HYPERFUNCTIONS ON A MAXIMALLY REAL SUBMANIFOLD OF COMPLEX SPACE

INTRODUCTION

Hyperfunctions in $\mathbb{R}^m$ are commonly thought as sums of "boundary values" of holomorphic functions in wedges $\mathcal{W} \subset \mathbb{C}^m \backslash \mathbb{R}^m$ whose edge lies on $\mathbb{R}^m$. That the same is true when $\mathbb{R}^m$ is replaced by a maximally real submanifold $\mathcal{X}$ of $\mathbb{C}^m$ is shown in the first three sections of Chapter II. The *boundary value* of a holomorphic function in the wedge $\mathcal{W}$ whose growth at the edge is unrestricted, is given a precise meaning in Theorem II.1.1. If the boundary value vanishes identically, the same is true of the holomorphic function (Theorem II.1.2).

Section II.2 introduces the FBI transform of an analytic functional carried by a small compact subset of $\mathcal{X}$, and then that of the germ of a hyperfunction on $\mathcal{X}$. Inverting the FBI transform is made possible by using the *real structure bundle* of $\mathcal{X}$, $\mathbb{R}T'_{\mathcal{X}}$ (beginning of section II.2) and by selecting the local embedding of $\mathcal{X}$ to ensure that $\mathcal{X}$ be *well–positioned* at the central point z_0 [see (II.2.3)]. The inversion formula is exploited to equate the hypo–analyticity of a germ of hyperfunction u at z_0 (ie., the fact that u is the restriction to $\mathcal{X}$ of the germ of a holomorphic function in $\mathbb{C}^m$) with the exponential decay of its FBI transform $\mathcal{F}u(z,\zeta)$ when (z,ζ) stays in a conic neighborhood of $\{z_0\}\times(\mathbb{R}^m\backslash\{0\})$ in $\mathbb{C}^m\times(\mathbb{C}^m\backslash\{0\})$ (Theorem II.2.5). And also to show that an arbitrary hyperfunction u on $\mathcal{X}$ can be represented, near z_0, as the sum of the boundary values of holomorphic functions h_i in wedges $\mathcal{W}_\delta(U,\Gamma_i)$ (Theorem II.3.1). In the present context U is an open neighborhood of z_0 and the Γ_i can always be thought as open convex cones, with vertex at the origin, in $\mathbb{R}^m\backslash\{0\}$. The fact that they can all be selected to lie in a half–space $\{ v \in \mathbb{R}^m;\ \xi\cdot v < 0 \}$ is equivalent to the exponential decay of $\mathcal{F}u(z,\zeta)$ when (z,ζ) stays in a conic neighborhood of (z_0,ξ) in $\mathbb{C}^m\times(\mathbb{C}^m\backslash\{0\})$ (Theorem II.4.1). If these properties are true one says that u is hypo–analytic at (z_0,ξ) or that (z_0,ξ) does not belong to the ***hypo–analytic wave–front set*** of u, $WF_{ha}(u)$ (Definition II.3.2).

Theorem II.4.1 is exploited to prove the Martineau's Edge of the Wedge Theorem, with the edge on the maximally real submanifold $\mathcal{X}$ (Theorem II.5.1). As already pointed out the result applies to all holo–morphic functions in the relevant wedges, regardless of the rate of growth at the edge of their absolute values.

The last section of Chapter II introduces the sheaf of ***microfunc–tions*** following the prescription of Sato: its stalk at a point (z_0,θ_0) of the cosphere bundle of $\mathcal{X}$ (or better, of the sphere bundle associated to $\mathbb{R}T'_{\mathcal{X}}$) is obtained by quotienting out the hyperfunctions whose hypo–analytic wave–front lies away from the ray corresponding to (z_0,θ_0). For a micro–function $\dot{u}$ at (z_0,θ_0) to be a ***microdistribution***, meaning that $\dot{u}$ can be represented in a neighborhood of z_0 by a distribution, it is necessary and sufficient that the FBI transform (at z_0) of any hyperfunction represent–ing $\dot{u}$ have tempered growth at infinity (Theorem II.6.2).

II.1 BOUNDARY VALUES OF HOLOMORPHIC FUNCTIONS IN WEDGES

Let $\mathcal{X}$ be a maximally real submanifold of $\mathbb{C}^m$. As in most of Chapter I we shall reason in the neighborhood of a point $z_0 \in \mathcal{X}$ which we take to be origin of $\mathbb{C}^m$; and we assume that the tangent space to $\mathcal{X}$ at 0 is $\mathbb{R}^m$. We select two open balls B_0, $B_1 \subset \mathbb{R}^m$ centered at 0 and a C^∞ function Φ: $B_0 \to B_1$ such that

$$\Omega = \{ z = x+iy \in B_{01};\ y = \Phi(x) \} = \mathcal{X} \cap (B_{01}) \quad (B_{01} = B_0+iB_1).$$

This entails $\Phi(0) = 0$, $D\Phi(0) = 0$, ie., Property (I.3.2) is valid; and in fact we shall assume throughout that (I.3.3) holds.

Let $U \subset\subset \Omega$ be an open neighborhood of 0 in $\mathcal{X}$. We also select an open cone Γ in $\mathbb{R}^m \backslash \{0\}$ (unless specified otherwise, all cones will have their vertices at the origin) and an arbitrary number $\delta > 0$. We define the *wedge* with *edge* U, *directrix* Γ and *height* δ,

$$\mathcal{W}_\delta(U,\Gamma) = \{ z+i\gamma \in \mathbb{C}^m;\ z \in U,\ \gamma \in \Gamma,\ |\gamma| < \delta \}. \tag{II.1.1}$$

Select arbitrarily a function $g \in \mathcal{O}(\mathcal{W}_\delta(U,\Gamma))$, an open subset $V \subset\subset U$ of Ω whose boundary ∂V in U is smooth, and a vector $\gamma \in \Gamma$, $|\gamma| < \delta$. Define, for any $h \in \mathcal{O}(\mathbb{C}^m)$,

$$<\mu^\gamma_{g,V}, h> = \int_{V+i\gamma} g(z)h(z)dz \tag{II.1.2}$$

($dz = dz_1 \wedge \cdots \wedge dz_m$). Formula (II.1.2) defines an analytic functional in $\mathbb{C}^m$, obviously carried by $\mathscr{Cl}V + i\gamma$ (as a translate of $\mathscr{Cl}V$, $\mathscr{Cl}V + i\gamma$ is polynomially convex). The following result will play an important role throughout Chapter II:

THEOREM II.1.1.— *Suppose* $V \subset\subset U$, *that* ∂V *is smooth and that the cone* Γ *is convex. There is an analytic functional* $\mu_{g,V} \in \mathcal{O}'(\mathscr{C}\ell V)$ *that has the following property:*

(II.1.3) *To every open neighborhood* $\mathscr{N}$ *of* ∂V *in* $\mathbb{C}^m$ *there is* ϵ, $0 < \epsilon < \delta$, *such that* $\mu_{g,V} - \mu^{\gamma}_{g,V} \in \mathcal{O}'(\mathscr{N})$ *if* $\gamma \in \Gamma$, $|\gamma| < \epsilon$.

If also the boundary ∂V_1 *of* $V_1 \subset V$ *in* $\mathscr{X}$ *is smooth and if* μ_{g,V_1} *is the analogue of* $\mu_{g,V}$ *when* V_1 *is substituted for* V, *then the analytic functional* $\mu_{g,V} - \mu_{g,V_1}$ *is carried by* $(\mathscr{C}\ell V)\backslash V_1$.

Proof: We begin with a remark. Suppose γ, $\gamma' \in \Gamma$, $|\gamma|$, $|\gamma'| < \delta$. The hypothesis that ∂V is C^∞ and that Γ is convex allows us to apply Stokes' formula, to get

(II.1.4) $$<\mu^{\gamma}_{g,V} - \mu^{\gamma'}_{g,V}, h> = \int_{\partial V + i[\gamma,\gamma']} g(z)h(z)dz,$$

where $[\gamma,\gamma']$ is the straight—line segment joining γ to γ' and the boundary ∂V is suitably oriented. For j = 1,2,..., set $K_j = \{ z \in \mathbb{C}^m;\ \mathrm{dist}(z,\partial V) \leq 3j^{-1} \}$. Select a vector $\gamma_0 \in \Gamma$, $|\gamma_0| \leq \delta$ (≤ 1); and set $\gamma_j = \gamma_0/j$. We are going to show that there is $\mu_{g,V} \in \mathcal{O}'(\mathscr{C}\ell V)$ such that, whatever j = 1,2,...,

(II.1.5) $$\mu_{g,V} - \mu^{\gamma_j}_{g,V} \in \mathcal{O}'(K_j).$$

Note that (II.1.5) implies (II.1.3). For given any open neighborhood $\mathscr{N}$ of

∂V, select $j_0 \geq 1$ and $\epsilon > 0$ ($\epsilon < \delta$) such that $K_{j_0} \subset \mathcal{N}$ and $\partial V + i[\gamma_{j_0},\gamma]$ $\subset \mathcal{N}$, $\forall\ \gamma \in \Gamma$, $0 < |\gamma| < \epsilon$. For those γ, $\mu_{g,V} - \mu^{\gamma}_{g,V} = (\mu_{g,V} - \mu^{\gamma_{j_0}}_{g,V}) + (\mu^{\gamma_{j_0}}_{g,V} - \mu^{\gamma}_{g,V})$ is carried by $\mathcal{N}$.

We proceed with the proof of (II.1.5). Given $h \in \mathcal{O}(\mathbb{C}^m)$ we define the function in $\mathbb{C}^m\times\mathbb{C}$, $H(z,w) = h(z+\imath w\gamma_0)$, and consider its Taylor expansion with respect to w of order $k \geq 1$,

$$\text{(II.1.6)}\quad H(z,w) = \sum_{p=0}^{k} \frac{w^p}{p!}\partial^p_w H(z,0) + \frac{1}{k!}\int_0^w (w-\tau)^k(\partial_w^{k+1}H)(z,\tau)d\tau.$$

We have, by virtue of (II.1.4),

$$\text{(II.1.7)}\quad <\mu^{\gamma_{j+1}}_{g,V},h> - <\mu^{\gamma_j}_{g,V},h> =$$
$$\int_{z\in\partial V}\int_{t=1/(j+1)}^{t=1/j} g(z+\imath t\gamma_0)h(z+\imath t\gamma_0)d(z+\imath t\gamma_0),$$

where $d(z+\imath t\gamma_0) = d(z_1+\imath t\gamma_{01})\wedge\cdots\wedge d(z_m+\imath t\gamma_{0m})$ [and $\gamma_0 = (\gamma_{01},\ldots,\gamma_{0m})$]: We take (II.1.6) into account in (II.1.7), thus getting

$$<\mu^{\gamma_{j+1}}_{g,V},h> - <\mu^{\gamma_j}_{g,V},h> = <\nu^j_k,h> + <\rho^j_k,h>,$$

with

$$<\nu^j_k,h> = \sum_{p=0}^{k}\frac{1}{p!}\int_{z\in\partial V}\int_{t=1/(j+1)}^{t=1/j} t^p g(z+\imath t\gamma_0)(\partial^p_w H)(z,0)d(z+\imath t\gamma_0),$$

$$<\rho^j_k,h> =$$
$$\frac{1}{k!}\int_{z\in\partial V}\int_{t=1/(j+1)}^{t=1/j} g(z+\imath t\gamma_0)\left[\int_0^t (\partial_w^{k+1}H)(z,\tau)(t-\tau)^k d\tau\right]d(z+\imath t\gamma_0).$$

It is clear that $\nu^j_k \in \mathcal{O}'(\partial V)$ whatever j, k $\geq$ 1. On the other hand, the Cauchy inequalities yield, for all $z \in \partial V$,

$$\operatorname*{Max}_{0\leq\tau\leq 1/j} |(\partial_w^{k+1}H)(z,\tau)| \leq \tfrac{3}{2}\frac{(k+1)!}{(2j^{-1})^{k+1}} \operatorname*{Max}_{|w|\leq 3/j} |H(z,w)|,$$

whence

$$|<\rho_k^j,h>| \leq C_j \operatorname*{Max}_{z\in\partial V,\ 0\leq\tau\leq 1/j} |(\partial_w^{k+1}H)(z,\tau)| \operatorname*{Max}_{0\leq t\leq 1/j} \int_0^t \frac{\tau^k}{k!} d\tau \leq$$

$$\tfrac{3}{2}C_j 2^{-k-1} \operatorname*{Max}_{z\in\partial V,\ 0\leq|w|\leq 3/j} |H(z,w)| \leq C_j 2^{-k} \operatorname*{Max}_{K_j} |h|.$$

If we select k large enough that $2^{-k}C_j \leq 2^{-j}$ and if then we write $\alpha_j = \nu_k^j$, we reach the following conclusion: to each $j = 1,2,\ldots$, there is $\alpha_j \in \mathcal{O}'(\partial V)$ such that, for all $h \in \mathcal{O}'(\mathbb{C}^m)$,

$$\text{(II.1.8)} \qquad |<\mu_{g,V}^{\gamma_{j+1}} - \mu_{g,V}^{\gamma_j},h> - <\alpha_j,h>| \leq 2^{-j} \operatorname*{Max}_{K_j} |h|.$$

Set $\beta_j = \mu_{g,V}^{\gamma_j} - \alpha_1 - \cdots - \alpha_{j-1}$ ($j = 1,2,\ldots;\ \alpha_0 = 0$); (II.1.8) is equivalent to

$$|<\beta_{j+1} - \beta_j,h>| \leq 2^{-j} \operatorname*{Max}_{K_j} |h|.$$

The analytic functional β_j is carried by $[(\mathcal{C}\ell\, V)+i\frac{1}{j}\gamma_0]\cup(\partial V)$. As a consequence any compact neighborhood $\mathcal{K}$ of $\mathcal{C}\ell V$ in $\mathbb{C}^m$ will carry every analytic functional β_j, $j \geq j_0$, and will contain K_{j_0} — provided j_0 is large enough. The preceding estimate implies that $\{\beta_j\}_{j=j_0,j_0+1,\ldots}$ is a Cauchy sequence in the Fréchet space $\mathcal{O}'(\mathcal{K})$; we define $\mu_{g,V}$ to be its limit (it does not depend on the choice of $\mathcal{K}$). By letting $\mathcal{K}$ contract to $\mathcal{C}\ell V$ (and j_0 go to $+\infty$) we conclude that $\mu_{g,V}$ is carried by $\mathcal{C}\ell V$. Then consider

$$<\mu_{g,V},h> - <\mu_{g,V}^{\gamma_{j_0}},h> =$$

$$\lim_{j_0\leq j\to+\infty} \left\{<\mu_{g,V}^{\gamma_{j+1}},h> - <\mu_{g,V}^{\gamma_{j_0}},h> - \sum_{p=0}^{j} <\alpha_p,h>\right\} =$$

$$\lim_{j_0 \leq j \to +\infty} \left\{ \sum_{i=j_0}^{j} (<\mu_{g,V}^{\gamma_{i+1}},h> - <\mu_{g,V}^{\gamma_i},h> - <\alpha_i,h>) - \sum_{p=0}^{j_0-1} <\alpha_p,h> \right\},$$

whence, by (II.1.8):

$$|<\mu_{g,V},h> - <\mu_{g,V}^{\gamma_{j_0}},h>| \leq const. \operatorname{Max}_{K_{j_0}} |h|,$$

which proves (II.1.5).

Finally we prove the last part of the statement. Let $\mathcal{N}$ be an arbitrary neighborhood of $K = (\mathcal{C}\ell V)\setminus V_1 \supset (\partial V)\cup(\partial V_1)$ in $\mathbb{C}^m$. We can select $\gamma \in \Gamma$, $|\gamma| < \epsilon$, such that both $\mu_{g,V} - \mu_{g,V}^{\gamma}$ and $\mu_{g,V_1} - \mu_{g,V_1}^{\gamma}$ are carried by $\mathcal{N}$. We then note that

$$<\mu_{g,V}^{\gamma} - \mu_{g,V_1}^{\gamma},h> = \int_{K+\imath\gamma} g(z)h(z)dz, \quad \forall\, h \in \mathcal{O}(\mathbb{C}^m),$$

which shows that $\mu_{g,V}^{\gamma} - \mu_{g,V_1}^{\gamma}$ is carried by $\mathcal{N}$, whence the claim. □

Notice that if also $\mu_V' \in \mathcal{O}'(\mathcal{C}\ell V)$ has Property (II.1.3) then $\mu_{g,V} - \mu_V'$ is carried by every neighborhood of ∂V, that is to say, is carried by ∂V. The analytic functionals $\mu_{g,V}$, μ_V' define one and the same hyperfunction $b_V g$ in V. The last part in the statement of Theorem II.1.1 means that that the restriction to $V_1 \subset V$ of $b_V g$ coincides with $b_{V_1} g$. It follows that, if we let V expand to fill U, the hyperfunctions $b_V g$ define a hyperfunction $b_U g$ in U, which is called the *boundary value* of g in U.

Actually, in order to define the boundary value $b_U g$ of a holomorphic function in $\mathcal{W}_\delta(U,\Gamma)$ the cone Γ need not be convex: that Γ is connected suffices. Indeed, consider an open covering of Γ by convex and open cones $\{\Gamma_\alpha\}_{\alpha \in A}$. To $g \in \mathcal{O}(\mathcal{W}_\delta(U,\Gamma))$ and to each $\alpha \in A$ Formula

(II.1.2) (and Theorem II.1.1) assigns the boundary value $b_U g_\alpha$ of the restriction of g to $\mathcal{W}_\delta(U,\Gamma_\alpha)$. But one of the main points of Theorem II.1.1 is that this boundary value is independent of the vector $\gamma \in \Gamma_\alpha$. As a consequence, if $\Gamma_\alpha \cap \Gamma_\beta \neq \emptyset$ (α, $\beta \in A$) then necessarily $b_U f_\alpha = b_U f_\beta$. This entails that all the hyperfunctions in U, $b_U f_\alpha$, are equal when Γ is connected, since in this case any two cones Γ_α and Γ_β (α, $\beta \in A$) can be linked by a chain $\{\Gamma_{\alpha_j}\}$ ($\alpha_j \in A$, $j = 0,1,\ldots,r$) such that $\alpha_0 = \alpha$, $\alpha_r = \beta$ and $\Gamma_{\alpha_{j-1}} \cap \Gamma_{\alpha_j} \neq \emptyset$ for each $j = 1,\ldots,r$.

THEOREM II.1.2.— *Assume that* Γ *is connected. The mapping* $g \to b_U g$ *is an injection of* $\mathcal{O}(\mathcal{W}_\delta(U,\Gamma))$ *into* $\mathcal{B}(U)$.

Proof: Suppose $b_U g = 0$ and let $V \subset\subset U$ have a C^∞ boundary in U. Since $b_V g = 0$, to each open neighborhood $\mathcal{N}$ of ∂V in $\mathbb{C}^m$ there is ϵ, $0 < \epsilon < \delta$, such that $\mu^\gamma_{g,V} \in \mathcal{O}'(\mathcal{N})$, $\forall\, \gamma \in \Gamma$, $|\gamma| < \epsilon$. Consider, for large $\tau > 0$,

$$G^\gamma_{V,\tau}(z) = (\tau/\pi)^{m/2} \langle (\mu^\gamma_{g,V})_w, e^{-\tau\langle z-w\rangle^2} \rangle.$$

Given any $\gamma \in \Gamma$, $|\gamma| < \epsilon$, there is $C_\gamma > 0$ such that

$$\text{(II.1.9)} \qquad |G^\gamma_{V,\tau}(z)| \le C_\gamma \tau^{m/2} \sup_{w\in\mathcal{N}} e^{-\tau\, \mathcal{R}e\langle z-w\rangle^2}, \ \forall\, z \in \mathbb{C}^m.$$

We select $\mathcal{N}$ sufficiently "thin" around ∂V to ensure that there is an open subset of $\mathcal{X}$, $V_1 \subset V$, such that $\mathcal{R}e\langle z-w\rangle^2 \ge d > 0$, $\forall\, z \in V_1$, $w \in \mathcal{N}$. Then $G^\gamma_{V,\tau} \to 0$ uniformly in V_1 as $\tau \to +\infty$ (for fixed γ). On the other hand, by the definition of $\mu^{\gamma}_{g,\mathrm{V}}$,

$$G^{\gamma}_{\mathrm{V},\tau}(z) = (\tau/\pi)^{\mathrm{m}/2}\int_{\mathrm{V}+\imath\gamma} e^{-\tau<z-w>^2} g(w)\mathrm{d}w =$$

$$(\tau/\pi)^{\mathrm{m}/2}\int_{\mathrm{V}_0} e^{-\tau<z-Z(u)-\imath\gamma>^2} g(Z(u)+\imath\gamma)\mathrm{d}[Z(u)],$$

where we call V_0 the projection of V under the map $w = u+\imath v \to u$, and denote by $u \to Z(u) = u + \imath\Phi(u)$ the inverse map. We are in a position to take advantage of the approximation scheme described in Section II.2, [TREVES, 1992], relative to the first integrals $Z'_{\mathrm{j}}(u) = Z_{\mathrm{j}}(u)+\imath\gamma_{\mathrm{j}}$ ($\gamma = (\gamma_1,\ldots,\gamma_{\mathrm{m}})$). By Lemma II.2.3, *loc. cit.*, we conclude that $g(z+\imath\gamma) = 0$ for all $z \in \mathrm{V}_1$ and all $\gamma \in \Gamma$, $|\gamma| < \epsilon$, hence $g \equiv 0$ in $\mathcal{W}_\delta(\mathrm{V}_1,\Gamma)$. Since the sets such as V_1 form a covering of U we conclude that $g \equiv 0$ in $\mathcal{W}_\delta(\mathrm{U},\Gamma)$. □

Distribution boundary values

In this subsection we consider a holomorphic function g in $\mathcal{W}_\delta(\mathrm{U},\Gamma)$ with "slow growth at the edge", ie., endowed with the following property:

(II.1.10) *There are constants $C > 0$, $\mathrm{N} \in \mathbb{Z}_+$, such that*

$$|g(z+\imath\eta)| \le C|\eta|^{-\mathrm{N}},\ \forall\, z \in \mathrm{U},\ \eta \in \Gamma,\ |\eta| < \delta.$$

THEOREM II.1.3.— *Suppose* Γ *is connected and* $g \in \mathcal{O}(\mathcal{W}_\delta(\mathrm{U},\Gamma))$ *satisfies Condition* (II.1.10). *Then* $b_{\mathrm{U}}g \in \mathcal{D}'(\mathrm{U})$, *and for all* $\chi \in C^\infty_c(\mathrm{U})$ *and* $\gamma \in \Gamma$, $|\gamma| < \delta$,

(II.1.11) $$<b_{\mathrm{U}}g,\chi> = \lim_{t \to +0} \int_{\mathcal{X}} g(z+\imath t\gamma)\chi(z)\mathrm{d}z.$$

Proof: Set, for $0 < t \leq 1$,

$$f_\gamma(t) = \int_{\mathrm{U}} g(z+\imath t\gamma)\chi(z)\mathrm{d}z = \int_{\mathrm{U}_0} g(Z(u)+\imath t\gamma)\chi(Z(u))\mathrm{d}Z(u),$$

where $\mathrm{U}_0 = \{ u \in \mathbb{R}^m;\ Z(u) = u+\imath\Phi(u) \in \mathrm{U} \}$. We have

$$\partial_t[g(Z(u)+\imath t\gamma)] = \gamma\cdot(\partial_y g)(Z(u)+\imath t\gamma) =$$

$$\imath\gamma\cdot(\partial_x g)(Z(u)+\imath t\gamma) = \imath \mathrm{M}_\gamma g(Z(u)+\imath t\gamma),$$

where

$$\mathrm{M}_\gamma = \sum_{j=1}^{m} \gamma_j \mathrm{M}_j,$$

with M_j the vector field in u–space defined by the relations $\mathrm{M}_j[Z_k(u)] = \delta_{jk}$ (Kronecker's index; see Section II.1, [TREVES, 1992]). According to Lemma II.1.1, *loc. cit.*, the transpose of $-\mathrm{M}_j$ with respect to the m–form $\mathrm{d}Z_1(u)\wedge\cdots\wedge\mathrm{d}Z_m(u)$ is equal to M_j, hence, for any $p \in \mathbb{Z}_+$,

$$f_\gamma^{(p)}(t) = (-\imath)^p \int_{\mathrm{U}_0} g(Z(u)+\imath t\gamma)\mathrm{M}_\gamma^p[\chi(Z(u))]\mathrm{d}Z(u),$$

and thus

(II.1.12) $$|f_\gamma^{(p)}(t)| \leq C_p t^{-N}, \quad \forall\, p \in \mathbb{Z}_+.$$

After N+1 integrations from, say, 1 to t we conclude that every derivative of f_γ is bounded in the semiclosed interval $[0,\delta)$, and therefore $f_\gamma \in \mathcal{C}^\infty([0,\delta))$, which shows that the right–hand side of (II.1.11) defines a distribution g_0 in U.

We are now going to show that g_0 and $b_{\mathrm{U}}g$ agree as hyperfunctions. It will suffice to show that each point $z_0 \in \mathrm{U}$ has an open neighborhood $\mathrm{V} \subset\subset \mathrm{U}$ in $\mathcal{X}$, with a $\mathcal{C}^\infty$ boundary, in which there is a compactly supported $\mathcal{C}^{\infty}$ function χ, equal to 1 in some open neighborhood $V' \subset\subset V$ of z_0 in $\mathcal{X}$, such that $\mu_{g,V} - \chi g_0 \in \mathcal{O}'((\mathscr{C}\ell V)\backslash V')$ ($\mu_{g,V}$ is the analytic functional in Theorem II.1.1).

Let $\mu^{\gamma}_{g,V}$ be the analytic functional defined in (II.1.2); here as before, $\gamma \in \Gamma$, $|\gamma| < \delta$. Define $\nu^{\gamma} = \mu^{\gamma}_{g,V} - \chi g_0$. We have, for arbitrary $h \in \mathcal{O}(\mathbb{C}^m)$,

$$<\nu^{\gamma},h> = \int_V g(z+\imath\gamma)h(z+\imath\gamma)dz - \lim_{t\to+0} \int_V g(z+\imath t\gamma)\chi(z)h(z)dz.$$

Hypothesis (II.1.10) combined with the Taylor's expansion of h of order N,

$$h(z) = h(z+\imath t\gamma) - \sum_{1\leq|\alpha|\leq N} \frac{1}{\alpha!}(\imath t\gamma)^{\alpha} h^{(\alpha)}(z)dz + O(t^{N+1}),$$

shows that

$$\int_V g(z+\imath t\gamma)\chi(z)h(z)dz = \int_V g(z+\imath t\gamma)h(z+\imath tz)\chi(z)dz -$$

$$\sum_{1\leq|\alpha|\leq N} \frac{1}{\alpha!}(\imath t\gamma)^{\alpha}\int_V g(z+\imath t\gamma)\chi(z)h^{(\alpha)}(z)dz + O(t).$$

Since we know that each integral $\int_V g(z+\imath t\gamma)\chi(z)h^{(\alpha)}(z)dz$ has a finite limit as $t \to +0$ we conclude that

$$\lim_{t\to+0} \int_V g(z+\imath t\gamma)\chi(z)h(z)dz = \lim_{t\to+0} \int_V g(z+\imath t\gamma)h(z+\imath tz)\chi(z)dz,$$

and therefore

$$<\nu^{\gamma},h> = \int_V g(z+\imath\gamma)h(z+\imath\gamma)[1-\chi(z)]dz +$$

$$\lim_{t\to+0} \int_V [g(z+\imath\gamma)h(z+\imath\gamma)-g(z+\imath t\gamma)h(z+\imath tz)]\chi(z)dz.$$

The first integral at the right defines an analytic functional carried by $(\mathscr{C}\ell V\backslash V') + \imath\gamma$. Regarding the second term in the right–hand side we observe that, if we write $F = gh$, then, by the argument in the first part of the proof we have

$$\int_V [F(z+\imath\gamma)-F(z+\imath t\gamma)]\chi(z)dz =$$

$$\imath(1-t)\int_V \int_0^1 (\gamma\cdot\partial_z F)(z+\imath(1-(1-t)\theta)\gamma)\ \chi(z)\mathrm{d}z\mathrm{d}\theta =$$

$$\imath(1-t)\int_{V_0}\int_0^1 M_\gamma[F(Z(u)+\imath(1-(1-t)\theta)\gamma)]\chi(Z(u))\mathrm{d}Z(u)\mathrm{d}\theta =$$

$$-\imath(1-t)\int_{V_0}\int_0^1 F(Z(u)+\imath(1-(1-t)\theta)\gamma)\ M_\gamma[\chi(Z(u))]\mathrm{d}Z(u)\mathrm{d}\theta,$$

where $V_0 = \{\ u \in \mathbb{R}^m;\ Z(u) = u + \imath\Phi(u) \in V\ \}$. If $t,\ \theta \in [0,1]$, perforce $1-(1-t)\theta \geq t$, and the argument in the first part of the proof easily shows that

$$h \to \lim_{t\to+0} \int_V \int_0^1 F(Z(u)+\imath(1-(1-t)\theta)\gamma)\ M_\gamma[\chi(Z(u))]\mathrm{d}Z(u)$$

defines an analytic functional carried by $\mathrm{supp}(\mathrm{d}\chi) + \imath[0,\gamma]$ ($[0,\gamma] \subset \Gamma$ is the straight–line segment joining 0 to γ).

Take $\gamma = c\gamma_0$, $\gamma_0 \in \Gamma$, $|\gamma_0| < \delta$, $0 < c < 1$. It is quite evident that the definition of g_0 is independent of c (a priori g_0 could depend on the choice of γ_0; of course, it will be seen not to be so when we will have established that $g_0 = b_U g$). Let then $\mathcal{N}$ be an arbitrary open neighborhood of $(\mathscr{C}\ell V)\backslash V'$. For c suitably small, we conclude that $\mu^\gamma_{g,V} - \chi g_0$ is carried by $\mathcal{N}$. Thanks to Theorem II.1.1, by taking c small we can also ensure that $\mu_{g,V}-\mu^\gamma_{g,V}$ be carried by $\mathcal{N}$, and thus that the same be true of $\mu_{g,V}-\chi g_0$. But the latter is independent of c and therefore must be carried by $(\mathscr{C}\ell V)\backslash V'$, which is what we wanted to prove. □

II.2 FBI TRANSFORM OF A HYPERFUNCTION. INVERSION OF THE FBI TRANSFORM. HYPO–ANALYTICITY CHARACTERIZED BY THE EXPONENTIAL DECAY OF THE FBI TRANSFORM

In this section we extend to hyperfunctions on the maximally real submanifold $\mathcal{X}$ the Fourier—Brós—Iagolnitzer transform defined in Section IX.1, [TREVES, 1992], for distributions. In the following section II.3 the inversion formula for the FBI transform of an analytic functional will be used to show that, locally, every hyperfunction is the sum of boundary values of holomorphic functions in wedges, as defined in section II.1.

We are going to make use of concepts introduced in Section IX.2, [TREVES, 1992], and first of all, of the *real structure bundle* of $\mathcal{X}$, $\mathbb{R}T'_{\mathcal{X}}$. We recall its definition. Call $T'^{1,0}$ the cotangent structure bundle of $\mathbb{C}^m$, ie., the vector subbundle of $\mathbb{C}T^*\mathbb{C}^m$ spanned by $dz_1,...,dz_m$; that $\mathcal{X}$ is maximally real means that the natural pullback map $\mathbb{C}T^*\mathbb{C}^m\big|_{\mathcal{X}} \to \mathbb{C}T^*\mathcal{X}$ induces a bijection $T'^{1,0}\big|_{\mathcal{X}} \to \mathbb{C}T^*\mathcal{X}$; $\mathbb{R}T'_{\mathcal{X}}$ is the pre–image of the real cotangent bundle of $\mathcal{X}$, $T^*\mathcal{X}$, under this bijection.

If we use complex coordinates $z_1,...,z_m$ in $\mathbb{C}^m$ we may identify to $\mathbb{C}^m\times\mathbb{C}^{2m}$ the complexified cotangent bundle $\mathbb{C}T^*\mathbb{C}^m$, via the map $(z,(\zeta,\zeta'))$ $\to (z,\sum_{i=1}^{m}(\zeta_i dz_i+\zeta'_i d\bar{z}_i))$. Then $T'^{1,0}$ gets identified to $\mathbb{C}^{2m}$ regarded as the vector subbundle of $\mathbb{C}^m\times\mathbb{C}^{2m}$ consisting of the points $(z,(\zeta,0))$. Thus $\mathbb{R}T'_{\mathcal{X}}$ can be viewed as a *real* vector subbundle of $T'^{1,0}$. If we equip $T'^{1,0} \cong \mathbb{C}^{2m}$ with its natural complex structure $\mathbb{R}T'_{\mathcal{X}}$ becomes a maximally real submanifold of $T'^{1,0}$.

We continue to avail ourselves of the usual parametric representation of $\mathcal{X}$ in a neighborhood of one of its points z_0: $\mathrm{B}_0 \ni x \to Z(x) = x + \imath\Phi(x) \in \mathrm{B}_{01} = \mathrm{B}_0 + \imath\mathrm{B}_1$. As before we write $\Omega = Z(\mathrm{B}_0) \subset \mathcal{X}$. And as before it is convenient to take the point $z_0 \in \mathcal{X}$ to be the origin in $\mathbb{C}^m$ and, possibly after a change of the complex coordinates, to assume that $\mathcal{X}$ is tangent to $\mathbb{R}^m$ at 0; and as a matter of fact, that the order of contact of $\mathcal{X}$ with $\mathbb{R}^m$ at 0 is as high as we wish. This can always be achieved by a holomorphic change of the local coordinates; it enables us, here, to assume

$$|\Phi(x)| \leq const.|x|^3.$$

After further contraction of B_0 we may assume that (I.3.3) holds and also, for later use, the following:

$$|\Phi(x)| + 2|\Phi(x)|^2 \leq |x|^2, \ \forall\, x \in \mathrm{B}_0. \tag{II.2.1}$$

Under these hypotheses $Z_x(0) = I_m$, the m×m identity matrix; and $Z_x(x)$ is nonsingular for all $x \in \mathrm{B}_0$. The submanifold $\mathbb{R}\mathrm{T}'_{\mathcal{X}}|_{\Omega}$ of $\mathrm{T}'^{1,0}$ is defined, near z_0, by the equations $z = Z(x)$, $\zeta = {}^tZ_x(x)^{-1}\xi$ ($x \in \mathrm{B}_0$, $\xi \in \mathbb{R}^m$). If we identify $\mathbb{C}^m$ to an arbitrary fibre of $\mathrm{T}'^{1,0}$ by means of the coordinates ζ_i with respect to the basis $dz_1,\ldots,dz_m$, the fibre of $\mathbb{R}\mathrm{T}'_{\mathcal{X}}$ at 0 gets identified to $\mathbb{R}^m$. If $\mathcal{C}$ is any open cone in $\mathbb{C}^m\backslash\{0\}$ such that $\mathbb{R}^m\backslash\{0\} \subset \mathcal{C}$, the neighborhood Ω of 0 in $\mathcal{X}$ can be taken sufficiently small that the set of points $(z,\zeta) \in \Omega\times\mathcal{C}$ will be an open and conic neighborhood of $\mathbb{R}\mathrm{T}'_{\mathcal{X}}\backslash 0|_{\Omega}$ in $\mathrm{T}'^{1,0}|_{\Omega}$.

All the preceding hypotheses combined entail that $\mathcal{X}$ is *well–positioned* at z_0 (Definition IX.2.2, [TREVES, 1992]): it is possible to choose the neighborhood Ω and the number κ, $0 < \kappa < 1$, in such a way that

(II.2.2) $$|\mathcal{Im}\,\zeta| \leq \kappa|\mathcal{Re}\,\zeta|,\ \forall\, z \in \Omega,\ \zeta \in \mathbb{R}T'_{\mathcal{X}}\big|_z;$$

(II.2.3) $$\mathcal{Im}[\zeta\cdot(z-z')+\imath\langle\zeta\rangle\langle z-z'\rangle^2] \geq (1-\kappa)|\zeta|\,|z-z'|^2,$$
$$\forall\, z, z' \in \Omega, \forall\, \zeta \in (\mathbb{R}T'_{\mathcal{X}}\big|_z)\cup(\mathbb{R}T'_{\mathcal{X}}\big|_{z'}).$$

We start by considering an arbitrary analytic functional μ carried by a compact subset K of Ω. By definition (cf. Definition IX.1.1, [TREVES, 1992]) the FBI transform of μ will be the function

(II.2.4) $$\mathcal{F}\mu(z,\zeta) = \langle \mu_w, e^{\imath\zeta\cdot(z-w)-\langle\zeta\rangle\langle z-w\rangle^2}\Delta(z-w,\zeta)\rangle,$$

holomorphic with respect to $(z,\zeta) \in \mathbb{C}^m\times\mathcal{C}_\kappa$ if

$$\mathcal{C}_\kappa = \{\ \zeta \in \mathbb{C}^m;\ |\mathcal{Im}\,\zeta| < \kappa|\mathcal{Re}\,\zeta|\ \}.$$

In (II.2.4) $\Delta(z,\zeta)$ stands for the Jacobian determinant of the map $\zeta \to \zeta + \imath\langle\zeta\rangle z$ ($\langle\zeta\rangle$ stands for the main square root of $\langle\zeta\rangle^2 = \zeta_1^2+\cdots+\zeta_m^2$). From (II.2.4) we derive, for any $\delta > 0$,

(II.2.5) $$|\mathcal{F}\mu(z,\zeta)| \leq C_\delta \sup_w \Big\{|\Delta(z-w,\zeta)|\exp(-\mathcal{Q}(z-w,\zeta)|\zeta|)\Big\},$$

where

$$\mathcal{Q}(z,\zeta) = \mathcal{Im}(\zeta\cdot z+\imath\langle\zeta\rangle\langle z\rangle^2)/|\zeta|,$$

and the supremum is taken over all points $w \in \mathbb{C}^m$ with dist$(w,K) < \delta$. Finding upper and lower bounds for the quantity $\mathcal{Q}$ is the key to all the estimates that follow. Notice that $\mathcal{Q}$ is positive–homogeneous of degree zero with respect to ζ. As a rule we shall establish a bound (either from below or from above) on $\mathcal{Q}$ at a given point (or points) of $\mathbb{R}T'_{\mathcal{X}}$, and then use continuity to obtain a (slightly weaker) bound in a conic neighborhood of those points in $T'^{1,0}$ ($\cong \mathbb{C}^{2m}$).

THEOREM II.2.1.— *Suppose $\mathcal{X}$ is well–positioned at z_0. If the open neigh–borhood Ω of z_0 in $\mathcal{X}$ is sufficiently small, then to every number $\epsilon > 0$ there is an open and conic neighborhood $\mathcal{U}^\epsilon$ of $\mathbb{R}T'_{\mathcal{X}}\backslash 0\big|_\Omega$ in $T'^{1,0}$ that has the following property:*

To every analytic functional μ carried by Ω there is a constant $C_\epsilon > 0$ such that

$$|\mathcal{F}\mu(z,\zeta)| \leq C_\epsilon e^{\epsilon|\zeta|}, \ \forall\,(z,\zeta) \in \mathcal{U}^\epsilon. \tag{II.2.6}$$

Proof: We assume that (II.2.2) and (II.2.3) hold. Consider a point $w \in \mathbb{C}^m$ such that there is a point $z' \in \operatorname{supp}\mu$ such that $|w-z'| < \delta$; then $\mathcal{Q}(z-w,\zeta) = Q(z-z',\zeta)+O(\delta)$. We avail ourselves of (II.2.3): by taking $(z,\zeta) \in \mathbb{R}T'_{\mathcal{X}}\backslash 0\big|_{\Omega}$, δ sufficiently small and w such that $\operatorname{dist}(w,\operatorname{supp}\mu) < \delta$, we get $\mathcal{Q}(z-w,\zeta) \geq -\frac{1}{2}\epsilon$. We can therefore let (z,ζ) vary in a conic neigh–borhood of $\mathbb{R}T'_{\mathcal{X}}\big|_\Omega$ in $\mathbb{C}^{2m}$ while still retaining the estimate $\mathcal{Q}(z-w,\zeta) \geq -\epsilon$. The result follows then from (II.2.5). □

THEOREM II.2.2.— *Suppose $\mathcal{X}$ is well–positioned at z_0. If the open neigh–borhood Ω of z_0 in $\mathcal{X}$ is sufficiently small, then every open subset U of $\mathcal{X}$ with compact closure contained in Ω has the following property:*

Given an arbitrary compact subset K of U and another one, K′, of $(\mathcal{C}\ell\,U)\backslash K$ there are an open and conic neighborhood $\mathcal{U}$ of $\mathbb{R}T'_{\mathcal{X}}\backslash 0\big|_K$ in $T'^{1,0}$ and a number $R > 0$ such that to each analytic functional $\mu \in \mathcal{O}(K')$ there is a constant $C > 0$ such that

$$|\mathcal{F}\mu(z,\zeta)| \leq C\,e^{-|\zeta|/R}, \ \forall\,(z,\zeta) \in \mathcal{U}. \tag{II.2.7}$$

Proof: Same notation as in the proof of Theorem II.2.1. Let K′ be a compact subset of $(\mathscr{C}\ell U)\backslash K$ which carries the analytic functional μ. Given any $\delta > 0$ we derive from (II.2.3):

$$\mathcal{Q}(z-w,\zeta) \geq (1-\kappa)|z-z'|^2 + O(\delta),$$

for all $(z,\zeta) \in \mathbb{R}T'_{\mathcal{X}}\backslash 0|_K$ and all $w \in \mathbb{C}^m$ such that $\mathrm{dist}(w,\mathrm{supp}\ \mu) < \delta$. But $z \in K$, $z' \in K' \Rightarrow |z-z'| > d > 0$. This implies that if δ is sufficiently small and if the open and conic neighborhood $\mathscr{U}$ of $\mathbb{R}T'_{\mathcal{X}}\backslash 0|_K$ in $T'^{1,0}$ ($\cong \mathbb{C}^{2m}$) suitably "thin", then

$$\mathcal{Q}(z-w,\zeta) \geq c > 0,$$

for all $(z,\zeta) \in \mathbb{R}T'_{\mathcal{X}}\backslash 0|_K$ and all $w \in \mathbb{C}^m$ such that $\mathrm{dist}(w,\mathrm{supp}\ \mu) < \delta$. □

We are now in a position to define the FBI transform of a hyper-function $u \in \mathcal{B}(U)$. First we must introduce the space $\mathfrak{E}(U)$ of the holo-morphic functions h in some conic neighborhood $\mathscr{U}$ of $\mathbb{R}T'_{\mathcal{X}}\backslash 0|_U$ in $T'^{1,0}$ ($\mathscr{U}$ may depend on h) that have the following property:

(II.2.8) *Given any number $\epsilon > 0$ there is an open and conic neighborhood $\mathscr{U}^\epsilon \subset \mathscr{U}$ of $\mathbb{R}T'_{\mathcal{X}}\backslash 0|_U$ in $T'^{1,0}$ such that*

$$\sup_{(z,\zeta)\in\mathscr{U}^\epsilon}\left[e^{-\epsilon|\zeta|}\,|h(z,\zeta)|\right] < +\infty.$$

A function $h \in \mathfrak{E}(U)$ will belong to the subspace $\mathfrak{E}_0(U)$ if it is defined and holomorphic in a conic and open neighborhood $\mathscr{U}$ of $\mathbb{R}T'_{\mathcal{X}}\backslash 0|_U$ and is such that

(II.2.9) *to each compact set $K \subset U$ there are an open and conic neighborhood $\mathscr{U}_K \subset \mathscr{U}$ of $\mathbb{R}T'_{\mathcal{X}}\backslash 0|_K$ in $T'^{1,0}$ and a constant*

$c > 0$ *such that*

$$\sup_{(z,\zeta)\in\mathcal{U}_K}\left[e^{c|\zeta|}\,|h(z,\zeta)|\right] < +\infty.$$

Theorems II.2.1 and II.2.2 imply that the FBI transform of analytic functionals defines a linear map $\mathcal{F} : \mathcal{B}(U) \to \mathfrak{E}(U)/\mathfrak{E}_0(U)$, by associating to a representative $\mu \in \mathcal{O}'(\mathscr{C}\ell U)$ of u the equivalence class mod $\mathfrak{E}_0(U)$ of $\mathcal{F}\mu$. By (II.2.6) the latter belongs to $\mathfrak{E}(U)$; and by (II.2.7) the equivalence class of $\mathcal{F}\mu$ does not depend on the choice of the representative μ.

Inversion of the FBI transform of an analytic functional

In this subsection we obtain various inversion formulas for the FBI transform of an analytic functional carried by the compact closure K $= \mathscr{C}\ell U$ of an open subset U of $\mathcal{X}$ when $K \subset \Omega = \mathcal{Z}(B_0) \subset \mathcal{X}$. We continue to assume that (II.2.2) and (II.2.3) hold. Our starting point will be the proof of Lemma I.3.1, more precisely the following

LEMMA II.2.1.— *Given any function* $\chi \in C_c^\infty(\Omega)$, $\chi \equiv 1$ *in some neighborhood of* K *in* Ω, *there is an open neighborhood* $\mathcal{U}$ *of* K *in* $\mathbb{C}^m$ *such that, whatever* $h \in \mathcal{O}(\mathbb{C}^m)$, *as* $\nu \to +\infty$ *the functions* $\mathcal{E}_\nu(\chi h)$ [see (I.3.4)] *converge uniformly to* h *in* $\mathcal{U}$.

Using the Fourier transform of the Gaussian shows that, if $\epsilon = 1/4\nu$, $\mathcal{E}_\nu(\chi h)(w)$ is equal to the function

$$(2\pi)^{-m}\int_{z\in\Omega}\int_{\xi\in\mathbb{R}^m} e^{i\xi\cdot(w-z)-\epsilon|\xi|^2}\chi(z)h(z)dzd\xi.$$

Notice that for the conclusion in Lemma II.2.1 to be valid there is no need for χ to be smooth. We may as well let χ be the characteristic function of an open neighborhood $V \subset\subset \Omega$ of K in $\mathcal{X}$. We shall consider the functions

$$H^{\epsilon}(w) = (2\pi)^{-m} \int_{z \in V} \int_{\xi \in \mathbb{R}^m} e^{i\xi \cdot (w-z) - \epsilon |\xi|^2} h(z) dz d\xi.$$

Let us select a compact subset K′ of $\mathcal{U} \cap \Omega \subset V$ and an open ball $B' = \{ \gamma \in \mathbb{R}^m;\ |\gamma| < r \}$ with $r < 1$ to be chosen more precisely below, but in any case such that $K' + iB' \subset \mathcal{U}$. We restrict the variation of w to $K' + iB'$: we take $w = w_* + i\gamma$ with $w_* = Z(s) \in K'$ ($s \in B_0$) and $\gamma \in B'$; and we carry out a change of variables $\xi \to \theta = \xi + i|\xi|(w-z)$. By choosing diam B_0 and diam B' sufficiently small, we may ensure that $|\mathcal{I}m\,\theta| \leq \frac{1}{2} |\mathcal{R}e\,\theta|$ is valid, whatever $(s,\gamma) \in B_0 \times B'$, $z \in \Omega$, $\xi \in \mathbb{R}^m$. We conclude that, as $\epsilon \to +0$,

$$H^{\epsilon}(w_* + i\gamma) =$$

$$(2\pi)^{-m} \int_{z \in V} \int_{\xi \in \mathbb{R}^m} e^{i\theta \cdot (w_* + i\gamma - z) - \epsilon \langle \theta \rangle^2} h(z) \Delta(w_* + i\gamma - z, \xi) dz d\xi$$

converges uniformly to $h(w_* + i\gamma)$ in $K' + iB'$. We recall (cf. (II.2.4)) that $\Delta(z,\xi)$ is the Jacobian determinant of the map $\xi \to \theta = \xi + i|\xi| z$.

Next we deform the domain of z–integration, somewhat as in the proof of Lemma I.3.1. We select $\psi \in C_c^{\infty}(V)$, $\psi \equiv 1$ in a neighborhood V′ of K′ in $\mathcal{X}$, $0 \leq \psi \leq 1$ everywhere, and $\lambda > 0$. Then we move the domain of z–integration from V to the image $\tilde{V}$ of the map $V \ni z_* \to z = z_* - i\lambda\psi(z_*)\dot{\xi}$ ($\dot{\xi} = \xi/|\xi|$).

After this is done we deform the domain of ξ–integration from $\mathbb{R}^m$ to the image of $\mathbb{R}^m$ under the map $\xi \to \zeta = {}^t Z_x(x_*)^{-1}\xi$ if $z_* = Z(x_*)$ (thus $\zeta \in \mathbb{R}T'_{\mathcal{X}}|_{z_*}$). We focus on the quantity

$$\text{(II.2.10)} \qquad \mathcal{Q} = -\ \mathcal{R}e[\imath\zeta\cdot(w_*+\imath\gamma-z_*+\imath\lambda\psi(z_*)\dot{\zeta}) - <\zeta><w_*+\imath\gamma-z_*+\imath\lambda\psi(z_*)\dot{\zeta}>^2]/|\zeta|,$$

where $\dot{\zeta} = \zeta/<\zeta>$. By contracting Ω about 0 we can select the number κ in (II.2.2) and (II.2.3) as small as we wish, and therefore small enough to ensure that

$$|<\zeta>| \leq |\zeta| \leq 2\ \mathcal{R}e<\zeta>,\ \forall\ z \in \Omega,\ \zeta \in \mathbb{R}\mathrm{T}'_{\mathcal{X}}\big|_{z},$$

and thus obtain

$$\mathcal{Q} \geq \mathcal{I}m[(w_*-z_*)\cdot\zeta + \imath<\zeta><w_*-z_*>^2]/|\zeta| - |\gamma| + \tfrac{1}{2}\lambda\psi(z_*) - 2|w_*-z_*|\cdot[|\gamma|+\lambda\psi(z_*)] - [|\gamma|+\lambda\psi(z_*)]^2,$$

whence, whatever the number $\tau > 0$,

$$\mathcal{Q} \geq \mathcal{I}m[\zeta\cdot(w_*-z_*) + \imath<\zeta><w_*-z_*>^2]/|\zeta| + \tfrac{1}{2}\lambda\psi(z_*) - |\gamma| - \tau^{-1}|w_*-z_*|^2 - 2(1+\tau)[|\gamma|^2+\lambda^2\psi(z_*)^2].$$

At this point we exploit the fact that $\zeta \in \mathbb{R}\mathrm{T}'_{\mathcal{X}}\big|_{z_*}$ and the hypothesis that $\mathcal{X}$ is well positioned at the origin, more precisely we exploit the inequality (II.2.3), getting

$$\mathcal{Q} \geq (1-\kappa-\tau^{-1})|w_*-z_*|^2 + [\tfrac{1}{2}-2(1+\tau)\lambda]\lambda\psi(z_*) - [1+2(1+\tau)|\gamma|]|\gamma|.$$

First we select $\tau = 2/(1-\kappa)$ and keep it fixed. Next we require $0 < \lambda \leq 1/8(1+\tau) = \frac{1}{8}(1-\kappa)/(3-\kappa)$. Since $r \leq 1$ we conclude that, if $C > 0$ is suitably large,

$$\text{(II.2.11)} \qquad \mathcal{Q} \geq \tfrac{1}{2}(1-\kappa)|w_*-z_*|^2 + \tfrac{1}{4}\lambda\psi(z_*) - Cr,$$
$$\forall\ w_* \in \mathrm{K}',\ z_* \in \Omega,\ \zeta \in \mathbb{R}\mathrm{T}'_{\mathcal{X}}\big|_{z_*},\ \gamma \in \mathrm{B}'.$$

From now on λ is kept fixed. Recall that $\psi(z_*) = 1$ if $z_* \in \mathrm{V}'$. On the other hand there is a number $d > 0$ such that $|w_* - z_*| \geq d$ whatever $w_* \in \mathrm{K}'$, $z_* \in \Omega \backslash \mathrm{V}'$. It follows that there is a number $c > 0$ such that

$$\tfrac{1}{2}(1-\kappa)|w_*-z_*|^2 + \tfrac{1}{4}\lambda\psi(z_*) \geq 2c, \ \forall\ w_* \in \mathrm{K}', z_* \in \Omega.$$

If we require $r \lesssim c/C$ (II.2.11) entails

$$\text{(II.2.12)} \qquad \mathcal{Q} \geq c, \forall\ w_* \in \mathrm{K}', z_* \in \Omega, \zeta \in \mathbb{R}\mathrm{T}'_{\mathcal{X}}\big|_{z_*}, \gamma \in \mathrm{B}'.$$

By the Lebesgue dominated convergence theorem we derive from (II.2.12) that, if $w = w_*+\imath\gamma$, with $w_* \in \mathrm{K}'$ and $\gamma \in \mathrm{B}'$, the limit, as $\epsilon \to +0$, of the integral $H^{\epsilon}(w)$ is equal to

$$(2\pi)^{-\mathrm{m}}\int_{z_*\in\mathrm{V}}\int_{\xi\in\mathbb{R}^{\mathrm{m}}} e^{\imath\theta\cdot(w-z)}h(z)\Delta(w-z,\zeta)\mathrm{d}z\mathrm{d}\zeta$$

[recall that $z = z_* - \imath\lambda\psi(z_*)\xi/|\xi|$]. Requiring that $\mathrm{K} \subset \mathcal{I}nt\mathrm{K}' \subset \mathcal{U}\cap\mathrm{V}$ we see that $\mathcal{U}' = \mathcal{I}nt\mathrm{K}'+\imath\mathrm{B}'$ is an open neighborhood of K contained in $\mathcal{U}$. Setting $\lambda_0 = \frac{1}{8}(1-\kappa)/(3-\kappa)$ we may state:

LEMMA II.2.2.— *If* $\Omega = Z(\mathrm{B}_0) \subset \mathcal{X}$ *is sufficiently small then, given any open neighborhood* $\mathrm{V} \subset\subset \Omega$ *of* $\mathrm{K} \subset\subset \Omega$ *in* $\mathcal{X}$ *and any number* λ, $0 < \lambda < \lambda_0$, *there is an open neighborhood* $\mathcal{U}'$ *of* K *in* $\mathbb{C}^{\mathrm{m}}$, $\mathcal{U}'\cap\Omega \subset\subset \mathrm{V}$, *such that, for all* $h \in \mathcal{O}(\mathbb{C}^{\mathrm{m}})$ *and* $w \in \mathcal{U}'$,

$$\text{(II.2.13)} \qquad h(w) =$$

$$(2\pi)^{-\mathrm{m}}\int_{z_*\in\mathrm{V}}\int_{\zeta\in\mathbb{R}\mathrm{T}'_{\mathcal{X}}|_{z_*}} e^{\imath\zeta\cdot(w-z)-<\zeta><w-z>^2}\Delta(w-z,\zeta)h(z)\mathrm{d}z\mathrm{d}\zeta,$$

where $z = z_* - i\lambda\psi(z_*)\zeta/<\zeta>$, $\psi \in C_c^\infty(\mathrm{V})$, $\psi \equiv 1$ *in* $\mathcal{U}'\cap\Omega$, $0 \leq \psi \leq 1$ *everywhere.*

By letting $\mu \in \mathcal{O}'(\mathrm{K})$ act with respect to w under the integral sign, in the right–hand side of (II.2.13), changing variable $\zeta \to -\zeta$ and taking the definition (II.2.4) into account, we obtain

THEOREM II.2.3.— *Suppose* $\mathcal{X}$ *is well–positioned at* z_0. *Let* Ω *be a sufficiently small open neighborhood of* z_0 *in* $\mathcal{X}$, K *a compact subset of* Ω *and* V $\subset\subset \Omega$ *an open neighborhood of* K *in* $\mathcal{X}$. *For all* $\mu \in \mathcal{O}'(\mathrm{K})$ *and* $h \in \mathcal{O}(\mathbb{C}^m)$,

$$(\mathrm{II}.2.14)\qquad <\mu,h> = (2\pi)^{-m}\int_{z_*\in \mathrm{V}}\int_{\zeta\in \mathbb{R}\mathrm{T}'_{\mathcal{X}}|_{z_*}} \mathcal{F}\mu(z,\zeta)\, h(z)\mathrm{d}z\mathrm{d}\zeta$$

where $z = z_* + i\lambda\psi(z_*)\zeta/<\zeta>$, $0 < \lambda < \lambda_0$.

Formula (II.2.14) will be used in the last subsection of the present section.

From (II.2.13) we can also derive an approximate inversion formula for the FBI transform of an analytic functional carried by K $\subset\subset \Omega$, generalizing Theorem IX.2.2, [TREVES, 1992]. Again we use dominated convergence, now to replace the convergence factor $\exp(-\epsilon<\theta>^2)$ by $\exp(-\epsilon<\zeta>^2)$. In this manner we see that the limit, as $\epsilon \to +0$, of the integral $H^\epsilon(w)$ is the same as that of

$$(2\pi)^{-m}\int_{z_*\in \mathrm{V}}\int_{\zeta\in \mathbb{R}\mathrm{T}'_{\mathcal{X}}|_{z_*}} e^{i\theta\cdot(w-z)-\epsilon<\zeta>^2} h(z)\Delta(w-z,\zeta)\mathrm{d}z\mathrm{d}\zeta,$$

where, as before, $z = z_* - \imath\lambda\psi(z_*)\dot\zeta$. But in this last integral we have the right to "undeform" the domain of z–integration back to Ω (to be precise we must switch back from ζ to $\xi \in \mathbb{R}^m$, only then undeform the z–integration back to Ω and, once this is done, revert from ξ to ζ). We reach the following conclusion:

LEMMA II.2.3.— *If the open neighborhood Ω of z_0 in $\mathcal{X}$ is sufficiently small then, given any open neighborhood* V $\subset\subset \Omega$ *of* K *in $\mathcal{X}$, there is an open neighborhood $\mathcal{U}'$ of* K *in $\mathbb{C}^m$ such that, whatever $h \in \mathcal{O}(\mathbb{C}^m)$, as $\epsilon \to +0$ the function*

$$(2\pi)^{-m}\int_{z\in V}\int_{\zeta\in \mathbb{R}T'_{\mathcal{X}}|_z} e^{\imath\zeta\cdot(w-z)-<\zeta><w-z>^2-\epsilon<\zeta>^2}\Delta(w-z,\zeta)h(z)dzd\zeta$$

converges uniformly to h in $\mathcal{U}'$.

We let $\mu \in \mathcal{O}'(K)$ act under the integral sign with respect to the variable w. In the notation (II.2.4) and after a change of variable $\xi \to -\xi$, Lemma II.2.3 entails

THEOREM II.2.4.— *If $\mathcal{X}$ is well–positioned at z_0 and if the neighborhood Ω of z_0 in $\mathcal{X}$ is sufficiently small then, given any open neighborhood* V $\subset\subset \Omega$ *of* K *in $\mathcal{X}$, any analytic functional $\mu \in \mathcal{O}'$*(K) *and any function $h \in \mathcal{O}(\mathbb{C}^m)$,*

(II.2.15)
$$<\mu,h> = \lim_{\epsilon\to+0} (2\pi)^{-m}\int_{z\in V}\int_{\zeta\in \mathbb{R}T'_{\mathcal{X}}|_z} \mathcal{F}\mu(z,\zeta)e^{-\epsilon<\zeta>^2}h(z)dzd\zeta.$$

Theorems II.2.2 and II.2.4 lead to other "inversion formulas". These can be used to prove that, locally, any hyperfunction can be represented as the sum of finitely many boundary values of holomorphic functions in wedges.

It is checked at once that the whole argument of the preceding pages remains valid if we replace the definition of the FBI transform (II.2.4) by

(II.2.16)
$$\mathcal{F}^{\varrho}\mu(z,\zeta) = < \mu_w, e^{i\zeta\cdot(z-w)-\rho<\zeta><z-w>^2}\Delta(\rho(z-w),\zeta)>,$$

where ρ is any number > 0. In particular, the analogue of (II.2.14) is valid: if V is a sufficiently small open subset of Ω such that $K \subset\subset V \subset\subset \Omega$ is sufficiently small (the meaning of "sufficiently small" will depend on ρ), then

$$\text{(II.2.17)} \qquad <\mu,h> = (2\pi)^{-m}\int_{z_*\in V}\int_{\zeta\in\mathbb{R}T'_{\mathcal{X}}|_{z_*}} \mathcal{F}^{\varrho}\mu(z,\zeta)h(z)dzd\zeta$$

for all $\mu \in \mathcal{O}'(K)$ and all $h \in \mathcal{O}(\mathbb{C}^m)$ [z_* has the same meaning as in (II.2.14)]. We make use of the following identity (p. 431, [TREVES, 1992]):

$$(2<\zeta>/\pi)^{-m'/2}\Delta(\tfrac{1}{2}(z-w),\zeta)e^{i\zeta\cdot(z-w)-\frac{1}{2}<\zeta><z-w>^2} =$$

$$\int_{\mathbb{R}^m} e^{i\zeta\cdot(z-t)+i\zeta\cdot(t-w)-<\zeta>[<z-t>^2+<t-w>^2]}\Delta(t-w,\zeta)dt.$$

Letting μ act with respect to w on both sides of this identity we get

(II.2.18)
$$\mathcal{F}^{\frac{1}{2}}\mu(z,\zeta) = (2<\zeta>/\pi)^{m'/2}\int_{\mathbb{R}^m} e^{i\zeta\cdot(z-t)-<\zeta><z-t>^2}\mathcal{F}\mu(t,\zeta)dt.$$

Combining (II.2.17), where $\rho = \frac{1}{2}$, with (II.2.18) yields

(II.2.19) $$<\mu,h> =$$

$$(2\pi^3)^{-m/2}\int_{z_*\in V}\int_{t\in\mathbb{R}^m}\int_{\zeta\in\mathbb{R}T'_{\mathcal{X}}|_{z_*}} e^{i\zeta\cdot(z-t)-<\zeta><z-t>^2}\mathcal{F}\mu(t,\zeta)h(z)\cdot$$

$$<\zeta>^{m/2}dzdtd\zeta$$

for all $\mu \in \mathcal{O}'(K)$, $h \in \mathcal{O}(\mathbb{C}^m)$ (cf. (IX.4.1), [TREVES, 1992]).

At this stage we return to the concept of a very well positioned manifold ([TREVES, 1992], p. 422). Recall that Ω is the image of the open ball B_0 under the map $x \to Z(x) = x + i\Phi(x)$. Let $\chi \in C_c^\infty(B_0)$, $\chi(x) \equiv 1$ if $|x| < r$, $|\chi^{(\alpha)}(x)| \leq C_\alpha r^{-|\alpha|}$ for all $\alpha \in \mathbb{Z}_+^m$ and all $x \in \mathbb{R}^m$. Set then $\hat{\Phi} = \chi\Phi$. Since Φ vanish to order 3 at the origin if $|\alpha| \leq 2$ then

$$|\hat{\Phi}^{(\alpha)}(x)| \leq \sum_{\beta+\gamma=\alpha} \binom{\alpha}{\gamma}|\chi^{(\beta)}(x)||\Phi^{(\gamma)}(x)| \leq$$

$$C\sum_{\beta+\gamma=\alpha} r^{-|\beta|}r^{3-|\gamma|} \leq C'r.$$

The crucial remark here is that, given any κ, $0 < \kappa < 1$, there is $r > 0$ such that, if we substitute $\hat{Z}(x) = x + i\hat{\Phi}(x)$ for $Z(x)$, the properties analogous to (II.2.2) and (II.2.3) are now valid in the whole of $\hat{\Omega} = \hat{Z}(\mathbb{R}^m)$. We may assume $V \subset\subset \{ x \in \mathbb{R}^m;\ |x| < r \}$ and, in (II.2.19), move the integration with respect to t from $\mathbb{R}^m$ to $\hat{\Omega}$:

$$(2\pi^3)^{m/2}<\mu,h> =$$

$$\int_{z_*\in V}\int_{z'\in\hat{\Omega}}\int_{\zeta\in\mathbb{R}T'_{\mathcal{X}}|_{z_*}} e^{i\zeta\cdot(z-z')-<\zeta><z-z'>^2}\mathcal{F}\mu(z',\zeta)\cdot$$

$$h(z)<\zeta>^{m/2}dzdz'd\zeta.$$

In this last integral we can move the integration with respect to ζ from $\mathbb{R}T'_{\mathcal{X}}\big|_{z_*}$ to $\mathbb{R}T'_{\mathcal{X}}\big|_{z'}$. We obtain

(II.2.20) $$(2\pi^3)^{m/2}<\mu,h> =$$

$$\int_{z_*\in V}\int_{z'\in\hat{\Omega}}\int_{\zeta\in\mathbb{R}T'_{\mathcal{X}}|_{z'}} e^{i\zeta\cdot(z-z')-<\zeta><z-z'>^2}\mathcal{F}\mu(z',\zeta)\cdot$$
$$h(z)<\zeta>^{m/2}dzdz'd\zeta.$$

The same argument leads to the following variant of (II.2.15):

(II.2.21) $$(2\pi^3)^{m/2}<\mu,h> =$$

$$\lim_{\epsilon\to+0}\int_{z\in V}\int_{z'\in\hat{\Omega}}\int_{\zeta\in\mathbb{R}T'_{\mathcal{X}}|_{z'}} e^{i\zeta\cdot(z-z')-<\zeta><z-z'>^2-\epsilon<\zeta>^2}\cdot$$
$$\mathcal{F}\mu(z',\zeta)h(z)<\zeta>^{m/2}dzdz'd\zeta.$$

Hypo–analyticity of a hyperfunction characterized by the exponential decay of its FBI transform

THEOREM II.2.5.— *Suppose $\mathcal{X}$ is well–positioned at z_0. For a hyperfunction u in $\mathcal{X}$ to be equal, in an open neighborhood of z_0 in $\mathcal{X}$, to the restriction of a holomorphic function in some open neighborhood of z_0 in $\mathbb{C}^m$, it is necessary and sufficient that there be a representative $\mu \in \mathcal{O}'(\mathcal{C}\ell U)$ of u in an open neighborhood* U of *z_0 in $\mathcal{X}$ whose FBI transform $\mathcal{F}\mu$ is exponentially decaying, ie., there is an open conic neighborhood $\mathcal{U}$ of $\mathbb{R}T'_{\mathcal{X}}\backslash 0\big|_{z_0}$ in $T'^{1,0}$ such that*

(II.2.22) $$|\mathcal{F}\mu(z,\zeta)| \leq Ce^{-|\zeta|/R}, \ \forall\ (z,\zeta) \in \mathcal{U}.$$

Proof: If $u = h$ in some open neighborhood U of z_0 in $\mathcal{X}$, with h defined and holomorphic in an open neighborhood $\mathcal{N}$ of $\mathcal{C}\ell$U in $\mathbb{C}^m$, then $\mu = \chi_U h \in \mathcal{O}'(\mathcal{C}\ell\,\mathrm{U})$ represents u in U. By the result for functions (Theorem IX.3.1, [TREVES, 1992]) $\mathcal{F}\mu$ will satisfy (II.2.22).

Conversely suppose there are an open neighborhood $\mathcal{N}$ of z_0 in $\mathbb{C}^m$ and a conic open neighborhood $\mathcal{C}$ of $\mathbb{R}^m$ in $\mathbb{C}^m$ such that, for suitably large constants C, $R > 0$,

(II.2.23) $$|\mathcal{F}\mu(z,\zeta)| \leq Ce^{-|\zeta|/R}, \ \forall\ z \in \mathcal{N},\ \zeta \in \mathcal{C}.$$

We select an open neighborhood W of z_0 in $\mathcal{X}$, having a smooth boundary ∂W, and a number λ, $0 < \lambda < \lambda_0 = \frac{1}{8}(1-\kappa)/(3-\kappa)$ such that, if $z_* \in$ W and $0 < \lambda' \leq \lambda$, then $z_* - \imath\lambda'\psi(z_*)\zeta/<\zeta> \in \mathcal{N}$. We subdivide the integral with respect to z_* at the right in (II.2.14) into two integrals, one over W and the second one over U\W. In the integral over W we first deform the domain of ζ–integration from $\mathbb{R}\mathrm{T}'_{\mathcal{X}}|_{z_*}$ to $\mathbb{R}^m$. Once this is done we deform the domain of z–integration from the image of W under the map $z_* \to z = z_* - \imath\lambda\psi(z_*)\zeta/<\zeta>$ back to W. We get

$$(2\pi)^m<\mu,h> = \int_{z_*\in \mathrm{V}\setminus\mathrm{W}} \int_{\zeta\in\mathbb{R}\mathrm{T}'_{\mathcal{X}}|_{z_*}} \mathcal{F}\mu(z,\zeta)\ h(z)\mathrm{d}z\mathrm{d}\zeta +$$

$$\int_0^\lambda \int_{z_*\in\partial\mathrm{W}} \int_{\xi\in\mathbb{R}^m} \mathcal{F}\mu(z,\xi)\ h(z)\mathrm{d}z\mathrm{d}\xi\mathrm{d}\lambda' + \int_{z\in\mathrm{W}} \int_{\xi\in\mathbb{R}^m} \mathcal{F}\mu(z,\xi)\ h(z)\mathrm{d}z\mathrm{d}\xi.$$

In the second integral at the right, $z = z_* - \imath\lambda'\psi(z_*)\xi/|\xi|$. The first and second integrals define analytic functionals carried by $(\mathcal{C}\ell\,U)\backslash W$. The reader will recall that

$$|\mathcal{F}\mu(z_* - \imath\lambda\psi(z_*)\zeta/\langle\zeta\rangle,\zeta)| \leq const.e^{-c|\zeta|}$$

thanks to (II.2.12). In the second integral, where $z_* \in \partial W$, we avail ourselves of (II.2.23) which also implies that $\int_{\xi\in\mathbb{R}^m} \mathcal{F}\mu(z,\xi)d\xi$ is a holo–morphic function of z in $\mathcal{N}$, whose restriction to W is therefore equal to that of u. □

Rather than looking at the sheaf $\mathcal{B}$ of the germs of hyperfunc–tions in $\mathcal{X}$ we may look at the quotient $\mathcal{B}/\mathcal{O}_{\mathcal{X}}$ of $\mathcal{B}$ by the restriction $\mathcal{O}_{\mathcal{X}}$ to $\mathcal{X}$ of the sheaf $\mathcal{O}$ of the germs of holomorphic functions in $\mathbb{C}^m$. The FBI transform (in Ω) defines a sheaf map from $\mathcal{B}/\mathcal{O}_{\mathcal{X}}$ (over Ω) into the sheaf over Ω defined by the linear spaces $\mathfrak{E}(U)/\mathfrak{E}_0(U)$ and their natural restric–tion mappings [see (II.2.8) and (II.2.9)]. Theorem II.2.5 states that the FBI map is *well defined and injective.*

II.3 LOCAL REPRESENTATION OF A HYPERFUNCTION AS A SUM OF BOUNDARY VALUES OF HOLOMORPHIC FUNCTIONS IN WEDGES. HYPO–ANALYTIC WAVE FRONT SET OF A HYPERFUNCTION

We shall call $\mathfrak{C}_m$ the family of all convex and open cones in $\mathbb{R}^m$. If $\Gamma \in \mathfrak{C}_m$ then either $\Gamma \subset \mathbb{R}^m\backslash\{0\}$ or else $\Gamma = \mathbb{R}^m$. If Γ_1 and Γ_2 belong to $\mathfrak{C}_m$ then so does their vector sum $\Gamma_1+\Gamma_2$; $\Gamma_1+\Gamma_2$ is also the *convex hull* of $\Gamma_1\cup\Gamma_2$. Thus $\mathfrak{C}_m$ is the smallest family of subsets of $\mathbb{R}^m$ that contains all convex and open cones $\Gamma \subset \mathbb{R}^m\backslash\{0\}$ and that is stable under the vector sum (or convex hull) operation.

The set–up is exactly the same as in sections II.1 and II.2: $\mathcal{X} \subset \mathbb{C}^m$ is a maximally real submanifold of $\mathbb{C}^m$ and $\Omega = \mathcal{Z}(B_o)$. Condition (I.3.3) is satisfied, as well as (II.2.2) and (II.2.3); in particular $\mathcal{X}$ is well positioned at the central point, which we take provisionally to be the origin in $\mathbb{C}^m$.

In the sequel U will denote an open subset of $\mathcal{X}$ contained in Ω. We shall deal with the wedges $\mathcal{W}_\delta(U,\Gamma)$ with edge U, directrix $\Gamma \in \mathfrak{C}_m$ and arbitrary height $\delta > 0$ (see (II.1.1)). Consider then the direct sum

$$\mathbf{E}(U) = \bigoplus_{\Gamma\in\mathfrak{C}_m} \bigoplus_{\delta>0} \mathcal{O}(\mathcal{W}_\delta(U,\Gamma)).$$

Its elements are the finite sums $f_1+\cdots+f_r$ with $f_j \in \mathcal{O}(\mathcal{W}_{\delta_j}(U,\Gamma_j))$, $\Gamma_j \in \mathfrak{C}_m$, $\delta_j > 0$. The set $\mathbf{E}(U)$ can be regarded, in a natural manner, as a complex vector space. Theorem II.1.1 allows us to define the boundary value map b_U: $\mathbf{E}(U) \to \mathcal{B}(U)$, as follows:

$$b_U\left[\sum_{j=1}^r f_j\right] = \sum_{j=1}^r b_U f_j.$$

We are going to show that, if U is sufficiently small, *the boundary value map* b_U: $\mathbf{E}(U) \to \mathcal{B}(U)$ *is surjective.* Actually we shall prove a more precise statement. We select a finite subfamily $\Gamma_1,\dots,\Gamma_r \in \mathfrak{C}_m$ that has the following property:

There are open cones in $\mathbb{R}^m\backslash\{0\}$, $\mathscr{C}_1,\dots,\mathscr{C}_r$, satisfying the following properties:

(II.3.1) $\mathscr{C}_1,\dots,\mathscr{C}_r$ *are disjoint and* $\text{meas}[\mathbb{R}^m\backslash(\mathscr{C}_1\cup\cdots\cup\mathscr{C}_r)] = 0$;

(II.3.2) *for some constant* $c_0 > 0$ *and every* $j = 1,\dots,r$,

$$\xi\cdot v \geq c_0|\xi|\,|v|,\ \forall\, \xi \in \mathscr{C}_j,\ v \in \Gamma_j.$$

THEOREM II.3.1.— *There is an open neighborhood* $U \subset \Omega$ *of* 0 *in* $\mathcal{X}$ *and a number* $\delta > 0$ *such that every hyperfunction* u *in* $\mathcal{X}$ *is equal in* U *to a sum* $\sum_{j=1}^{r} b_U f_j$, *with* $f_j \in \mathcal{O}(\mathcal{W}_\delta(U,\Gamma_j))$, $j = 1,\dots,r$.

Proof: We begin by selecting an open neighborhood $V' \subset\subset \Omega$ of 0 in $\mathcal{X}$ such that

$$\text{(II.3.3)}\quad Z(x) \in V',\ v \in \Gamma_j,\ \xi \in \mathscr{C}_j \Rightarrow \mathscr{R}e({}^tZ_x(x)^{-1}\xi\cdot v) \geq \tfrac{1}{2}c_0|v|\,|\xi|.$$

We take $U \subset\subset V'$ (U open in $\mathcal{X}$, $0 \in U$) and let $\mu \in \mathcal{O}'(\mathscr{C}\ell U)$ represent the hyperfunction u in U. For each $j = 1,\dots,r$, and for any $\epsilon > 0$, we define

$$F_j^\epsilon(z) =$$
$$(2\pi^3)^{-m/2}\int_{z'\in V'}\int_{\xi\in\mathscr{C}_j} e^{i\zeta\cdot(z-z')-<\zeta><z-z'>^2-\epsilon<\zeta>^2}\mathcal{F}\mu(z',\zeta)\cdot$$
$$<\zeta>^{m/2}dz'd\zeta,$$

where $z' = Z(x')$ and $\zeta = {}^tZ_x(x')^{-1}\xi$. Each F_j^ϵ is an entire function in $\mathbb{C}^m$. Let us take $z = z_* + iv$, $z_* \in \Omega$, $v \in \Gamma_j$. We look at the quantity [cf. (II.2.10)]

$$\mathcal{Q} = -\ \mathcal{R}e[\imath\zeta\cdot(z-z')-<\zeta><z-z'>^2]/|\zeta|.$$

There is a constant $C > 0$ such that, for any number $\tau > 1$,

$$\mathcal{Q} \geq \mathcal{I}m[\zeta\cdot(z_*-z')+\imath<\zeta><z_*-z'>^2]/|\zeta| + \mathcal{R}e(\zeta\cdot v)/|\zeta| -$$
$$\tau^{-1}|z_*-z'|^2 - C\tau|v|^2.$$

Thanks to (II.2.3) and (II.3.3) we obtain, for a suitable choice of the constant $c > 0$,

$$\mathcal{Q} \geq (1-\kappa-\tau^{-1})|z_*-z'|^2 + (c-C\tau|v|)|v|.$$

If we choose $\tau^{-1} < 1-\kappa$ and select $|v| < \delta < c/C\tau$, we obtain, for some $c' > 0$,

(II.3.4) $\quad \mathcal{Q} \geq c'(|z_*-z'|^2+|v|),\ \forall\ z_* \in \Omega,\ z' \in \mathrm{V}',\ v \in \Gamma_j.$

This implies that, as $\epsilon \to +0$, F_j^ϵ converges in $\mathcal{O}(\mathcal{W}_\delta(\Omega,\Gamma_j))$ to a holomorphic function F_j.

We select an open neighborhood V of $\mathcal{C}\ell$U in $\mathcal{X}$, whose boundary in $\mathcal{X}$ is $\mathcal{C}^\infty$ and whose closure is compact and contained in V$'$. For each $j = 1,\ldots,r$, we select a vector $v_j \in \Gamma_j$, $|v_j| < \delta$. We form, for any $h \in \mathcal{O}(\mathbb{C}^m)$ (cf. (II.1.2)),

$$\sum_{j=1}^{r} <\mu_{F_j,V}^{v_j},h> = \sum_{j=1}^{r} \int_{V+\imath v_j} F_j(z)h(z)dz =$$
$$\lim_{\epsilon\to+0} \sum_{j=1}^{r} \int_{V+\imath v_j} F_j^\epsilon(z)h(z)dz =$$
$$\lim_{\epsilon\to+0} \sum_{j=1}^{r} \left\{\int_V F_j^\epsilon(z)h(z)dz + \int_{\partial V+\imath[0,v_j]} F_j^\epsilon(z)h(z)dz\right\}$$

by the Stokes' theorem. In view of the definition of the functions F_j^ϵ we get

$$(2\pi^3)^{m/2} \sum_{j=1}^{r} <\mu_{F_j,V}^{v_j},h> =$$

$$\lim_{\epsilon\to+0}\sum_{j=1}^{r}\Big\{\int_{z\in V}\int_{z'\in V'}\int_{\xi\in\mathscr{C}_j} e^{\imath\zeta\cdot(z-z')-<\zeta><z-z'>^2-\epsilon<\zeta>^2}\cdot$$
$$\mathcal{F}\mu(z',\zeta)h(z)<\zeta>^{m/2}dz\,dz'\,d\zeta+\int_{\partial V+\imath[0,v_j]}F_j^{\epsilon}(z)h(z)dz\Big\}.$$

If we compare with (II.2.21) we obtain

(II.3.5)
$$<\mu-\sum_{j=1}^{r}\mu_{F_j,V}^{v_j},h>=$$
$$\lim_{\epsilon\to+0}\Big\{\sum_{j=1}^{r}\int_{z\in V}\int_{z'\in\hat\Omega\setminus V'}\int_{\xi\in\mathscr{C}_j} e^{\imath\zeta\cdot(z-z')-<\zeta><z-z'>^2-\epsilon<\zeta>^2}\cdot$$
$$\mathcal{F}\mu(z',\zeta)h(z)<\zeta>^{m/2}dz\,dz'\,d\zeta/(2\pi^3)^{m/2}+\int_{\partial V+\imath[0,v_j]}F_j^{\epsilon}(z)h(z)dz\Big\}.$$

Here, as in (II.2.21), $\hat\Omega=\hat Z(\mathbb{R}^m)$. If $z\in V$, $z'\in\hat\Omega$ and $\zeta\in\mathbb{R}T'_{\mathcal{X}}\big|_{z'}$,

$$\mathcal{Q}=-\mathscr{R}e[\imath\zeta\cdot(z-z')-<\zeta><z-z'>^2]\geq(1-\kappa)|z-z'|^2|\zeta|$$

by (II.2.3). If moreover $z'\in\hat\Omega\setminus V'$ then $\mathcal{Q}\geq d>0$ and we may let ϵ go to zero in the first integral in the right—hand side of (II.3.5); each integral

$$\int_{z'\in\hat\Omega\setminus V'}\int_{\xi\in\mathscr{C}_j} e^{\imath\zeta\cdot(z-z')-<\zeta><z-z'>^2}\mathcal{F}\mu(z',\zeta)\cdot$$
$$<\zeta>^{m/2}dz'\,d\zeta/(2\pi^3)^{m/2}$$

defines a holomorphic function G in an open neighborhood $\tilde V$ of $\mathscr{C}\ell V$ in $\mathbb{C}^m$. We obtain:

(II.3.6)
$$<\mu-\sum_{j=1}^{r}\mu_{F_j,V}^{v_j},h>=\int_V G(z)h(z)dz+\lim_{\epsilon\to+0}\sum_{j=1}^{r}\int_{\partial V+\imath[0,v_j]}F_j^{\epsilon}(z)h(z)dz.$$

Lastly we must look at the integrals $\int_{\partial V+\imath[0,v_j]}F_j^{\epsilon}(z)h(z)dz$. Take

$z = z_* + \imath v$, with $z_* \in \partial V$ and $v = \lambda v_j \in \Gamma_j$ $(0 < \lambda < 1,\ |v| < \delta)$. Here also we can avail ourselves of (II.3.4). Moreover we take advantage of the fact that the analytic functional μ is carried by a compact subset of V, $\mathscr{C}\ell U$. We obtain, for suitably large positive constants C, R, and all $z = z_* + \imath v$, $z_* \in \partial V$, $v \in \Gamma_j$, $|v| < \delta$, $z' \in V'$,

$$|e^{\imath\zeta\cdot(z-z') - <\zeta><z-z'>^2 - \epsilon<\zeta>^2}\mathcal{F}\mu(z',\zeta)| \leq Ce^{-|\zeta|/R}.$$

The estimate is valid when z' is away from ∂V, by virtue of (II.3.4) and of Theorem II.2.1. It is also valid when z' is near ∂V as one sees by applying Theorem II.2.2 with ∂V substituted for K and $\mathscr{C}\ell U$ for K$'$ and, here also, by taking advantage of (II.3.4). This gives us the right to let ϵ go to zero: the function F_j^ϵ converges uniformly in an open subset $\mathscr{N}_j$ of $\mathbb{C}^m$ which contains the set of points $z = z_* + \imath v$, $z_* \in \partial V$, $v \in \Gamma_j$, $|v| < \delta$. Its limit can also be denoted by F_j. We have finally shown

(II.3.7)

$$<\mu - \sum_{j=1}^{r} \mu_{F_j,V}^{v_j}, h> - \int_V G(z)h(z)dz = \sum_{j=1}^{r} \int_{\partial V + \imath[0,v_j]} F_j(z)h(z)dz.$$

Given an open neighborhood $\mathscr{N}$ of ∂V in $\mathbb{C}^m$ we can select a number $\epsilon > 0$ such that, if $|v_j| < \epsilon$ for every j = 1,...,r, the analytic functional defined by the right—hand side of (II.3.7) is carried by $\mathscr{N}$. It suffices to apply Theorem II.1.1: if ε is sufficiently small we can also achieve that every analytic functional $\mu_{F_j,V} - \mu_{F_j,V}^{v_j}$ is carried by $\mathscr{N}$. If we denote by $\mu_{G,V}$ the analytic functional $h \to \int_V G(z)h(z)dz$ we have reached the conclusion that $\mu - \mu_{G,V} - \sum_{j=1}^{r} \mu_{F_j,V}$ is carried by $\mathscr{N}$, hence by ∂V. This means that the hyperfunction defined by μ in V is equal to the sum of the restriction of G to V and of $\sum_{j=1}^{r} b_V F_j$. Restricting to U and setting (possibly after having decreased $\delta > 0$) $f_j = F_j + \frac{1}{r}G \in \mathcal{O}(\mathcal{W}_\delta(U,\Gamma_j))$ ($j = 1,\ldots,r$) yields the sought conclusion. □

Naturally, at this stage, one would like to know the kernel of the map b_U: $\mathbf{E}(U) \to \mathcal{B}(U)$. The answer to this question is related to the celebrated *Edge of the Wedge Theorem* (see section II.5). But without waiting for this result we avail ourselves of Theorem II.3.1 to introduce the important concept of the hypo–analytic wave–front set of a hyperfunction. We shall give right–away an invariant definition of the wave–front set, based on the notion of germ of wedge.

Let z_0 be an arbitrary point of $\mathcal{X}$. Select arbitrarily complex coordinates $z_1,\ldots, z_m$ in $\mathbb{C}^m$ vanishing at z_0 and such that the tangent space to $\mathcal{X}$ at z_0 is defined by the equations $y_1 = \cdots = y_m = 0$. In a neighborhood $\mathcal{N}$ of 0 in $\mathbb{C}^m$ the submanifold $\mathcal{X}$ is now defined by equations $y = \Phi(x)$, with Φ real–valued, $\Phi(0) = 0$, $\Phi_x(0) = 0$. We may select open balls B_0, B_1 in $\mathbb{R}^m$ such that $B_0 \times B_1 \subset\subset \mathcal{N}$ and $\Phi(B_0) \subset B_1$. By contracting B_0 about z_0 (now the origin) we can ensure that all the hypotheses in sections II.1 and II.2 are satisfied. In this situation we can define the wedges with edges $U \subset \Omega = Z(B_0)$ [as usual, $Z(x) = x + \imath\Phi(x)$] according to Formula (II.1.1).

Suppose we change complex coordinates, from z_j to $z_j^\# = x_j^\# + \imath y_j^\#$, with $z_j^\# = 0$ at z_0, and the tangent space to $\mathcal{X}$ at z_0 defined by the equations $y_j^\# = 0$ ($j = 1,\ldots,m$). An equation for $\mathcal{X}$ near z_0 will be $y^\# = \Phi^\#(x^\#)$, giving rise to the diffeomorphism $x^\# \to Z^\#(x^\#)$ from an open neighborhood of 0 in $\mathbb{R}^m$ onto one of z_0 in $\mathcal{X}$; we take the former to be an open ball B_0 centered at 0. Call $\mathcal{W}^{\#}_{\delta}(U,\Gamma)$ the analogue of (II.1.1) in the coordinates $z^{\#}_j$ (they determine the meaning of $\imath\eta$, $\eta \in \mathbb{R}^m$). The wedges $\mathcal{W}_{\delta}(U,\Gamma)$ and $\mathcal{W}^{\#}_{\delta}(U^{\#},\Gamma^{\#})$ can be nested one inside the other, a fact whose proof is easy, and is left to the reader:

PROPOSITION II.3.1.— *Let* $\boldsymbol{v}_0 \in \mathbb{R}^m\backslash\{0\}$. *Given any open neighborhood* $U \subset \Omega\cap\Omega^{\#}$ *in* $\mathcal{X}$ $[\Omega = Z(B_0),\ \Omega^{\#} = Z^{\#}(B_0)]$, *any cone* $\Gamma \in \mathfrak{C}_m$ *containing* $\boldsymbol{v}_0$ *and any number* $\delta > 0$, *there is an open neighborhood* $U' \subset U$ *of* 0 *in* $\mathcal{X}$, *a cone* $\Gamma' \in \mathfrak{C}_m$ *such that* $\boldsymbol{v}_0 \in \Gamma' \subset \Gamma$, *and a positive number* $\delta' \leq \delta$ *such that*

$$\mathcal{W}^{\#}_{\delta'}(U',\Gamma') \subset \mathcal{W}_{\delta}(U,\Gamma).$$

We can also select U', Γ' *and* δ' *as indicated, such that*

$$\mathcal{W}_{\delta'}(U',\Gamma') \subset \mathcal{W}^{\#}_{\delta}(U,\Gamma).$$

Now let $\boldsymbol{v}_0$ be a tangent vector to $\mathcal{X}$ at z_0, and $\mathfrak{W}(z_0,\boldsymbol{v}_0)$ denote the collection of all subsets of $\mathbb{C}^m$ that contain some wedge $\mathcal{W}_{\delta}(U,\Gamma)$ with U an open neighborhood of z_0 in $\mathcal{X}$, Γ an open cone in $\mathbb{R}^m\backslash\{0\}$ such that $\boldsymbol{v}_0 \in \Gamma$, δ a suitably small number > 0. The family of these wedges $\mathcal{W}_{\delta}(U,\Gamma)$ forms the *basis of a filter*, and $\mathfrak{W}(z_0,\boldsymbol{v}_0)$ forms a *filter*: any subset of $\mathbb{C}^m$ that contains a set $\Sigma \in \mathfrak{W}(z_0,\boldsymbol{v}_0)$, as well as the intersection of any two sets that belong to $\mathfrak{W}(z_0,\boldsymbol{v}_0)$, also belong to $\mathfrak{W}(z_0,\boldsymbol{v}_0)$. It follows at once from Proposition II.3.1 that the filter $\mathfrak{W}(z_0,\boldsymbol{v}_0)$ does not depend on the choice of the coordinates $z_1,\ldots, z_m$ such that $z_0 = 0$ and $T_{z_0}\mathcal{X} = \mathbb{R}^m$, used in the definition of the wedges.

DEFINITION II.3.1.— *We shall refer to the filter* $\mathfrak{W}(z_0,\boldsymbol{v}_0)$ *as the* ***germ of wedge at*** z_0 ***with edge*** $(\mathcal{X},z_0)$ *and* ***direction*** $\boldsymbol{v}_0$.

The notation $(\mathcal{X},z_0)$ of course refers to the germ of set $\mathcal{X}$ at the point z_0.

REMARK II.3.1.— It is obvious that the germ of wedge $\mathfrak{W}(z_0,\boldsymbol{v}_0)$ only depends on the *ray* $\dot{\boldsymbol{v}}_0 = \{\lambda\boldsymbol{v}_0\}_{\lambda>0}$ spanned by $\boldsymbol{v}_0$ in $\mathrm{T}_{z_0}\mathcal{X}$, and not on the vector $\boldsymbol{v}_0$ itself. The set of rays $\dot{\boldsymbol{v}}_0$ is the *sphere* S_{z_0} associated to the real vector space $\mathrm{T}_{z_0}\mathcal{X}$. If $\mathrm{T}_{z_0}\mathcal{X}$ is equipped with a norm, S_{z_0} can be identified to the unit sphere in $\mathrm{T}_{z_0}\mathcal{X}$. As z_0 ranges over $\mathcal{X}$ the sets S_{z_0} form the *tangent sphere bundle* of $\mathcal{X}$, $\mathrm{S}\mathcal{X}$. Clearly $\mathrm{S}\mathcal{X} \ni (z,\dot{\boldsymbol{v}}) \to \mathfrak{W}(z,\dot{\boldsymbol{v}})$ is a bijection of $\mathrm{S}\mathcal{X}$ onto the set of germs of wedges in $\mathcal{X}$. In a way germs of wedges provide us with a special interpretation of the tangent sphere bundle. Even more suggestive is to think of $\mathfrak{W}(z,\boldsymbol{v})$ as the *germ of* $\mathrm{S}\mathcal{X}$ *at the point* $(z,\boldsymbol{v})$. □

We may now talk of the germ $\dot{f}$ of a holomorphic function in the germ of wedge $\mathfrak{W}(z_0,\boldsymbol{v}_0)$: this is an equivalence class of pairs $(\mathcal{W}_\delta(\mathrm{U},\Gamma),f)$ with $f \in \mathcal{O}(\mathcal{W}_\delta(\mathrm{U},\Gamma))$, z_0 $(= 0) \in \mathrm{U}$, $\boldsymbol{v}_0 \in \Gamma$ $(\Gamma \in \mathfrak{C}_\mathrm{m})$ for the equivalence relation

$$(\mathcal{W}_\delta(\mathrm{U},\Gamma),f) \approx (\mathcal{W}_{\delta'}(\mathrm{U}',\Gamma'),f')$$

defined as follows:

$\exists\ \mathrm{U}'' \subset \mathcal{X}$ *open*, $z_0 \in \mathrm{U}''$, $\Gamma'' \in \mathfrak{C}_\mathrm{m}$, $\boldsymbol{v}_0 \in \Gamma''$, $\delta'' > 0$, *such that* $\mathcal{W}_{\delta''}(\mathrm{U}'',\Gamma'') \subset \mathcal{W}_\delta(\mathrm{U},\Gamma) \cap \mathcal{W}_{\delta'}(\mathrm{U}',\Gamma')$ *and* $f = f'$ in $\mathcal{W}_{\delta''}(\mathrm{U}'',\Gamma'')$.

Assuming that the open cone Γ is connected we define the *boundary value* $b\dot{f}$ of $\dot{f}$ as the germ of $b_\mathrm{U}f$ at z_0, if $(\mathcal{W}_\delta(\mathrm{U},\Gamma),f)$ is an arbitrary representative of $\dot{f}$.

DEFINITION II.3.2.— *We shall say that the germ of a hyperfunction* $\dot{u}$ *at* z_0

is ***hypo—analytic at a point*** $(z_0,\xi) \in T^*\mathcal{X}$ *if there are finitely many germs of wedges* $\mathfrak{W}(z_0,\mathbf{v}_1),\ldots,\mathfrak{W}(z_0,\mathbf{v}_r)$ *and, for each* $j = 1,\ldots,r$, *the germ of a holomorphic function* $\dot{f}_j$ *in* $\mathfrak{W}(z_0,\mathbf{v}_j)$, *satisfying the following two conditions:*

(II.3.8) $$<\xi,\mathbf{v}_j> < 0,\ j = 1,\ldots,r;$$

(II.3.9) $$\dot{u} = \sum_{j=1}^{r} b\dot{f}_j.$$

We shall say that a hyperfunction u in $\mathcal{X}$ *is hypo—analytic at* (z_0,ξ) *if its germ at* z_0 *is hypo—analytic at* (z_0,ξ). *The subset of* $T^*\mathcal{X}$ *consisting of the points* (z_0,ξ) *at which u is not hypo—analytic will be called the* ***hypo—analytic wave front set*** *of u and denoted by* $WF_{ha}(u)$.

In (II.3.8) < , > represents the bracket of the duality between tangent and cotangent vectors to $\mathcal{X}$ (at z_0).

In the analytic theory the name "*essential support*", and the con—comitant notation ess supp *u*, are sometimes used in place of "wave—front set" and of $WF_a(u)$.

The subset of $T^*\mathcal{X}\backslash 0\big|_\Omega$ in which $u \in \mathcal{B}(\Omega)$ is hypo—analytic is open and conic; $WF_{ha}(u)$ is a closed and conic subset of $T^*\mathcal{X}\backslash 0\big|_\Omega$.

II.4 DELIMITATION OF THE HYPO–ANALYTIC WAVE–FRONT SET OF A HYPERFUNCTION BY THE DECAY OF ITS FBI TRANSFORM

In Definition II.3.2 the hypo–analytic wave front set of a hyper–function on the maximally real submanifold $\mathcal{X}$ has been defined as a (conic and closed) subset of the cotangent bundle $T^*\mathcal{X}$ of $\mathcal{X}$, more precisely of the complement of the zero section, $T^*\mathcal{X}\backslash 0$. In the present section we shall regard it as a subset of $\mathbb{R}T'_{\mathcal{X}}\backslash 0$. The equivalence of the two approaches derives, of course, from the natural isomorphism $T^*\mathcal{X} \cong \mathbb{R}T'_{\mathcal{X}}$. What makes the embedding in $\mathbb{R}T'_{\mathcal{X}}$ preferable is the fact that the structure bundle of complex space $\mathbb{C}^m$, $T'^{1,0}$, is a ready made complex vector space (of com–plex dimension 2m) in which $\mathbb{R}T'_{\mathcal{X}}$ is embedded as a maximally real sub–manifold. In contrast to this, note that $\mathbb{C}T^*\mathcal{X} \supset T^*\mathcal{X}$ is not a complex manifold.

Section II.6 shall present a systematic microlocalization of hyper–function theory. Here microlocalization means focusing one's attention on the behaviour of the hyperfunction u under study in a conic neighborhood, in $T'^{1,0}$, of some point (z_0,ζ_0) of $\mathbb{R}T'_{\mathcal{X}}\backslash 0$; and one disregards (ie., "quo–tients out") those hyperfunctions that are hypo–analytic in such a neigh–borhood. A powerful method of analyzing the behaviour of u in a conic neighborhood of (z_0,ζ_0), in particular of determining whether u is hypo–analytic at (z_0,ζ_0) (Definition II.3.2), is to analyze its FBI transform. In accordance with the local results (Theorem II.2.5) one expects that the microlocal hypo–analyticity of u is equivalent to the exponential decay of its FBI transform, $\mathcal{F}u$. To show this we must begin by establishing some facts about the FBI transform of those special (germs of) hyperfunctions that are the boundary values of holomorphic functions in wedges whose edge is an open neighborhood of z_0 in $\mathcal{X}$.

We continue to reason in the set—up of the preceding sections, as described at the beginning of section II.1. The submanifold $\mathcal{X}$ will be well—positioned at z_0. We shall make use of the customary parametric representation of $\mathcal{X}$, $B_0 \ni x \to Z(x) = x + \imath\Phi(x)$. We take $z_0 = Z(0) = 0$, $\Omega = Z(B_0)$; and we assume that Properties (I.3.3), (II.2.1), (II.2.2) and (II.2.3) are valid. We systematically identify the fibre of $\mathbb{R}T'_{\mathcal{X}}$ at z_0 to $\mathbb{R}^m$ and we regard the unit sphere S^{m-1} as a subset of $\mathbb{R}T'_{\mathcal{X}}\big|_{z_0}$.

Let $U \subset\subset \Omega$ be an open neighborhood of z_0 in $\mathcal{X}$, $\Gamma \subset \mathbb{R}^m\backslash\{0\}$ an open and convex cone and δ a number > 0; V shall denote an open neighborhood of z_0 in $\mathcal{X}$ whose closure $\mathcal{C}\ell V$ is compact and contained in U, and whose boundary in $\mathcal{X}$, ∂V, is $\mathcal{C}^\infty$. In such a situation Theorem II.1.1 associates to an arbitrary holomorphic function g in $\mathcal{W}_\delta(U,\Gamma)$ an analytic functional $\mu_{g,V}$ carried by $\mathcal{C}\ell V$. We recall that $\mu_{g,V}$ represents $b_U g$ in V.

PROPOSITION II.4.1.— *Suppose $\theta \in S^{m-1}$ verifies, for some $c_0 > 0$,*

$$\theta\cdot v \leq -c_0|v|,\ \forall\, v \in \Gamma. \tag{II.4.1}$$

Then there exist an open and conic neighborhood $\mathcal{U}_0$ of (z_0,θ) in $T'^{1,0}$ and numbers C, $R > 0$ such that

$$|\mathcal{F}\mu_{g,V}(z,\zeta)| \leq Ce^{-|\zeta|/R},\ \forall\, (z,\zeta) \in \mathcal{U}_0. \tag{II.4.2}$$

Proof: Arguing as in the proof of Theorem II.2.2 one can show that there are an open neighborhood $\mathcal{N}$ of ∂V in $\mathbb{C}^m$, an open neighborhood V_1

$\subset\subset$ V of z_0 in $\mathcal{X}$, and a constant $c > 0$ such that, whatever $\beta \in \mathcal{O}'(\mathcal{N})$,

$$\text{(II.4.3)} \qquad \sup \left[e^{c|\zeta|} \, |\mathcal{F}\beta(z,\zeta)| \right] < +\infty,$$

with the supremum taken over an open and conic neighborhood $\mathcal{U}_1$ of $\mathbb{R}T'_{\mathcal{X}}\backslash 0\big|_{V_1}$ in $T'^{1,0}$.

If $\gamma \in \Gamma$, $|\gamma| < \delta$, we can define the analytic functional $\mu^{\gamma}_{g,V}$ by the formula (II.1.2). We avail ourselves of Theorem II.1.1, which enables us to take $\beta = \mu_{g,V} - \mu^{\gamma}_{g,V}$. Proposition II.4.1 will then follows from (II.4.3) and from an appropriate estimate of $|\mathcal{F}\mu^{\gamma}_{g,V}|$. Set $V_0 = \{ x \in \mathbb{R}^m;$ $Z(x) \in V \}$. By (II.1.2) we have

$$\text{(II.4.4)} \qquad \mathcal{F}\mu^{\gamma}_{g,V}(z,\zeta) = \int_{V_0} e^{i\zeta\cdot(z-Z(x)-i\gamma)-<\zeta><z-Z(x)-i\gamma>^2} g(Z(x)+i\gamma)\cdot \Delta(z-Z(x)-i\gamma,\zeta)\,dZ(x).$$

We have

$$\mathcal{Q}(0,Z(x)+i\gamma,\theta) = -\theta\cdot(\Phi(x)+\gamma) + |x|^2 - |\Phi(x)+\gamma|^2 \geq$$
$$|x|^2 - |\Phi(x)| - 2|\Phi(x)|^2 - \theta\cdot\gamma - 2|\gamma|^2 \geq -\theta\cdot\gamma - 2|\gamma|^2$$

by (II.2.1). Now take (II.4.1) into account: $|\gamma| \leq \mathrm{Min}(\delta, c_0/4)$ entails $\mathcal{Q}(0,Z(x)+i\gamma,\theta) \geq (c_0 - 2|\gamma|)|\gamma| \geq c_0|\gamma|/2$. If this estimate is valid for $z = 0$, $\zeta = \theta$, we can find an open neighborhood $\mathcal{U}_0$ of $(0,\theta)$ in $T'^{1,0}$ such that $\mathcal{Q}(z,Z(x)+i\gamma,\zeta) \geq c_0|\gamma|/4$ holds for all $(z,\zeta) \in \mathcal{U}_0$ and all $x \in \mathcal{R}e\,V$. This entails, for suitable constants C, $R > 0$,

$$\text{(II.4.5)} \qquad |\mathcal{F}\mu^{\gamma}_{g,V}(z,\zeta)| \leq Ce^{-|\zeta|/R}, \ \forall\, (z,\zeta) \in \mathcal{U}_0.$$

As we said earlier Proposition II.4.1 follows at once from (II.4.3) and (II.4.5). □

Proposition II.4.1 implies easily the necessity of the condition in the following statement:

THEOREM II.4.1.— *Let u be a hyperfunction in $\mathcal{X}$. For u to be hypo–analytic at the point $(z_0,\theta) \in \mathbb{R}T'_{\mathcal{X}}\backslash 0$ it is necessary and sufficient that the FBI transform of u decay exponentially at infinity, in an open and conic neighborhood of (z_0,θ) in $T'^{1,0}\backslash 0$.*

Proof: I. *Necessity.* Let u be a hyperfunction in $\mathcal{X}$ such that $(z_0,\theta) \notin WF_{ha}(u)$ $(\theta \in S^{m-1})$. According to Definition II.3.2 there are an open subset $U \subset\subset \Omega$ of $\mathcal{X}$, cones $\Gamma_j \in \mathfrak{C}_m$ contained in the open half–space $\{\, v \in \mathbb{R}^m;\ <\theta,v> \, < 0 \,\}$ and a number $\delta > 0$, such that the restriction of u to U is equal to a sum $\sum_{j=1}^{r} b_U f_j$, with $f_j \in \mathcal{O}(\mathcal{W}_\delta(U,\Gamma_j))$, $j = 1,\dots,r$. After contracting each Γ_j about one of its rays we may assume that every cone Γ_j satisfies condition (II.4.1). If $\mu \in \mathcal{O}'(\mathcal{C}\ell U)$ represents u in U and $\mu_j \in \mathcal{O}'(\mathcal{C}\ell U)$ represents $b_U f_j$ in U, for each $j = 1,\dots,r$, $\mu - \sum_{j=1}^{r} \mu_j$ is carried by ∂U. We derive from Theorem II.2.2 that, given any compact neighborhood $K \subset U$ of z_0 in $\mathcal{X}$, there are constants C, $R > 0$ such that

$$\left| \mathcal{F}\mu(z,\zeta) - \sum_{j=1}^{r} \mathcal{F}\mu_j(z,\zeta) \right| \leq C e^{-|\zeta|/R},$$

provided (z,ζ) stays in a suitably "thin" conic neighborhood $\mathcal{U}$ of $\mathbb{R}T'_{\mathcal{X}}\backslash 0 \big|_K$ in $T'^{1,0}\backslash 0$. Proposition II.4.1 implies that each analytic functional μ_j satisfies (II.4.2) for some choice of the conic neighborhood $\mathcal{U}_0$, and so therefore does μ. By Theorem II.2.2 this remains true if we change the representative μ of u.

II. *Sufficiency.* As before we use the parametric representation $B_0 \ni x \to Z(x) = x+\imath\Phi(x) \in \Omega$ (and we identify z_0 to the origin of $\mathbb{C}^m$, $\mathbb{R}T'_{\mathcal{X}}|_{z_0}$ to $\mathbb{R}^m$, $T'^{1,0}_{z_0}$ to $\mathbb{C}^m$). Let U_0 be an open neighborhood of z_0 in $\mathcal{X}$, $U_0 \subset\subset \Omega$, and let $\mu_0 \in \mathcal{O}'(\mathcal{C}\mathcal{L}U_0)$ represent u in U_0. Here our hypothesis is that there are an open and conic neighborhood $\mathcal{U}_0$ of (z_0,θ) in $T'^{1,0}$ and constants C, $R > 0$, such that

$$|\mathcal{F}\mu_0(z,\zeta)| \le Ce^{-|\zeta|/R},\ \forall\,(z,\zeta) \in \mathcal{U}_0. \tag{II.4.6}$$

We can select open neighborhoods of z_0, $\mathcal{N}_0$ and V_0 in $\mathbb{C}^m$ and in $\mathcal{X}$ respectively, and an open cone $\mathcal{C}_0$ in $\mathbb{R}^m\backslash\{0\}$, with $\theta \in \mathcal{C}_0$ and such that

$$z \in \mathcal{N}_0,\ Z(x) \in V_0,\ \xi \in \mathcal{C}_0 \Rightarrow (z,{}^tZ_x(x)^{-1}\xi) \in \mathcal{U}_0.$$

We select additional open cones $\mathcal{C}_1,\ldots,\mathcal{C}_r$ in $\mathbb{R}^m\backslash\{0\}$ such that the following properties are valid:

(II.4.7) $\mathcal{C}_0, \mathcal{C}_1,\ldots,\mathcal{C}_r$ *are pairwise disjoint and*

$$\text{meas}[\mathbb{R}^m\backslash(\mathcal{C}_0 \cup \mathcal{C}_1 \cdots \cup \mathcal{C}_r)] = 0;$$

(II.4.8) *for some number* $c_0 > 0$ *and each* $j = 1,\ldots,r$, *there is* $\boldsymbol{v}_j \in \mathbb{R}^m$ *satisfying* $\theta\cdot\boldsymbol{v}_j < 0$ *and*

$$\xi\cdot\boldsymbol{v}_j \ge c_0|\xi|\,|\boldsymbol{v}_j| \text{ for all } \xi \in \mathcal{C}_j.$$

To achieve this it suffices to require that θ not belong to the closed convex hull of any of the cones $\mathcal{C}_j$ $(1 \le j \le r)$, and to apply the Hahn—Banach theorem. It follows from (II.4.8) that, for each $j = 1,\ldots,r$, there is a cone $\Gamma_j \in \mathfrak{C}_m$, $\boldsymbol{v}_j \in \Gamma_j$, such that

(II.4.9) $$\forall\, \boldsymbol{v} \in \Gamma_j,\ \ \theta\cdot\boldsymbol{v} < 0;$$

(II.4.10) $$\forall\, \xi \in \mathscr{C}_j,\ \boldsymbol{v} \in \Gamma_j,\ \ \xi\cdot\boldsymbol{v} \geq \tfrac{1}{2}c_0|\xi|\,|\boldsymbol{v}|.$$

Thanks to (II.4.10) we can find an open neighborhood $V' \subset\subset U_0$ of z_0 in $\mathcal{X}$ such that

(II.4.11)
$$\forall\, Z(x) \in V',\ \xi \in \mathscr{C}_j,\ \boldsymbol{v} \in \Gamma_j,\ \ \mathscr{R}e\,{}^tZ_x(x)^{-1}\xi\cdot\boldsymbol{v} \geq \tfrac{1}{4}c_0|\xi|\,|\boldsymbol{v}|.$$

Now let μ be a representative of u in an open neighborhood $U \subset\subset V'$ of z_0 in $\mathcal{X}$. By Theorem II.2.2 we know that, possibly after increasing C and R, there is an open and conic neighborhood $\mathscr{U}_1$ of $\mathbb{R}T_{\mathcal{X}}\backslash 0\big|_{z_0}$ in $T'^{1,0}\backslash 0$ such that

(II.4.12) $$|F(\mu_0-\mu)(z,\zeta)| \leq Ce^{-|\zeta|/R},\ \forall\, (z,\zeta) \in \mathscr{U}_1.$$

The conjunction of (II.4.6) and (II.4.12) allows us to select an an open neighborhood $\mathcal{N}$ of z_0 in $\mathbb{C}^m$ such that

(II.4.13) $$|\mathcal{F}\mu(z,\zeta)| \leq Ce^{-|\zeta|/R},$$
$$\forall\, z \in \mathcal{N},\ \zeta = {}^tZ_x(x)^{-1}\xi,\ Z(x) \in \mathcal{N}\cap\mathcal{X},\ \xi \in \mathscr{C}_0.$$

At this point we refer the reader to the proof of Theorem II.3.1. We select an open subset V of $\mathcal{X}$, with a smooth boundary ∂V relative to $\mathcal{X}$ and such that $U \subset\subset V \subset\subset V'$ (keep in mind that μ is carried by $\mathscr{C}\ell U$). Define the holomorphic function F_j in the wedge $\mathcal{W}_\delta(\Omega,\Gamma_j)$ ($j = 1,\ldots,r$) exactly as in the proof of Theorem II.3.1. The reasoning leading to (II.3.7) is valid here, except for the fact that the integration with respect to ξ over $\mathscr{C}_0$ must be accounted for. We get, in the notation of the proof of Theorem II.3.1,

$$(II.4.14) \qquad <\mu - \sum_{j=1}^{r} \mu_{F_j,V}^{v_j}, h> -$$

$$\int_V G(z)h(z)dz - \sum_{j=1}^{r} \int_{\partial V + \imath[0,v_j]} F_j(z)h(z)dz =$$

$$\lim_{\epsilon \to +0} \left\{ \int_{z\in V} \int_{z'\in V'} \int_{\xi \in \mathscr{C}_0} e^{\imath\zeta\cdot(z-z') - <\zeta><z-z'>^2 - \epsilon<\zeta>^2} \mathcal{F}\mu(z',\zeta)\cdot \right.$$

$$\left. h(z)<\zeta>^{m/2} dz dz' d\zeta/(2\pi^3)^{m/2} \right\}$$

[in the integral at the right $z' = Z(x') \in V'$ and $\zeta = {}^tZ_x(x')^{-1}\xi$].

We avail ourselves of (II.4.13). Call W′ an open neighborhood of z_0 in $\mathcal{X}$ such that $W' \subset\subset V' \cap (\mathcal{N} \cap \mathcal{X})$;(II.2.3) and (II.4.13) imply that, as $\epsilon \to +0$, the integral

$$\int_{z'\in W'} \int_{\xi \in \mathscr{C}_0} e^{\imath\zeta\cdot(z-z') - <\zeta><z-z'>^2 - \epsilon<\zeta>^2} \mathcal{F}\mu(z',\zeta)\cdot$$

$$<\zeta>^{m/2} dz' d\zeta/(2\pi^3)^{m/2}$$

converges uniformly to a holomorphic function H in a neighborhood of $\mathscr{C}\ell V$. It remains to study the analogous integral but with the z'–integration carried out over $V' \backslash W'$. First we look at that integral when the variation of z is restricted to an open neighborhood $W \subset\subset W'$ of z_0 in $\mathcal{X}$, with a smooth boundary ∂W relative to $\mathcal{X}$. By virtue of (II.2.3) and of Theorem II.2.1 we see that

$$\left| e^{\imath\zeta\cdot(z-z') - <\zeta><z-z'>^2} \mathcal{F}\mu(z',\zeta)<\zeta>^{m/2} \right| \leq Ce^{-d|\zeta|},$$

where $d > 0$ depends on dist(W,V′\W′). As a consequence of this, as $\epsilon \to +0$, the integral

$$\int_{z'\in V'\backslash W'}\int_{\zeta\in\mathscr{C}_0} e^{\imath\zeta\cdot(z-z')-<\zeta><z-z'>^2-\epsilon<\zeta>^2}\mathcal{F}\mu(z',\zeta)\cdot$$
$$<\zeta>^{m'2}dz'd\zeta/(2\pi^3)^{m'2}$$

converges to a holomorphic function H_1 in a neighborhood of $\mathscr{C}\ell$W.

Lastly we look at the analytic functionals defined by

(II.4.15)
$$\int_{z\in V\backslash W}\int_{z'\in V'\backslash W'}\int_{\zeta\in\mathscr{C}_0} e^{\imath\zeta\cdot(z-z')-<\zeta><z-z'>^2-\epsilon<\zeta>^2}\mathcal{F}\mu(z',\zeta)\cdot$$
$$h(z)<\zeta>^{m'2}dzdz'd\zeta.$$

In dealing with these integrals we deform the domain of z–integration from V\W to the image Σ of V\W under the map $z \to z+\imath\lambda\dot{\zeta}$ ($\dot{\zeta} = \zeta/<\zeta>$), somewhat as in the proof of Theorem II.2.5. This deformation leads to three integrals with respect to the variable z: one over Σ, one over $\partial W+\imath[0,\lambda]\dot{\zeta}$ and the third one over $\partial V+\imath[0,\lambda]\dot{\zeta}$. In each one of these integrals we have the right to let ϵ go to zero. Over Σ this is due to the fact that the quantity

$$\mathcal{Q} = -\ \mathcal{R}e[\imath\dot{\zeta}\cdot(z+\imath\lambda\dot{\zeta}-z')-<\dot{\zeta}><z+\imath\lambda\dot{\zeta}-z'>^2] \geq$$
$$(1-\kappa-\tau^{-1})|z-z'|^2 + c\lambda - C\tau\lambda^2$$

is bounded from below if $\tau^{-1} \leq \frac{1}{2}(1-\kappa)$ and $C\tau\lambda < \frac{1}{2}c$. The integral over $\partial W+\imath[0,\lambda]\dot{\zeta}$ is seen to converge by exploiting the fact that $|z-z'| \geq d$ if $z \in \partial W$ and $z' \in V'\backslash W'$. The integral over $\partial V+\imath[0,\lambda]\dot{\zeta}$ is seen to converge by noting that either $z' \in V'\backslash W'$ lies near ∂V, in which case $|\mathcal{F}\mu(z',\zeta)|$ is exponentially decaying, by virtue of Theorem II.2.2; or else z' lies away from ∂V, in which case $|z-z'|$ is bounded away from zero. We leave the details to the reader. The upshot is that the limit, as $\epsilon \to +0$, of the integral (II.4.15) defines an analytic functional ν carried by $(\mathscr{C}\ell V\backslash W) + \imath[0,\lambda]B'$, where $B' = \{\ \zeta \in \mathbb{C}^m;\ |\zeta| < 2\ \}$. Since we can select $\lambda > 0$ as small as we wish we see that ν will be carried by $(\mathscr{C}\ell V)\backslash W$. Returning to

(II.4.14) and observing that the analytic functionals

$$h \to \int_{\partial V + i[0,v_j]} F_j(z)h(z)dz \ (j = 1,\dots,r)$$

are carried by compact sets that do not intersect $\mathscr{C}\ell W$ we conclude that, in W,

$$u = \sum_{j=1}^{r} b_W F_j + G + H + H_1,$$

which shows that $(z_0,\theta) \notin WF_{ha}(u)$. □

Below, if Γ and Γ' are two cones in $\mathbb{R}^m\backslash\{0\}$ we write $\Gamma' \subset\subset \Gamma$ to mean that the closure of Γ' in $\mathbb{R}^m\backslash\{0\}$ is contained in Γ, ie. $\Gamma' \cap S^{m-1} \subset\subset \Gamma \cap S^{m-1}$. As for Γ^0 it denotes the *polar* of Γ, ie.,

$$\Gamma^0 = \{\ \xi \in \mathbb{R}^m;\ \forall\ v \in \Gamma,\ \xi \cdot v \geq 0\ \}.$$

THEOREM II.4.2.— *Let u be a hyperfunction in $\mathcal{X}$ and $\Gamma_1,\dots,\Gamma_r$ be open and convex cones in $\mathbb{R}^m\backslash\{0\}$. The following properties are equivalent:*

(II.4.16) *Let $\{\Gamma'_1,\dots,\Gamma'_r\}$ be any set of* r *open and convex cones in $\mathbb{R}^m\backslash\{0\}$ such that $\Gamma'_j \subset\subset \Gamma_j$ for each* j = 1,...,r. *There exist an open neighborhood* $U \subset \Omega$ *of z_0 in $\mathcal{X}$, a number $\delta > 0$ and holomorphic functions f_j in $\mathcal{W}_\delta(U,\Gamma'_j)$* (j = 1,...,r) *such that $u = \sum_{j=1}^{r} b_U f_j$ in* U.

(II.4.17) $$WF_{ha}(u) \cap (\mathbb{R}T'_{\mathcal{X}}|_{z_0}) \subset \Gamma_1^0 \cup \cdots \cup \Gamma_r^0.$$

Proof: (II.4.16) ⇒ (II.4.17). If $\theta \notin \Gamma_1^0 \cup \cdots \cup \Gamma_r^0$ and if the cones Γ'_j are as in (II.4.16), then

$$v \in \bigcup_{j=1}^{r} \Gamma_j' \Rightarrow \theta \cdot v < 0.$$

Definition II.3.2 implies that $(z_0, \theta) \notin \mathrm{WF}_{ha}(u)$.

(II.4.17) ⇒ (II.4.16). For each j = 1,...,r, select open and convex cones Γ_j' and Γ_j'' such that $\Gamma_j' \subset\subset \Gamma_j'' \subset\subset \Gamma_j$. Moreover we require the boundary of Γ_j'' in $\mathbb{R}^m \backslash \{0\}$ to be smooth. For each j = 1,...,r we define

$$\mathscr{C}_1 = \mathscr{Int}\, \Gamma_1''^{\,o},\ \mathscr{C}_2 = \mathscr{Int}\,\{\Gamma_2''^{\,o} \backslash (\mathscr{C}_1 \cap \Gamma_2''^{\,o})\},...,$$

$$\mathscr{C}_r = \mathscr{Int}\,\{\Gamma_r''^{\,o} \backslash [(\mathscr{C}_1 \cup \cdots \cup \mathscr{C}_{r-1}) \cap \Gamma_r''^{\,o}]\}.$$

Finally call $\mathscr{C}_0$ the interior of $(\mathbb{R}^m \backslash \{0\}) \backslash (\mathscr{C}_1 \cup \cdots \cup \mathscr{C}_r)$. Notice that Conditions (II.4.7) and (II.4.10) are satisfied if we substitute Γ_j' for Γ_j and if c_0 is sufficiently small. By the necessity part in Theorem II.4.1 there is an open neighborhood $\mathscr{N}$ of z_0 in $\mathbb{C}^m$ and positive constants C, $R > 0$ such that (II.4.13) hold for some representative μ of u. As a consequence the whole construction in the sufficiency part of the proof of Theorem II.4.1 applies, with the above choice of $\mathscr{C}_0, \mathscr{C}_1,..., \mathscr{C}_r$. We conclude that, in a suitably small open neighborhood U of z_0 in $\mathscr{X}$, $u \cong \sum_{j=1}^{r} b_U F_j$ modulo holomorphic functions in a full neighborhood of z_0 in $\mathbb{C}^m$. Here the F_j are defined and holomorphic in the wedges $\mathscr{W}_\delta(U, \Gamma_j')$. □

COROLLARY II.4.1.— *Let u be a hyperfunction in $\mathscr{X}$ and Γ an open and convex cone in $\mathbb{R}^m \backslash \{0\}$. The following properties are equivalent:*

(II.4.18) *If Γ' is an open and convex cone in $\mathbb{R}^m \backslash \{0\}$ whose closure in $\mathbb{R}^m \backslash \{0\}$ is contained in Γ there exist an open neighborhood $U \subset \Omega$ of z_0 in $\mathscr{X}$, a number $\delta > 0$ and a holomorphic* *function f in $\mathcal{W}_\delta(\mathrm{U},\Gamma')$ such that $u = b_{\mathrm{U}} f$ in* U.

(II.4.19) $$\mathrm{WF}_{\mathrm{ha}}(u)\cap(\mathbb{R}\mathrm{T}'_{\mathcal{X}}\big|_{z_0}) \subset \Gamma^{\mathrm{o}}.$$

II.5 EDGE OF THE WEDGE

The theorems of section II.4 give us the means to describe with some precision the kernel of the composite map

$$\mathrm{E}(\mathrm{U}) \xrightarrow{b_{\mathrm{U}}} \mathcal{B}(\mathrm{U}) \xrightarrow{\rho^{\mathrm{U}}_{\mathrm{V}}} \mathcal{B}(\mathrm{V})$$

(see section II.3). Here U and V are open neighborhoods of z_0 in $\mathcal{X}$ such that $\mathrm{V} \subset\subset \mathrm{U} \subset\subset \Omega = \mathcal{Z}(\mathrm{B}_0)$; $\rho^{\mathrm{U}}_{\mathrm{V}}$ is the restriction map from U to V.

We continue to use the notation $\Gamma' \subset\subset \Gamma$ when Γ and Γ' are two cones in $\mathbb{R}^m\backslash\{0\}$, to indicate that the closure of Γ' in $\mathbb{R}^m\backslash\{0\}$ is contained in Γ, ie., $\Gamma' \cap S^{m-1} \subset\subset \Gamma \cap S^{m-1}$. For simplicity we shall write

$$\mathrm{WF}_{\mathrm{ha}}(u)\Big|_{z_0} = \mathrm{WF}_{\mathrm{ha}}(u) \cap \left[\mathbb{R}\mathrm{T}'_{\mathcal{X}}\Big|_{z_0}\right] \quad \text{if } u \in \mathcal{B}(\mathcal{X}).$$

The following consequence of Theorem II.4.2 can be regarded as a version of *Martineau's Edge of the Wedge Theorem:*

THEOREM II.5.1.— *Let* $\mathrm{U} \subset\subset \Omega = \mathcal{Z}(\mathrm{B}_0)$ *be an open neighborhood of* z_0 *in* $\mathcal{X}$, $\Gamma_1,\ldots,\Gamma_r$ *be open and convex cones in* $\mathbb{R}^m\backslash\{0\}$ *and* δ *be a number* > 0. *Given any open neighborhood* $\mathrm{V} \subset \mathrm{U}$ *of* z_0 *and any set of* r *open and convex cones* Γ'_j *such that* $\emptyset \neq \Gamma'_j \subset\subset \Gamma_j$ $(j = 1,\ldots,r)$ *there are an open neighborhood* $\mathrm{U}' \subset \mathrm{U}$ *of* z_0 *in* $\mathcal{X}$ *and a number* δ', $0 < \delta' < \delta$, *such that the following two properties of a set of* r *functions* $f_j \in \mathcal{O}(\mathcal{W}_\delta(\mathrm{U},\Gamma_j))$ $(j = 1,\ldots,r)$ *are equivalent:*

(II.5.1) *The boundary value* $\sum_{j=1}^{r} b_{\mathrm{U}} f_j \in \mathcal{B}(\mathrm{U})$ *vanishes identically in* V.

(II.5.2) *For each pair of indices* (j,k), $1 \leq j, k \leq r$, *there is a holomorphic function* g_{jk} *in the wedge* $\mathcal{W}_{\delta'}(\mathrm{U}',\Gamma'_j+\Gamma'_k)$ *such that* $g_{jk} \equiv$

$-g_{kj}$ *for all* j, k = 1,...,r (thus $g_{jj} \equiv 0$, $\forall$ j = 1,..,r) *and such moreover that, for each* j = 1,..., r, *the restrictions of* f_j *and of* $\sum_{k=1}^{r} g_{jk}$ *to* $\mathcal{W}_{\delta'}(U',\Gamma_j')$ *are equal.*

Proof: If the functions $g_{jk} \in \mathcal{O}(\mathcal{W}_{\delta'}(U',\Gamma_j'+\Gamma_k'))$ are as in (II.5.2) then

$$\sum_{j=1}^{r} b_{U'} f_j = b_{U'}\Big(\sum_{j,k=1}^{r} g_{jk}\Big)$$

and (II.5.1) with V ⊂ U′ follows directly from the skew symmetry of the g_{jk}. We shall therefore concentrate our efforts on the proof that (II.5.1) ⇒ (II.5.2). If r = 1 (II.5.1) ⇒ $f_1 \equiv 0$ in $\mathcal{W}_\delta(V,\Gamma_1)$ by Theorem II.1.2. Henceforth we assume $r \geq 2$.

For each i = 1,...,r, we select arbitrarily a sequence of cones $\Gamma_i^{(q)} \in \mathfrak{C}_m$ (q = 1,...,r) such that $\Gamma_i' \subset \Gamma_i^{(q+1)} \subset\subset \Gamma_i^{(q)} \subset\subset \Gamma_i$. For each q, $1 \leq q \leq r$, call $\mathcal{S}_q$ the family of all *ordered* subsets of q integers $i_1,\dots,i_q$ such that $1 \leq i_1 < \cdots < i_q \leq r$. To the subset $I = \{i_1,\dots,i_q\} \in \mathcal{S}_q$ and to $p \geq 1$ we associate the vector sum $\Gamma_I^{(p)} = \Gamma_{i_1}^{(p)} + \cdots + \Gamma_{i_q}^{(p)}$. Note that $\Gamma_I^{(p)}$ is a convex and open cone in $\mathbb{R}^m\backslash\{0\}$. We also introduce the set $\mathcal{S}_0$ whose only element is the empty subset $\emptyset$ of [1,...,r]; we set $\Gamma_\emptyset^{(p)} = \emptyset$.

Given q, $1 \leq q < r$, suppose there are an open neighborhood $U^{(q)} \subset V$ of z_0 in $\mathcal{X}$, a number $\delta_q > 0$ and, for each set $I \in \mathcal{S}_{q-1}$ and each integer $j \notin I$, $1 \leq j \leq r$, a function $F_{I,j} \in \mathcal{O}(\mathcal{W}_{\delta_q}(U^{(q)},\Gamma_I^{(q)}+\Gamma_j^{(q)}))$ such that

(II.5.3) $$\sum_{I\in\mathcal{S}_{q-1}} \sum_{j\notin I} b_{U^{(q)}} F_{I,j} \equiv 0.$$

According to our hypothesis such a system is provided, when q = 1, by taking $\Gamma_I^{(0)} = \emptyset$ and $\Gamma_j^{(1)}$ any cone $\subset\subset \Gamma_j$, $F_{\emptyset,j} = f_j$ (j = 1,...,r), $U^{(1)} = V$ and $\delta^{(1)} = \delta$.

For each $J \in \mathcal{S}_q$ we introduce the symmetrization

$$F_J^{\sigma} = \Sigma\, F_{I,j},$$

where the sum ranges over the set of all the pairs consisting of an ordered (q–1)–tuple I and of an integer $j \notin I$ such that J is equal to I∪{j} after re–ordering. [When q = 1 and $J = \{j\} \in \mathcal{S}_1$, $F_J^{\sigma} = f_j$.] We have $F_J^{\sigma} \in \mathcal{O}(\mathcal{W}_{\delta_q}(U^{(q)}, \Gamma_J^{(q)}))$ and

$$\text{(II.5.4)} \qquad \sum_{J \in \mathcal{S}_q} b_{U(q)} F_J^{\sigma} \equiv 0.$$

It follows from (II.5.4) and Definition II.3.2 that

$$WF_{ha}(b_{U(q)} F_J^{\sigma})\Big|_{z_0} \subset \Gamma_J^{(q)\,0}, \quad \forall J \in \mathcal{S}_q.$$

As a consequence of this and of (II.5.4) we have

$$WF_{ha}(b_{U(q)} F_J^{\sigma})\Big|_{z_0} \subset \Gamma_J^{(q)\,0} \cap \bigcup_{\substack{K \in \mathcal{S}_q \\ K \neq J}} \Gamma_K^{(q)\,0} = \bigcup_{\substack{K \in \mathcal{S}_q \\ K \neq J}} (\Gamma_K^{(q)} + \Gamma_J^{(q)})^0.$$

We apply Theorem II.4.2: there are an open neighborhood $U^{(q+1)} \subset U^{(q)}$ of z_0 in $\mathcal{X}$, a number δ_{q+1}, $0 < \delta_{q+1} < \delta_q$ (with $U^{(q+1)}$ and δ_q independent of the $F_{I,j}$) and functions

$$G_{JK} \in \mathcal{O}(\mathcal{W}_{\delta_{q+1}}(U^{(q+1)}, \Gamma_J^{(q+1)} + \Gamma_K^{(q+1)}))\ (J, K \in \mathcal{S}_q,\ J \neq K)$$

such that

$$b_{U(q+1)} F_J^{\sigma} = \sum_{\substack{K \in \mathcal{S}_q \\ K \neq J}} b_{U(q+1)} G_{JK}.$$

Keeping J $\in \mathcal{S}_q$ fixed let us call $\{k_1,...,k_{r-q}\}$ the complementary set $[1,...,r]\backslash J$ (with its natural order). Define F_{J,k_1} to be the restriction to $\mathcal{W}_{\delta_{q+1}}(U^{(q+1)},\Gamma_J^{(q+1)}+\Gamma_{k_1}^{(q+1)})$ of the sum ΣG_{JK} over the sets K $\in \mathcal{S}_q$ such that $k_1 \in K$; next define F_{J,k_2} to be the restriction to $\mathcal{W}_{\delta_{q+1}}(U^{(q+1)},\Gamma_J^{(q+1)}+\Gamma_{k_2}^{(q+1)})$ of the sum ΣG_{JK} over the sets K $\in \mathcal{S}_q$ such that $k_1 \notin K$, $k_2 \in K$; etc. We have

$$F_{J,k_\alpha} \in \mathcal{O}(\mathcal{W}_{\delta_{q+1}}(U^{(q+1)},\Gamma_J^{(q+1)}+\Gamma_{k_\alpha}^{(q+1)}))\ (\alpha = 1,...,r-q).$$

We derive

$$b_{U^{(q+1)}} F_J^\sigma = \sum_{\alpha=1}^{r-q} b_{U^{(q+1)}} F_{J,k_\alpha}. \tag{II.5.5}$$

Actually, according to the injectivity of the boundary value map (Theorem II.1.2) Property (II.5.5) is equivalent to the following:

$$F_J{}^\sigma = \sum_{\alpha=1}^{r-q} F_{J,k_\alpha} \ \text{in}\ \mathcal{W}_{\delta_{q+1}}(U^{(q+1)},\Gamma_J^{(q+1)}). \tag{II.5.6}$$

In summary, we have defined a holomorphic function $F_{J,k}$ in the wedge $\mathcal{W}_{\delta_{q+1}}(U^{(q+1)},\Gamma_J^{(q+1)}+\Gamma_k^{(q+1)})$, now for each pair (J,k), J $\in \mathcal{S}_q$, k $\in [1,...,r]\backslash J$. These functions obviously satisfy the analogue of (II.5.3):

$$\sum_{J\in\mathcal{S}_q}\sum_{k\notin J} b_{U^{(q+1)}} F_{J,k} \equiv 0, \tag{II.5.7}$$

and are related to the F_J^σ, $J \in \mathcal{S}_q$, by (II.5.6). For any (q+1)–tuple $K \in \mathcal{S}_{q+1}$ we can form the symmetrized F_K^σ; we have

$$\sum_{K \in \mathcal{S}_{q+1}} b_{U(q+1)} F_K^\sigma \equiv 0,$$

By repeating r times the procedure just described, starting with q = 1 we reach the value q = r–1. Because the only element of $\mathcal{S}_r$ is the interval [1,...,r], if $J \in \mathcal{S}_{r-1}$ and if $1 \leq k \leq r$, $k \notin J$, then perforce $\Gamma_J^{(r)}+\Gamma_k^{(r)} = \Gamma_1^{(r)}+\cdots+\Gamma_r^{(r)} = \Gamma^{(r)}$. Since all the functions $F_{J,k}$ are defined in one and the same wedge $\mathcal{W}_{\delta_r}(U^{(r)},\Gamma^{(r)})$ it follows from Theorem II.1.2 that, when q = r–1, Equation (II.5.7) is equivalent to

(II.5.8) $$\sum_{J \in \mathcal{S}_{r-1}} \sum_{k \notin J} F_{J,k} \equiv 0.$$

Note that (II.5.8) can be rewritten as

(II.5.9) $$F^\sigma_{[1,\dots,r]} \equiv 0.$$

With the notation introduced above we are going to define straight–away the functions g_{jk} in the wedges $\mathcal{W}_{\delta_r}(U^{(r)},\Gamma_j^{(r)}+\Gamma_k^{(r)})$. It suffices to consider the case j < k and to set $g_{kj} = -g_{jk}$. If k > 1, we define, in $\mathcal{W}_{\delta_r}(U^{(r)},\Gamma_1^{(r)}+\Gamma_k^{(r)})$,

$$g_{1k} = F_{1,k};$$

and if k > j > 1, we define, in $\mathcal{W}_{\delta_r}(U^{(r)},\Gamma_j^{(r)}+\Gamma_k^{(r)})$,

$$g_{jk} = F_{j,k} + \sum_{i_1<j} F_{i_1j,k} + \sum_{i_1<i_2<j} F_{i_1i_2j,k} + \cdots + F_{12\cdots j,k}.$$

We must show that, whatever j = 1,...,r, we have in $\mathscr{W}_{\delta_r}(U^{(r)},\Gamma_j^{(r)})$:

$$\text{(II.5.10)}\qquad \sum_{k=1}^{r} g_{jk} = \sum_{k<j} F_{j,k} + \sum_{k>j} F_{j,k} = f_j$$

The second equality in (II.5.10) follows from (II.5.6) applied with q = 1 [when J = {j} ∈ $\mathcal{S}_1$, $F_J^{\sigma} = f_J$]. We have in $\mathscr{W}_{\delta_r}(U^{(r)},\Gamma_j^{(r)})$:

$$\sum_{k=1}^{r} g_{jk} = \sum_{k>j} g_{jk} - \sum_{k<j} g_{kj} =$$

$$\sum_{k>j}\Big[F_{j,k} + \sum_{i_1<j} F_{i_1j,k} + \sum_{i_1<i_2<j} F_{i_1i_2j,k} + \cdots + F_{12\cdots j,k}\Big] -$$

$$\sum_{k<j}\Big[F_{k,j} + \sum_{i_1<k} F_{i_1k,j} + \sum_{i_1<i_2<k} F_{i_1i_2k,j} + \cdots + F_{12\cdots k,j}\Big].$$

In order to prove the first equality in (II.5.10) we must prove the validity in $\mathscr{W}_{\delta_r}(U^{(r)},\Gamma_j^{(r)})$ of the following equation (for each j)

$$\text{(II.5.11)}\qquad \sum_{k>j}\Big[\sum_{i_1<j} F_{i_1j,k} + \sum_{i_1<i_2<j} F_{i_1i_2j,k} + \cdots + F_{12\cdots j,k}\Big] -$$
$$\sum_{k<j}\Big[\sum_{i_1<k} F_{i_1k,j} + \sum_{i_1<i_2<k} F_{i_1i_2k,j} + \cdots + F_{12\cdots k,j}\Big] = \sum_{k<j} F_{kj}^{\sigma}.$$

Notice that (II.5.10) is immediate for j = 1; we may therefore assume j ≥ 2. Then (II.5.11) is the particular case $\nu = 1$ of the following identity:

$$\text{(II.5.12)}_\nu \qquad \sum_{k>j}\Big[\sum_{i_1<\cdots<i_\nu<j} F_{i_1\ldots i_\nu j,k} + \cdots + F_{12\ldots j,k}\Big] =$$

$$\sum_{i_1<\cdots<i_\nu<j} F^\sigma_{i_1\ldots i_\nu j} + \sum_{k<j}\left[\sum_{i_1<\cdots<i_\nu<k} F_{i_1\ldots i_\nu k,j} + \cdots + F_{12\cdots k,j}\right].$$

We are going to show that all the identities $(II.5.12)_\nu$, $\nu = 1,\ldots,r-1$, are equivalent. Since $(II.5.12)_{r-1}$ is a restatement of (II.5.9) this will prove $(II.5.12)_1$. We may therefore assume $r \geq 3$ and $\nu \leq r-2$. We apply (II.5.6) with $q = \nu+1$:

$$F^\sigma_{i_1\ldots i_\nu j} = \sum_k{}' F_{i_1\ldots i_\nu j,k}$$

with the sum Σ' restricted to the indices $k \notin \{i_1,\ldots,i_\nu,j\}$. Putting this into $(II.5.12)_\nu$ yields

$$\sum_{k>j}\left[\sum_{i_1<\cdots<i_{\nu+1}<j} F_{i_1\ldots i_{\nu+1}j,k} + \cdots + F_{12\ldots j,k}\right] =$$

$$\sum_{k<j}\Bigg[\sum_{\substack{i_1<\cdots<i_\nu<j\\ i_\alpha\neq k,\ \alpha=1,\ldots,\nu}} F_{i_1\ldots i_\nu j,k} +$$

$$\sum_{i_1<\cdots<i_\nu<k} F_{i_1\ldots i_\nu k,j} + \cdots + F_{12\cdots k,j}\Bigg].$$

We observe that

$$\sum_{k<j}\left[\sum_{\substack{i_1<\cdots<i_\nu<j\\ i_\alpha\neq k,\ \alpha=1,\ldots,\nu}} F_{i_1\ldots i_\nu j,k} + \sum_{i_1<\cdots<i_\nu<k} F_{i_1\ldots i_\nu k,j}\right] =$$

$$\sum_{i_1<\cdots<i_{\nu+1}<j} F_{i_1\ldots i_{\nu+1},j} + \sum_{k<j}\sum_{\substack{i_1<\cdots<i_\nu<j\\ k\neq i_\alpha,\ \alpha=1,\ldots,\nu}} F_{i_1\ldots i_\nu j,k} =$$

$$\sum_{i_1<\cdots<i_{\nu+1}<j} F^\sigma_{i_1\ldots i_{\nu+1},j}.$$

Putting this into $(II.5.12)_\nu$ transforms it into $(II.5.12)_{\nu+1}$. □

Let us denote by $\mathbf{N}(U)$ the subspace of $\mathbf{E}(U)$ consisting of the sums $\sum_{j=1}^{r} f_j$ with $f_j \in \mathcal{O}(\mathcal{W}_\delta(U,\Gamma_j))$, $\Gamma_j \in \mathfrak{C}_m$ $(1 \leq j \leq r)$, $\delta > 0$, that have the following property:

(II.5.13) *Given any set of* r *cones* $\Gamma'_j \in \mathfrak{C}_m$ *such that* $\Gamma'_j \subset\subset \Gamma_j$ $(j = 1,\dots,r)$ *there are an open neighborhood* $U' \subset U$ *of* z_0 *in* $\mathcal{X}$, *a number* δ', $0 < \delta' < \delta$, *and for each pair* (j,k), $1 \leq j, k \leq r$, *a function* $g_{jk} \in \mathcal{O}(\mathcal{W}_{\delta'}(U',\Gamma'_j+\Gamma'_k))$ *with the properties that* $g_{jk} + g_{kj} = 0$ *for all* $j, k = 1,\dots,r$, *and that for each* $j = 1,\dots,r$, *the restrictions of* f_j *and of* $\sum_{k=1}^{r} g_{jk}$ *to* $\mathcal{W}_{\delta'}(U',\Gamma'_j)$ *are equal.*

COROLLARY II.5.1.— *Given any* $\sum_{j=1}^{r} f_j \in \mathbf{E}(U)$, *with* $f_j \in \mathcal{O}(\mathcal{W}_\delta(U,\Gamma_j))$, $\Gamma_j \in \mathfrak{C}_m$, $\delta > 0$, *in order that* $\sum_{j=1}^{r} b_U f_j \equiv 0$ *in some open neighborhood* $V \subset U$ *of* z_0 *in* $\mathcal{X}$ *it is necessary and sufficient that* $\sum_{j=1}^{r} f_j \in \mathbf{N}(U)$.

REMARK II.5.1.— Consider a holomorphic function f in a wedge $\mathcal{W}_\delta(U,\Gamma)$, with $U \subset\subset \Omega$ an open neighborhood of 0 in $\mathcal{X}$, Γ an open cone in $\mathbb{R}^m \backslash \{0\}$ and δ a number > 0. Assume Γ to be connected but not necessarily convex; call $\hat{\Gamma}$ the convex hull of Γ. It follows from Theorem II.5.1 that given any open convex cone $\Gamma' \subset\subset \hat{\Gamma}$ there are an open neighborhood of 0 in $\mathcal{X}$, $U' \subset U$, and a number δ', $0 < \delta' < \delta$, such that f extends holomorphically to $\mathcal{W}_{\delta'}(U',\Gamma')$. Roughly (or rather, microlocally) speaking one can say that f extends holomorphically to $\{0\}+i\hat{\Gamma}$. This can be regarded as a microlocal version of the classical *Bochner Tube Theorem*, and justifies restricting one's attention to wedges $\mathscr{W}_\delta(U,\Gamma)$ with convex directrix Γ. □

II.6 THE SHEAF OF MICROFUNCTIONS

As before let $\mathcal{X}$ be a maximally real submanifold of $\mathbb{C}^m$ and let $\mathbb{R}T'_{\mathcal{X}}$ denote its real structure bundle, ie., the pre–image of $T^*\mathcal{X}$ under the isomorphism $T'^{1,0}\big|_{\mathcal{X}} \to \mathbb{C}T^*\mathcal{X}$ defined by the pullback to $\mathcal{X}$. We continue to view the hypo–analytic wave–front set of a hyperfunction u in $\mathcal{X}$, $WF_{ha}(u)$, as a closed and conic subset of $\mathbb{R}T'_{\mathcal{X}}\backslash 0$. Actually it will be convenient, below, to consider the projection of $WF_{ha}(u)$ into the *sphere bundle* $S'_{\mathcal{X}}$ associated to $\mathbb{R}T'_{\mathcal{X}}\backslash 0$: $S'_{\mathcal{X}}$ is the manifold of *rays* $\{z\}\times\{\rho\zeta;\ \rho > 0\}$ when (z,ζ) ranges over $\mathbb{R}T'_{\mathcal{X}}\backslash 0$; or, if one prefers, the quotient of $\mathbb{R}T'_{\mathcal{X}}\backslash 0$ for the equivalence relation

$$(z,\zeta) \approx (z',\zeta') \iff z = z',\ \zeta = \rho\zeta' \text{ for some } \rho > 0.$$

Clearly there is a one–to–one correspondence between conic subsets of $\mathbb{R}T'_{\mathcal{X}}\backslash 0$ and subsets of $S'_{\mathcal{X}}$. The standard Euclidean metric on $\mathbb{C}^m$ associated to the complex coordinates $z_1,\dots,z_m$ gives rise to norms on the fibres of the complex structure bundle $T'^{1,0}$ and therefore on those of its real subbundle $\mathbb{R}T'_{\mathcal{X}}$. In turn these norms allow us to identify each fibre $S'_{\mathcal{X}}\big|_z$ to the unit sphere in $\mathbb{R}T'_{\mathcal{X}}\big|_z$ $(z \in \mathcal{X})$.

The definition of the hypo–analytic wave–front set suggests that we "quotient out" the hypo–analytic functions at z_0; even better, that we quotient out the hyperfunctions which are hypo–analytic at a point (z_0,θ_0) (Definition II.3.2). There lies the motivation for introducing the concept of a microfunction. As a counterpart to this the use of the FBI transform involves holomorphic functions of (z,ζ) in conic neighborhoods in $T'^{1,0}$ (the cotangent structure bundle of $\mathbb{C}^m$) of points of the real structure bundle $\mathbb{R}T'_{\mathcal{X}}$ (off the zero section) and suggests that we quotient out those that decay exponentially as $|\zeta| \to +\infty$ [cf. remarks at the end of section II.2]. In the present section we formalize these twin microlocalizations.

It is obvious how to microlocalize at points of $S'_{\mathcal{X}}$ the notion of germ of a hyperfunction in $\mathcal{X}$. The wording itself in Definition II.3.2 defines the concept of hypo–analytic wave–front for the germ of a hyperfunction: if $\dot{u} \in \mathcal{B}_z$, ie., $\dot{u}$ is the germ at z of a hyperfunction, then $(z,\theta) \notin \mathrm{WF}_{\mathrm{ha}}(\dot{u})$ means that $(z,\theta) \notin \mathrm{WF}_{\mathrm{ha}}(u)$ for all or, equivalently, for some, representative u of $\dot{u}$.

Given any $(z,\theta) \in S'_{\mathcal{X}}$ we define $\mathcal{C}_{(z,\theta)}$ to be the quotient vector space

$$\mathcal{B}_z/\{\, \dot{u} \in \mathcal{B}_z\, ;\, (z,\theta) \notin \mathrm{WF}_{\mathrm{ha}}(\dot{u})\, \}.$$

As (z,θ) ranges over $S'_{\mathcal{X}}$ these vector spaces define a sheaf $\mathcal{C}$ on $S'_{\mathcal{X}}$.

DEFINITION II.6.1.— *The sheaf $\mathcal{C}$ is called the **sheaf of microfunctions** over $\mathcal{X}$.*

We shall refer to a section of $\mathcal{C}$ as a ***microfunction***. There is a natural multivalued "map" $\mathcal{B} \to \mathcal{C}$: for each $z \in \mathcal{X}$, it assigns to a germ $\dot{u} \in \mathcal{B}_z$ the family of its cosets $\dot{u}(\theta) \in \mathcal{C}_{(z,\theta)}$ as θ ranges over $S'_{\mathcal{X}}\big|_z$. This map $\mathcal{B} \to \mathcal{C}$ can also be defined as follows: denote by π the base projection $S'_{\mathcal{X}} \to \mathcal{X}$ and call $\pi^*\mathcal{B}$ the pullback of $\mathcal{B}$ to $S'_{\mathcal{X}}$ under π; then the multivalued map $\mathcal{B} \to \mathcal{C}$ is the compose of the pullback map $\mathcal{B} \to \pi^*\mathcal{B}$ with the natural quotient map $\pi^*\mathcal{B} \to \mathcal{C}$. The hypo–analytic wave front set of a continuous section of $\mathcal{B}$,ie., of a hyperfunction, is equal to the support of the corresponding microsection. This is why the wave–front set of a hyperfunction is sometimes called its ***microsupport*** (also called, sometimes, its *essential support*).

Next we must define the version of the FBI transform that maps the stalk of $\mathscr{C}$ at a point $(z_0,\theta) \in S'_{\mathcal{X}}$ into a stalk of germs of holomorphic functions in some conic neighborhood $\tilde{\mathcal{U}}_0$ in $T'^{1,0}$ of the ray defined by (z_0,θ). What the sheaf of those germs of holomorphic functions should be is suggested by Properties (II.2.8) and (II.2.9) — and by their evident microlocalizations.

Given any open and conic subset $\mathcal{U}$ of $\mathbb{R}T'_{\mathcal{X}}$ we shall denote by $\mathfrak{E}(\mathcal{U})$ the space of holomorphic functions h in some open and conic neigh–borhood $\tilde{\mathcal{U}}$ of $\mathcal{U}$ in $T'^{1,0}$ (with $\tilde{\mathcal{U}}$ depending on h) that have the following property [cf. (II.2.8)]:

$$\text{(II.6.1)} \qquad \forall\, \epsilon > 0,\ \sup_{\tilde{\mathcal{U}}} \left[e^{-\epsilon|\zeta|}\, |h(z,\zeta)| \right] < +\infty.$$

We shall denote by $\mathfrak{E}_0(\mathcal{U})$ the subspace of $\mathfrak{E}(\mathcal{U})$ consisting of those func–tions h, defined and holomorphic in some open and conic neighborhood $\tilde{\mathcal{U}}$ of $\mathcal{U}$ in $T'^{1,0}$ (with $\tilde{\mathcal{U}}$ depending on h) that have the following property [cf. (II.2.9)]:

$$\text{(II.6.2)} \qquad \exists\, c > 0 \textit{ such that } \sup_{\tilde{\mathcal{U}}} \left[e^{c|\zeta|}\, |h(z,\zeta)| \right] < +\infty.$$

The preceding definitions lead to two presheaves on $S'_{\mathcal{X}}$: to each open subset $\dot{\mathcal{U}}$ of $S'_{\mathcal{X}}$ we attach the vector spaces $\mathfrak{E}(\mathcal{U})$ and $\mathfrak{E}_0(\mathcal{U})$ respec–

tively, with $\mathcal{U}$ the (unique) conic subset of $\mathbb{R}T'_{\mathcal{X}}\backslash 0$ whose image under the quotient map $\mathbb{R}T'_{\mathcal{X}}\backslash 0 \to S'_{\mathcal{X}}$ is equal to $\dot{\mathcal{U}}$. The restriction mappings $\mathfrak{E}(\mathcal{U}_1) \to \mathfrak{E}(\mathcal{U}_2)$ and $\mathfrak{E}_0(\mathcal{U}_1) \to \mathfrak{E}_0(\mathcal{U}_2)$ are those naturally determined by the inclusion $\mathcal{U}_2 \subset \mathcal{U}_1$. In turn those presheaves define two sheaves $\tilde{\mathscr{E}}$ and $\tilde{\mathscr{E}}_0$ on $S'_{\mathcal{X}}$; clearly $\tilde{\mathscr{E}}_0$ is a subsheaf of $\tilde{\mathscr{E}}$.

Now we limit our attention to the vicinity of a point z_0 of $\mathcal{X}$. We assume that $\mathcal{X}$ is well–positioned at z_0 (Definition IX.2.2, [TREVES, 1992]; see also (II.2.3)) and we make use of our usual parametric representation of $\mathcal{X}$ near z_0, $B_0 \ni x \to Z(x) = x+i\Phi(x)$. We make all the needed hypotheses about the map Φ: $B_0 \to B_1$ (as usual, B_0 and B_1 are two open balls in $\mathbb{R}^m$ centered at the origin) and we write $\Omega = Z(B_0)$. This brings us into a position where we can use the FBI of a hyperfunction (defined in Ω) as described in the first part of section II.2. In turn the latter leads directly to the definition of the FBI transform of a germ $\dot{u} \in \mathscr{B}_{z_0}$ as an element of the stalk at z_0 of the sheaf over Ω defined by the quotient spaces $\mathfrak{E}(U)/\mathfrak{E}_0(U)$ — as implicitly pointed out in the closing remark of section II.2.

But now we can go further. Let us denote by $\mathscr{C}_{(z_0,\theta)}$, $\tilde{\mathscr{E}}_{(z_0,\theta)}$, etc., the stalks of the sheaves that have just been introduced, at an arbitrary point $(z_0,\theta) \in S'_{\mathcal{X}}$. Let $u \in \mathcal{B}(\Omega)$ be an arbitrary representative of a germ $\dot{u} \in \mathscr{C}_{(z_0,\theta)}$. The FBI transform $\mathcal{F}u$ defines an element of $\tilde{\mathscr{E}}_{(z_0,\theta)}$. If $v \in \mathcal{B}(\Omega)$ is another representative of the germ $\dot{u}$, $u—v$ is hypo–analytic at (z_0,θ) and Theorem II.4.1 tells us that $\mathcal{F}(u—v)$ decays exponentially in a conic neighborhood in $T'^{1,0}$ of the ray in $\mathbb{R}T'_{\mathcal{X}}\big|_{z_0}$ corresponding to θ. In other words, $\mathcal{F}(u—v)$ defines a germ that belongs to $\tilde{\mathscr{E}}_{0(z_0,\theta)}$, and the equivalence class in $(\tilde{\mathscr{E}}/\tilde{\mathscr{E}}_0)_{(z_0,\theta)}$ defined by $\mathcal{F}u$ does not depend on the representative u. It is legitimate to denote it by $\mathcal{F}_{(z_0,\theta)}\dot{u}$ and refer to it as the *FBI transform of the germ of microfunction* $\dot{u}$ at (z_0,θ). We may state:

THEOREM II.6.1.— *The FBI transform of the germs of microfunctions at* $(z_0,\theta) \in S'_{\mathcal{X}}$ *is an injective linear map* $\mathcal{F}_{(z_0,\theta)}: \mathcal{C}_{(z_0,\theta)} \to (\tilde{\mathcal{E}}/\tilde{\mathcal{E}}_0)_{(z_0,\theta)}$.

Microdistributions

DEFINITION II.6.2.— *We shall say that the germ of a microfunction* $\dot{u}$ *at* $(z_0,\theta) \in S'_{\mathcal{X}}$ *is the germ of a* ***microdistribution*** *if it can be represented by the germ of a distribution* u *at* z_0.

We can similarly say that the germ of a microfunction $\dot{u}$ at $(z_0,\theta) \in S'_{\mathcal{X}}$ is the germ of an H^s *microdistribution* if $\dot{u}$ can be represented by the germ of a H^s distribution at z_0. Here H^s is the sth Sobolev space; it can be replaced by any other space of distributions one wishes. We can also say that the germ of a microfunction $\dot{u}$ at $(z_0,\theta) \in S'_{\mathcal{X}}$ is the germ of a $\mathcal{C}^k$ *microfunction* if it can be represented by the germ of a $\mathcal{C}^k$ function at z_0 $(0 \leq k \leq +\infty)$. Here also $\mathcal{C}^k$ can be replaced by any other function space, such as L^p $(1 \leq p \leq +\infty)$. [Since we are dealing with germs of functions or distributions the notions of L^p, H^s, etc. are purely local and do not require the choice of a metric.] Provided the base manifold $\mathcal{X}$ is submitted to appropriate regularity hypotheses we could also consider classes of hyperfunctions more general than distributions, such as Gevrey ultradis— tributions, and define the corresponding microlocal objects.

Below we assume that $\mathcal{X}$ is well—positioned at z_0.

THEOREM II.6.2.— *For the germ of a microfunction* $\dot{u}$ *at* $(z_0,\theta) \in S'_{\mathcal{X}}$ *to be the germ of a microdistribution it is necessary and sufficient that, as* $|\zeta| \to +\infty$, *its FBI transform at* z_0 *grow slowly in a conic neighborhood of the ray defined by* (z_0,θ) *in* $\mathbb{R}T'_{\mathcal{X}}$.

The condition in Theorem II.6.2 means that there is an analytic functional μ carried by the closure of a suitably small open neighborhood Ω of z_0 in $\mathcal{X}$, representing a hyperfunction u in Ω that represents $\dot{u}$, such that the following is true:

(II.6.3) *There are an open and conic subset* $\mathcal{U}$ *of* $\mathbb{R}T'_{\mathcal{X}}|_{\Omega}$ *containing the ray defined by* (z_0,θ) *and positive constants* C, k *such that*

$$|\mathcal{F}\mu(z,\zeta)| \leq C(1+|\zeta|)^{k}, \forall (z,\zeta) \in \mathcal{U}.$$

Proof: The condition is necessary by Theorem IX.2.1, [TREVES, 1992]. In order to prove its sufficiency we duplicate to a large extent the proof of the sufficiency of the condition in Theorem II.4.1. We select the neighborhood V' of z_0 in $\mathcal{X}$ and the cone $\mathcal{C}_0$ in $\mathbb{R}^m\backslash\{0\}$ in such a way that $(z',\zeta) \in \mathcal{U}$ if $z' = Z(x') \in V'$ and if $\zeta = {}^tZ_x(x')^{-1}\xi$, $\xi \in \mathcal{C}_0$. We arrive at the same formula (II.4.14) but here we deal with an integral

$$\int_{z'\in V'}\int_{\xi\in\mathcal{C}_0} e^{i\zeta\cdot(z-z')-<\zeta><z-z'>^2-\epsilon<\zeta>^2}\mathcal{F}\mu(z',\zeta)<\zeta>^{m'2}dz'd\zeta$$

in which $\mathcal{F}\mu$ is submitted to Condition (II.6.3) instead of (II.4.13). The hypothesis that $\mathcal{X}$ is well–positioned at z_0 [cf. (II.2.2) & (II.2.3)] and integration by parts with respect to z' leads easily to the proof that the limit of the above integral as $\epsilon \to +0$ defines a distribution with respect to z in some open neighborhood of z_0 in $\mathcal{X}$. Thanks to this the desired con–clusion can be reached by essentially duplicating the end of the proof of Theorem II.4.1. □

Let a hyperfunction u in $\mathcal{X}$ define a microfunction $\dot{u}$ on $S'_{\mathcal{X}}$. The conic subset of $\mathbb{R}T'_{\mathcal{X}}$ corresponding to the microsupport of $\dot{u}$ is the hypo–analytic wave front set of u. One can define the *distribution wave front set* of u, $WF_{\mathcal{D}'}(u)$, as the complement of the conic set which projects onto the open subset of $S'_{\mathcal{X}}$ at whose points $\dot{u}$ is a microdistribution. Likewise one defines the $\mathcal{C}^\infty$ *wave front set* of u, $WF_{\mathcal{C}^\infty}(u)$, as the complement of the conic set which projects onto the open subset of $S'_{\mathcal{X}}$ at whose points $\dot{u}$ is a $\mathcal{C}^\infty$ microfunction. Note that

$$\text{(II.6.4)} \qquad WF_{\mathcal{D}'}(u) \subset WF_{\mathcal{C}^\infty}(u) \subset WF_{\mathrm{ha}}(u).$$

Naturally, one can replace $\mathcal{D}'$ or $\mathcal{C}^\infty$ by other functional spaces. The $\mathcal{C}^\infty$ wave front set can be characterized by the *rapid decay* of the FBI trans–form of u (cf. Theorem IX.4.1, [TREVES, 1992]). Similarly, many of the other kinds of wave front sets can be characterized by an appropriate rate of decay (or growth) of the FBI transform.

CHAPTER III

HYPERFUNCTION SOLUTIONS IN A HYPO–ANALYTIC MANIFOLD

INTRODUCTION

Section III.1 begins by recalling the main features of a hypo–analytic structure of rank m on a manifold $\mathcal{M}$, focusing on the case $m < \dim \mathcal{M}$ (ie., there are nonvanishing vector fields that annihilate all the hypo–analytic functions) and with special emphasis on the differential complex associated with such a structure. To define the sheaf of germs of hyper–function solutions in $\mathcal{M}$ we avail ourselves of local embeddings $\wp \to (Z(\wp), Z'(\wp)) \in \mathbb{C}^m \times \mathbb{C}^n$ such that the components $Z_1, \dots, Z_m$ are hypo–analytic in their domain of definition, an open subset $\mathcal{M}'$ of $\mathcal{M}$; the components $Z'_1, \dots, Z'_n$, are simply $\mathcal{C}^\infty$ functions. The map (Z, Z') is a assumed to be a diffeo–morphism of $\mathcal{M}'$ onto a *maximally real* submanifold Σ of $\mathbb{C}^m \times \mathbb{C}^n$. By push forward from $\mathcal{M}'$ to Σ we equip Σ with a hypo–analytic structure weaker than that induced by $\mathbb{C}^m \times \mathbb{C}^n$: the hypo–analytic functions in that struc–ture are not the restrictions of holomorphic functions in the ambient space but only of those that are locally independent of z'. It is induced by the *elliptic* structure on $\mathbb{C}^m \times \mathbb{C}^n$ associated with the differential operator $\bar{\partial}_z + d_{z'}$. It is therefore natural to take the hyperfunctions solutions on Σ to be those hyperfunctions u on Σ (well defined, according to Chapter I) such that $\partial_{z'} u \equiv 0$. By pulling back to $\mathcal{M}'$ we get a presheaf $U \to \mathfrak{Sol}(U)$ and thus a sheaf on $\mathcal{M}'$. It is not yet possible to patch up these local sheaves, to get a sheaf in the whole of $\mathcal{M}$. For this we must show that the local sheaves are hypo–analytically invariant, as a consequence of which they will agree on overlaps. But what can already be proved is that they contain the sheaves of germs of classical (ie., $\mathcal{C}^1$) solutions (Example III.1.1) and distribution solutions (Example III.1.2).

Section III.1. also introduces the cohomology space $\mathfrak{Sol}^{(q)}(U)$ of the differential complex $\partial_{z'}$: $\mathcal{B}(U;\Lambda^{q-1}) \to \mathcal{B}(U;\Lambda^q)$ (q = 1,...,n; Definition

III.1.2). The elements of $\mathcal{B}(U;\Lambda^q)$ are the currents $f = \sum_{|J|=q} f_J dz'_J$ whose coefficients f_J are hyperfunctions in U (U: open subset of the maximally real submanifold Σ of $\mathbb{C}^{m+n}$). The "holomorphic" differential operator $\partial_{z'}$ acts on those coefficients, and it acts in the customary manner on the current f.

The invariance of the spaces $\mathfrak{Sol}^{(q)}(U)$ is stated in section III.2 and proved in section III.3, for any q = 0,1,...,n [$\mathfrak{Sol}^{(0)}(U) = \mathfrak{Sol}(U)$]. It follows from the natural isomorphism of $\mathfrak{Sol}^{(q)}(U)$ with the relative co—homology space $H_U^{m,m+n+q}(\mathbb{C}^{m+n}\backslash\partial U)$ (Theorem III.2.2). The latter space pertains to the differential complex $\bar{\partial}_z + d_{z'}$ in the open set $\mathbb{C}^{m+n}\backslash\partial U$, of which U is a closed subset (∂U is the boundary of U with respect to Σ).

III.1 DEFINITION OF HYPERFUNCTION SOLUTIONS IN A HYPO–ANALYTIC MANIFOLD

The concept of *a hypo–analytic structure of rank* m on a $\mathcal{C}^\infty$ manifold $\mathcal{M}$ was recalled at the end of section I.5; for more details we refer the reader to Chapter III, [TREVES, 1992]. We recall some of the related concepts, such as the structure bundles and the associated differential complex. We shall make use of *hypo–analytic charts* $(\mathcal{M}',Z)$ in $\mathcal{M}$: $\mathcal{M}'$ is an open subset of $\mathcal{M}$ and Z is a map $\mathcal{M}' \to \mathbb{C}^m$ with hypo–analytic components Z_i such that $dZ_1\wedge\cdots\wedge dZ_m \neq 0$ at every point of $\mathcal{M}'$. It is convenient to enrich a bit the notion of hypo–analytic chart by requiring $\mathcal{M}'$ to be the domain of local coordinates $x_1,...,x_m,t_1,...,t_n$ [throughout the sequel we write $n = \dim \mathcal{M} - m \geq 0$] such that $\mathcal{M}' \ni (x,t) \to (Z(x,t),t) \in \mathbb{C}^m\times\mathbb{R}^n$ is a diffeomorphism onto a submanifold Σ of $\mathbb{C}^m\times\mathbb{R}^n$ [this subsumes that the Jacobian matrix $\partial Z/\partial x$ is nonsingular]. Automatically, the hypo–analytic structure on Σ transferred from $\mathcal{M}'$ by means of this diffeomorphism is the same as that inherited by Σ from the elliptic structure of the ambient space $\mathbb{C}^m\times\mathbb{R}^n$. We may view the hypo–analytic structure of $\mathcal{M}$ as a patch–work made up of these local maps $(\mathcal{M}',x,t,Z)$. On overlaps the change of coordinates x_j, t_k are $\mathcal{C}^\infty$; those of the "first integrals" Z_i must be biholo–morphic.

The differentials of the germs of the hypo–analytic functions in $\mathcal{M}$ span a vector subbundle T' of the complexified cotangent bundle $\mathbb{C}T^*\mathcal{M}$ to which we shall refer as the *cotangent structure bundle.* In the local chart $(\mathcal{M}',x,t,Z)$ the differentials $dZ_1,...,dZ_m$ make up a basis of T' at every point of $\mathcal{M}'$ (and $\{dZ_1,...,dZ_m,dt_1,...,dt_n\}$ is a basis of $\mathbb{C}T^*\mathcal{M}$ over $\mathcal{M}'$). For any pair of integers p, q ≥ 0 we define $T'^{p,q}$ as the vector subbundle of $\Lambda^{p+q}\mathbb{C}T^*\mathcal{M}$ whose fibre at a point $\wp \in \mathcal{M}$ is spanned by exterior products $\phi_1\wedge\cdots\wedge\phi_p\wedge\psi_1\wedge\cdots\wedge\psi_q$ with $\phi_j \in T'_\wp$ (j = 1,...,p) and $\psi_k \in \mathbb{C}T^*_\wp\mathcal{M}$ (k =

1,...,q). Clearly, $T'^{1,0} = T'$ and $T'^{p+1,q-1} \subset T'^{p,q}$. Note also that

(III.1.1) $$T'^{p,q} = T'^{p+1,q-1} \ \textit{if} \ q > n;$$

(III.1.2) $$T'^{p,q} = 0 \ \textit{if} \ p > m, \textit{ or if } p+q > m+n.$$

We can define the quotient vector bundles

(III.1.3) $$\Lambda_{\mathcal{M}}^{p,q} = T'^{p,q}/T'^{p+1,q-1};$$

and Properties (III.1.1) and (III.1.2) entail

(III.1.4) $$\Lambda_{\mathcal{M}}^{p,q} = 0 \ \textit{if} \ p > m, \textit{ or if } q > n.$$

Since T' is locally spanned by closed forms, the exterior derivative of a $\mathcal{C}^1$ section of T' is a $\mathcal{C}^0$ section of $T'^{1,1}$, a property sometimes referred to by saying that T' is closed. By the same token the exterior derivative of a $\mathcal{C}^1$ section of $T'^{p,q}$ is a $\mathcal{C}^0$ section of $T'^{p,q+1}$, a property we summarize by writing $dT'^{p,q} \subset T'^{p,q+1}$. Combining this with (III.1.3) shows that the exterior derivative d induces (for each p = 0,1,...,m) a sequence of differential operators

(III.1.5) $$d'^{p,q}: \mathcal{C}^{\infty}(\mathcal{M};\Lambda_{\mathcal{M}}^{p,q}) \to \mathcal{C}^{\infty}(\mathcal{M};\Lambda_{\mathcal{M}}^{p,q+1}), \quad q = 0,1,...,n.$$

In the sequel we shall often write d' rather than $d'^{p,q}$; of course $d'^2 = 0$. We may also look at the distribution sections of the vector bundles $\Lambda_{\mathcal{M}}^{p,q}$; this yields the differential complexes (for each p = 0,1,...,m)

(III.1.6) $$d' : \mathcal{D}'(\mathcal{M};\Lambda_{\mathcal{M}}^{p,q}) \to \mathcal{D}'(\mathcal{M};\Lambda_{\mathcal{M}}^{p,q+1}),\ q = 0,1,\dots,n.$$

And likewise we might want to look at the analogous differential complexes when the supports of the sections are required to be compact.

The orthogonal of $T' \subset \mathbb{C}T^*\mathcal{M}$ for the duality between tangent and cotangent vectors will be denoted by $\mathcal{V}$ and referred to as the *tangent structure bundle.* Over the domain $\mathcal{M}'$ of a local chart $(\mathcal{M}',x,t,Z)$ a linear basis of $\mathcal{V}$ consists of the vector fields $L_1,\dots,L_n$ defined by the following "orthonormality" conditions:

(III.1.7) $$L_j Z_i = 0,\ L_j t_k = \delta_{jk}\ (i = 1,\dots,m,\ j,\ k = 1,\dots,n).$$

In passing note that a basis of $\mathbb{C}T\mathcal{M}$ over $\mathcal{M}'$ is made up of the vector fields $L_1,\dots,L_n$ together with the vector fields $M_1,\dots,M_m$ defined by

(III.1.8) $$M_h Z_i = \delta_{hi},\ M_i t_j = 0\ \ (h,\ i = 1,\dots,m,\ j = 1,\dots,n).$$

An immediate consequence of (III.1.7) and (III.1.8) is that all the vector fields L_j, M_h pairwise commute. The fact that the vector $L_1,\dots,L_n$ commute among themselves proves that the vector subbundle $\mathcal{V}$ satisfies the Frobenius condition (of formal integrability): the commutation bracket of two smooth sections of $\mathcal{V}$ is a section of $\mathcal{V}$, which we abbreviate by writing $[\mathcal{V},\mathcal{V}] \subset \mathcal{V}$.

It is also easy to derive explicit expressions for these vector fields. Call m_{hi} the generic entry of the matrix $(\partial Z/\partial x)^{-1}$. Then, by (III.1.8),

(III.1.9) $$M_h = \sum_{i=1}^{m} m_{hi}(x,t)\partial/\partial x_i\ \ (h = 1,\dots,m).$$

Combining (III.1.8) and (III.1.9) shows that

(III.1.10) $$L_j = \partial/\partial t_j - \sum_{h=1}^{m} (\partial Z_h/\partial t_j)M_h \quad (j = 1,\dots,n).$$

This provides us with a convenient representation of the vector bundles $\Lambda_{\mathcal{M}}^{p,q}$ and of the differential operator d′ [see (III.1.5)] in the local chart $(\mathcal{M}',x,t,Z)$. Indeed, a section $\dot{f} \in C^{\infty}(\mathcal{M}';\Lambda_{\mathcal{M}}^{p,q})$ is represented by a unique differential form of the kind

(III.1.11) $$f = \sum_{|I|=p} \sum_{|J|=q} f_{I,J}(x,t)dZ_I \wedge dt_J,$$

where $f_{I,J} \in C^{\infty}(\mathcal{M}')$. Taking the coefficients in $\mathcal{D}'(\mathcal{M}')$ leads to a distri– bution section $\dot{f} \in \mathcal{D}'(\mathcal{M}';\Lambda_{\mathcal{M}}^{p,q})$; taking them in $C_c^{\infty}(\mathcal{M}')$ defines $\dot{f} \in C_c^{\infty}(\mathcal{M}';\Lambda_{\mathcal{M}}^{p,q})$ and taking them in $\mathcal{E}'(\mathcal{M}')$ defines $\dot{f} \in \mathcal{E}'(\mathcal{M}';\Lambda_{\mathcal{M}}^{p,q})$. The section $d'\dot{f}$ of $\Lambda_{\mathcal{M}}^{p,q+1}$ is represented by the differential form

(III.1.12) $$Lf = \sum_{|I|=p} \sum_{|J|=q} \sum_{k=1}^{n} L_k f_{I,J}(x,t)\, dt_k \wedge dZ_I \wedge dt_J.$$

Of course $L^2 = 0$. If we avail ourselves of (III.1.9) we can write

(III.1.13) $$Lf = d_t f - \sum_{h=1}^{m} d_t Z_h \wedge M_h f,$$

with the vector fields M_h acting coefficientwise on the differential form f.

The reader will note that the form (III.1.11) can be regarded as the pullback of the differential form on the image Σ of $\mathcal{M}'$ under the map $(x,t) \to (Z(x,t),t)$,

$$\tilde{f} = \sum_{|I|=p} \sum_{|J|=q} \tilde{f}_{I,J}(z,t)dz_I \wedge dt_J, \tag{III.1.14}$$

where $\tilde{f}_{I,J} \in C^{\infty}(\Sigma)$ is determined by the equation $f_{I,J}(x,t) = \tilde{f}_{I,J}(Z(x,t),t)$. The differential operator L acting on the forms (III.1.11) is the pullback of the operator $\bar{\partial}_z + d_t$ acting on the forms (III.1.14). The pushforward from $\mathcal{M}'$ to Σ of the vector bundle $\Lambda^{p,q}_{\mathcal{M}}$ is equal to the pullback to Σ of the vector bundle on $\mathbb{C}^m \times \mathbb{R}^n$ spanned by the exterior products $dz_I \wedge d\bar{z}_J \wedge dt_K$, $|I| = p$, $|J|+|K| = q$. As long as the analysis is limited to the hypo–analytic chart $(\mathcal{M}',x,t,Z)$, one may as well transfer it to the submanifold Σ equipped with the hypo–analytic structure induced by the elliptic structure of $\mathbb{C}^m \times \mathbb{R}^n$. Using the x_i and the t_j as coordinates in Σ we may regard $Z_h(x,t)$ as the expression in those coordinates of the restriction to Σ of the function z_h ($h = 1,\dots,m$). And the differential forms (or currents) (III.1.11), as well as the differential operator L, can be regarded as defined in Σ. This is the viewpoint more or less explicitely adopted in [TREVES, 1992] (for instance see p. 77).

Here we shall go one step further and also complexify t–space. In other words we shall think of the coarse local embedding (Section II.1, *loc. cit.*)

$$\mathcal{M}' \ni (x,t) \to (Z(x,t),t)$$

as valued in $\mathbb{C}^m \times \mathbb{C}^n$. This maps $\mathcal{M}'$ onto a *maximally real* submanifold of $\mathbb{C}^m \times \mathbb{C}^n$ which we continue to call Σ. As before we call $z_1,\dots,z_m$ the coordinates in the first factor $\mathbb{C}^m$. Provisionally let us call $z'_1,\dots,z'_n$ the coordinates in the second factor $\mathbb{C}^n$ and write $z' = (z'_1,\dots,z'_n)$. For the hypo–analytic structure inherited by Σ from the ambient space $\mathbb{C}^m \times \mathbb{C}^n$ the hypo–analytic functions in Σ are the restrictions of the holomorphic functions of (z,z'). Among these the functions that are hypo–analytic in the sense of the structure transferred from $\mathcal{M}'$ are those that are independent of z'.

As a matter of fact we can go one step further and equip $\mathcal{M}'$ with a maximal hypo–analytic structure (see section I.5) stronger than the hypo–analytic structure induced by $\mathcal{M}$. The latter simply means that if a function is hypo–analytic (at some point) in the sense of the structure of $\mathcal{M}$ it is *a fortiori* hypo–analytic in the sense of the maximal structure. This is equivalent to adjoining to the functions $Z_1,\dots,Z_m$ n more smooth functions Z'_j such that $dZ_1\wedge\cdots\wedge dZ_m\wedge dZ'_1\wedge\cdots\wedge dZ'_n \neq 0$; the hypo–analytic functions in the maximal structure will simply be the holomorphic functions of (Z,Z'). The difference between this more general situation and the earlier one is that the Z'_j are not necessarily real–valued, as the t_j were. We shall assume that the map $(x,t) \to (Z(x,t),Z'(x,t))$ is injective: it realizes an embedding of $\mathcal{M}'$ as a maximally real submanifold of $\mathbb{C}^{m+n}$, Σ. Here again the functions on Σ that are hypo–analytic for the structure transferred from $\mathcal{M}$ are the restrictions of holomorphic functions in $\mathbb{C}^m \times \mathbb{C}^n$ that are independent of z'.

We shall keep to this viewpoint and carry the analysis to Σ. Throughout the sequel we shall assume Σ to be sufficiently small that the construction of Chapter I can be carried out. In particular we can apply directly Definition I.5.1: if $U \subset\subset \Sigma$, a hyperfunction in U is an element of the quotient space $\mathcal{B}(U) = \mathcal{O}'(\mathcal{C}\ell U)/\mathcal{O}'(\partial U)$ [∂U shall always denote the boundary of U in Σ]. The linear spaces $\mathcal{B}(U)$ and the restriction mappings r^U_V: $\mathcal{B}(U) \to \mathcal{B}(V)$ (for any open subset V of U) define the sheaf of hyper–functions $\mathscr{B}_\Sigma$ on Σ. We may consider the linear spaces $\mathcal{B}(U) = \Gamma(U;\mathscr{B}_\Sigma)$

even when the closure of U is not a compact subset of Σ.

Of course the differential operator $f \to \partial_{z'} f = (\partial f/\partial z'_1,\dots,\partial f/\partial z'_n)$ acts on holomorphic functions and therefore, by transposition, on analytic functionals, and from there, on hyperfunctions in Σ.

DEFINITION III.1.1.— *By a **hyperfunction solution** in an open subset* U *of* Σ *we shall mean any element u of* $\mathcal{B}(U)$ *such that* $\partial_{z'} u \equiv 0$ *in* U. *The space of all hyperfunction solutions in* U *will be denoted by* $\mathfrak{Sol}(U)$.

If $U \subset\subset \Sigma$ and if $u \in \mathcal{B}(U)$ is a hyperfunction solution in U then each representative $\mu \in \mathcal{O}'(\mathcal{C}\ell U)$ of u has the property that $\partial_{z'}\mu \in \mathcal{O}'(\partial U)$.

EXAMPLE III.1.1 (***Classical solutions***).— Let U_0 be a relatively compact open subset of the ball $B_0 \subset \mathbb{R}^m$ with smooth boundary ∂U_0 and let Θ denote an open ball in t–spa– ce $\mathbb{R}^n$. Here we take $Z' = t$: we call U the image of $U_0 \times \Theta$ under the map $(x,t) \to (Z(x,t),t)$. We assume $U \subset\subset \Sigma$. In the sequel, whatever $t \in \Theta$ we call U_t the slice $\{ (z,z') \in U;\ z' = t \}$; U_t will often be regarded as a subset of $\mathbb{C}^m$.

Consider a "classical" solution $g(x,t) \in C^1(\mathcal{C}\ell U)$ of the system of equations

$$\text{(III.1.15)} \qquad L_j g = 0 \quad \textit{in } U,\ j = 1,\dots,n,$$

where $L_1,\dots,L_n$ are the basic vector fields (III.1.10). The function g defines an analytic functional $\nu_g(t)$ carried by $\mathcal{C}\ell U_t$ for each $t \in \Theta$, by the following formula, in which $h \in \mathcal{O}(\mathbb{C}^m)$,

(III.1.16)

$$<\nu_g(t),h> = \int_{U_t} gh dz_1\wedge\cdots dz_m = \int_{U_0} g(x,t)h(Z(x,t))[\det Z_x(x,t)]dx.$$

It is clear that $\mathscr{C}\ell\Theta \ni t \to <\nu_g(t),h>$ is a C^1 function. The integration $\mathcal{O}(\mathbb{C}^n) \ni h_0 \to \int_\Theta h_0(t)<\nu_g(t),h>dt$ defines an analytic functional in (z,z')—space $\mathbb{C}^{m+n}$, ν_g, carried by $\mathscr{C}\ell U$ [we are implicitly using the density in $\mathcal{O}(\mathbb{C}^{m+n})$ of polynomials, hence of finite sums of products $h(z)h_0(t)$]. Now let

$$\psi = \sum_{j=1}^n (-1)^{j-1}\psi_j(t)dt_1\wedge\cdots\wedge dt_{j-1}\wedge dt_{j+1}\wedge\cdots\wedge dt_n \in C_c^\infty(\Theta;\Lambda^{n-1})$$

be arbitrary; we have

$$\int_{\mathbb{R}^n} <\nu_g(t),h>d_t\psi = \int_{U_0\times\Theta} g(h\circ Z)d_t\psi\wedge dZ_1\wedge\cdots\wedge dZ_m.$$

We take advantage of the fact that $dZ_1,\ldots,dZ_m,dt_1,\ldots,dt_n$ span the whole cotangent bundle to $B_0\times\Theta$. If d denotes the total differential in (x,t)—space, according to (III.1.7) & (III.1.8) we have, for any function $f \in C^1(B_0\times\Theta)$,

$$df = \sum_{i=1}^m M_i f\, dZ_i + \sum_{j=1}^n L_j f\, dt_j. \tag{III.1.17}$$

We derive from (III.1.17):

$$d[g(h\circ Z)\psi\wedge dZ_1\wedge\cdots\wedge dZ_m] =$$

$$\sum_{j=1}^n \psi_j(t)L_j[g(h\circ Z)]dt_1\wedge\cdots\wedge dt_n\wedge dZ_1\wedge\cdots\wedge dZ_m +$$

$$g(h\circ Z)d_t\psi\wedge dZ_1\wedge\cdots\wedge dZ_m = g(h\circ Z)d_t\psi\wedge dZ_1\wedge\cdots\wedge dZ_m$$

by virtue of (III.1.15). We see that

$$\int_{\mathbb{R}^n} \psi \wedge d_t \langle \nu_g(t), h \rangle = \pm \int_{U_0 \times \Theta} d[g(h \circ Z)\psi \wedge dZ_1 \wedge \cdots \wedge dZ_m] =$$
$$\pm \int_{\partial U_0 \times \Theta} g(h \circ Z)\psi \wedge dZ_1 \wedge \cdots \wedge dZ_m$$

since supp $\psi \subset\subset \Theta$. And since ψ is otherwise arbitrary we obtain

(III.1.18)
$$(\partial/\partial t_j)\langle \nu_g(t), h \rangle =$$
$$\sum_{i=1}^{m} \pm \int_{\partial U_0} g(h \circ Z)(\partial Z_i/\partial t_j) dZ_1 \wedge \cdots \wedge dZ_{i-1} \wedge dZ_{i+1} \wedge \cdots \wedge dZ_m,$$

which shows that $d_t \nu_g(t)$ is carried by ∂U_t and therefore, that $\partial_{z'} \nu_g$ is carried by ∂U. We conclude that ν_g defines a hyperfunction solution in U.

□

EXAMPLE III.1.2 (***Distribution solutions***).— Same set—up as in Example III.1.1; let Ω be an open subset of Σ containing the image of $\mathscr{Cl}(B_0 \times \Theta)$ under the map $(x,t) \to (Z(x,t),t)$. Let $u \in \mathcal{D}'(\Omega)$ be a solution of the homogeneous equations $L_j u = 0$, $j = 1,\dots,n$, in Ω [see (III.1.10)]. According to Proposition I.4.3, [TREVES, 1992], u is a $\mathcal{C}^\infty$ function of t valued in the space of distributions of x. According to Theorem II.5.1, *loc. cit.* [see also (II.5.11), *ibid.*], if B_0 and Θ are sufficiently small there are a $\mathcal{C}^1$ solution v in an open neighborhood $\Omega' \subset \Omega$ of the image of $\mathscr{Cl}(B_0 \times \Theta)$ and an integer $\kappa \geq 0$ such that $u = \Delta_M^\kappa v$ in Ω'. Here $\Delta_M = M_1^2 + \cdots + M_m^2$ with M_j given by (III.1.9).

As before, let U denote the image of $U_0 \times \Theta$ under the map $(x,t) \to (Z(x,t),t)$ ($U_0 \subset\subset B_0$, U_0 open, ∂U_0 smooth); suppose $U \subset\subset \Omega'$. If χ_U stands for the characteristic function of U_0 the distribution $u - \Delta_M^\kappa(\chi_U v)$ vanishes identically in U, and therefore the element of $\mathcal{B}(U)$ defined by u

is the same as that defined by $\Delta_M^\kappa(\chi_U v)$. If then $h \in \mathcal{O}(\mathbb{C}^m)$ is arbitrary we have, in the sense of the duality between C^∞ functions in Ω'_t and compactly supported distributions in Ω'_t,

$$<\Delta_M^\kappa(\chi_U v), h> = <\chi_U v, \Delta_M^\kappa h> = \int_{U_0} v(x,t)\Delta_M^\kappa[h(Z(x,t))]\, dZ(x,t).$$

By virtue of (III.1.8) we have

$$\Delta_M^\kappa[h(Z(x,t))] = (\Delta_z^\kappa h)(Z(x,t)) \quad (\Delta_z = \sum_{j=1}^m \partial^2/\partial z_j^2).$$

We have already seen (Example III.1.1) that the integral

$$\int_{U_0} v(x,t)h(Z(x,t))dZ(x,t)$$

defines an element $\nu_v(t) \in \mathcal{O}'(\mathscr{C}\!\ell U_t)$ such that $d_t\nu_v$ is carried by ∂U_t. We conclude that the integral

$$\int_{U_0} v(x,t)\Delta_M^\kappa[h(Z(x,t))]\, dZ(x,t)$$

defines the analytic functional $\Delta_z^\kappa \nu_v(t)$; $d_t\Delta_z^\kappa\nu_v(t)$ is carried by ∂U_t. In turn, integration with respect to t over Θ produces out of $\nu_v(t)$ an analytic functional ν_v in $\mathbb{C}^{m+n}$ such that $\partial_{z'}\Delta_z^\kappa\nu_v$ is carried by ∂U. If then $\tilde{v} \in \mathcal{B}(U)$ is the hyperfunction solution in U defined by v (and represented by ν_v), $\Delta_z^\kappa\tilde{v}$ is the hyperfunction solution in U defined by the distribution u.

Now consider a distribution u in Ω; assume that u is a C^∞ function of t valued in the space of distributions of x. By using the ellipticity of the differential operator Δ_M in x–space, it is possible to find, in an open subset Ω' of Ω, a C^∞ function of t, v, valued in the space of C^1 functions of x, and an integer $\kappa \geq 0$ such that $u = \Delta_M^\kappa v$ in Ω'. Exactly as before, u and $\Delta_M^\kappa(\chi_U v)$ define one and the same element of $\mathcal{B}(U)$, $\tilde{u}$, represented by the analytic functional $\Delta_z^\kappa\nu_v$. Let us posit that $\tilde{u}$ is a hyperfunction solution in U. Then the analytic functionals defined by

$$h\otimes h_0 \to \int_\Theta h_0(t)(\partial/\partial t_j)<\nu_v(t),\Delta^\kappa_z h>dt \quad (j = 1,\dots,n)$$

are carried by ∂U and consequently, by repeating the preceding argument, we conclude that the analytic functionals

$$\mathcal{O}(\mathbb{C}^{m+n}) \ni H \to \int_U (L_j v)(\Delta^\kappa_z H)dZ\wedge dt$$

are also carried by ∂U. Let now $\varphi \in C^\infty_c(U)$ be arbitrary and form $\mathcal{E}_\nu\varphi$ according to (I.3.4) but in the maximal hypo—analytic structure defined by the functions $Z_1,\dots, Z_m,t_1,\dots,t_n$. If we apply the commutation formulas (II.2.8)—(II.2.9) in [TREVES, 1992] we obtain that, as $\nu \to +\infty$, $\Delta^\kappa_z\mathcal{E}_\nu\varphi$ converges uniformly to $\Delta^\kappa_M\varphi$ in U, and also that $\Delta^\kappa_z\mathcal{E}_\nu\varphi$ converges uniformly to zero in some neighborhood of ∂U in $\mathbb{C}^{m+n}$. We reach the conclusion that

$$\int_U (L_j v)(\Delta^\kappa_M\varphi)dZ\wedge dt = 0.$$

Since φ is arbitrary this means that $L_j\Delta^\kappa_M v = L_j u = 0$ for every $j = 1,\dots,n$. We have proved:

THEOREM III.1.1.— *For a distribution u in an open subset Ω of Σ to be a solution it is necessary and sufficient that u define a hyperfunction solution in Ω.*

□

We are also going to deal with "currents" whose coefficients are hyperfunctions. For any integer $q \geq 1$ we shall denote by $\mathcal{B}(U;\Lambda^q)$ the space of currents

(III.1.19) $$f = \sum_{|J|=q} f_J dz'_J$$

with coefficients $f_J \in \mathcal{B}(U)$. We set $\mathcal{B}(U;\Lambda^0) = \mathcal{B}(U)$; and of course, $\mathcal{B}(U;\Lambda^q) = \{0\}$ if $q > n$. Since the partial differentiations $\partial^{\alpha}_{z'}$ define endomorphisms of $\mathcal{B}(U)$ the following is a differential complex

$$\text{(III.1.20)} \qquad \partial_{z'}: \mathcal{B}(U;\Lambda^q) \to \mathcal{B}(U;\Lambda^{q+1}), \quad q = 0,1,\ldots,$$

with the standard action of $\partial_{z'}$:

$$\partial_{z'} f = \sum_{|J|=q} \sum_{k=1}^{n} (\partial f_J/\partial z'_k) dz'_k \wedge dz'_J.$$

DEFINITION III.1.2.— *Let* U *be an open subset of* Σ, q *an integer*, $0 \leq q \leq n$. *We shall denote by* $\mathfrak{Sol}^{(q)}(U)$ *the* q^{th} *cohomology space of the differential complex* (III.1.20).

The space $\mathfrak{Sol}^{(0)}(U)$ is naturally isomorphic to the space $\mathfrak{Sol}(U)$ of hyperfunction solutions in U (Definition III.1.1).

EXAMPLE III.1.3.— Suppose $1 \leq q \leq n$ and let $f \in C^{\infty}(\Sigma;\Lambda^{m,q}_{\Sigma})$ be given by (III.1.11) (with $p = m$) and satisfy $Lf \equiv 0$ [see (III.1.12)]. Let U be an open subset of Σ as in Example III.1.1; in particular, suppose ∂U_0 is smooth. We can use the integration in z–space,

$$\mathcal{O}(\mathbb{C}^m) \ni h \to \int_{U_t} hf = \langle \mu_f(t), h \rangle$$

to define a "current" in Θ, $\mu_f(t) = \sum_{|J|=q} \mu_J(t) dt_J$, whose coefficients are analytic functionals in z–space $\mathbb{C}^m$ depending on $t \in \Theta$ (and are carried by $\mathcal{C}\ell U_t$ for each t). The same integration by parts as in Example III.1.1 shows that

$$d_t\langle\mu_f(t),h\rangle = \int_{\partial U_t} h\omega_f$$

for a suitable form

$$\omega_f = \sum_{|I|=m-1}\ \sum_{|J|=q+1} \omega_{I,J}(z,t)dz_I \wedge dt_J$$

with smooth coefficients. Thus $d_t\mu_f(t)$ is carried by ∂U_t. We can identify each coefficient μ_J of μ_f to an analytic functional in $\mathbb{C}^{m+n}$ by means of integrals

$$\int_\Theta h_0(t)\langle\mu_J(t),h\rangle dt,\quad h \in \mathcal{O}(\mathbb{C}^m),\ h_0 \in \mathcal{O}(\mathbb{C}^n),$$

as in Example III.1.1. In this manner μ_f defines an element $\tilde{f}$ of $\mathcal{B}(U;\Lambda^q)$ such that $\partial_{z'}\tilde{f} \equiv 0$ in U. To say that the cohomology class $[\tilde{f}] \in \mathfrak{Sol}^{(q)}(U)$ vanishes is the same as saying that $\partial_{z'}\tilde{u} = \tilde{f}$ in U for some $\tilde{u} \in \mathcal{B}(U;\Lambda^{q-1})$. This is the correct version of the statement that $Lu = f$ in U [see (III.1.12); in general, the action of the vector fields L_j on elements of $\mathcal{B}(U)$ is not defined]. □

If $U \subset V$ are two open subsets of Σ the restriction mapping $\mathcal{B}(U) \to \mathcal{B}(V)$ induces a restriction mapping

(III.1.21) $$r_V^U\colon \mathfrak{Sol}^{(q)}(U) \to \mathfrak{Sol}^{(q)}(V).$$

The presheaf $(\mathfrak{Sol}^{(q)}(U), r_V^U)$ defines a sheaf on Σ which shall be denoted by $\mathcal{Sol}_\Sigma^{(q)}$.

Definitions III.1.1 and III.1.2 do not yet yield analogous definitions in the hypo–analytic manifold $\mathcal{M}$, since it is not clear how much they depend on the special embedding of the neighborhood $\mathcal{M}'$ as the submanifold Σ of $\mathbb{C}^{m+n}$. We are going to address this important question in the next section.

III.2 STATEMENT OF THE MAIN THEOREM ON THE INVARIANCE OF THE SHEAVES $\mathcal{Sol}^{(q)}$

We continue to make use of a local embedding

$$\mathcal{M}' \ni (x,t) \to (Z(x,t), Z'(x,t)) \in \Sigma \subset \mathbb{C}^m \times \mathbb{C}^n. \tag{III.2.1}$$

Recall that in $\mathcal{M}'$ the hypo–analytic structure of $\mathcal{M}$ is defined by the map $Z\colon \mathcal{M}' \to \mathbb{C}^m$; that $dZ_1 \wedge \cdots \wedge dZ_m \wedge dZ'_1 \wedge \cdots \wedge dZ'_n \neq 0$ at every point of $\mathcal{M}'$ and that the map (III.2.1) is injective, ie., it is a diffeomorphism of $\mathcal{M}'$ onto the maximally real submanifold Σ of $\mathbb{C}^{m+n}$.

The (local) transformations of $\mathbb{C}^m \times \mathbb{C}^n$ that preserve the hypo–analytic structure of Σ transferred from $\mathcal{M}'$ are given by

$$\tilde{z} = H(z),\ \tilde{z}' = K(z,z'), \tag{III.2.2}$$

where H is holomorphic, K is C^∞ and $\det \partial H/\partial z \neq 0$, $\det \partial(K,\overline{K})/\partial(z',\bar{z}') \neq 0$. These are the transformations that preserve the locally integrable structure of $\mathbb{C}^{m+n}$ defined by the combined action of the Cauchy–Riemann operator $\bar{\partial}_z$ in z–space $\mathbb{C}^m$ and of the total exterior derivative $d_{z'} = \partial_{z'} + \bar{\partial}_{z'}$ in z'–space $\mathbb{C}^n$. These transformations preserve the vector bundles $\tilde{\Lambda}^{p,q} \subset \Lambda^{p+q}\mathbb{C}T^*(\mathbb{C}^{m+n})$ spanned by the exterior products

$$dz_I \wedge d\bar{z}_J \wedge dz'_K \wedge d\bar{z}'_L,\ |I| = p,\ |J|+|K|+|L| = q$$

($0 \leq p \leq m$, $1 \leq q \leq m+2n$). In particular we shall pay special attention to the differential complex

$$\bar{\partial}_z + d_{z'}\colon C^\infty(\mathcal{U};\tilde{\Lambda}^{m,q}) \to C^\infty(\mathcal{U};\tilde{\Lambda}^{m,q+1}),\quad q = 0,1,\ldots,m+2n, \tag{III.2.3}$$

where $\mathcal{U}$ is an open subset of $\mathbb{C}^{m+n}$.

We shall denote by ${}_m\mathcal{O}$ the sheaf of germs of holomorphic functions of z alone, ie. independent of z'; ${}_m\mathcal{O}$ is viewed as a sheaf over $\mathbb{C}^{m+n}$. We denote by ${}_m\mathcal{O}^{(m)}$ the sheaf (over $\mathbb{C}^{m+n}$) of germs of differential forms $h dz_1\wedge\cdots\wedge dz_m$, with h a holomorphic function of z locally independent of z'. The differential complex (III.2.3) leads to the sheaf resolution

$$\text{(III.2.4)}\quad 0 \to {}_m\mathcal{O}^{(m)} \xrightarrow{i} \mathcal{C}^\infty\tilde{\Lambda}^{m,0} \xrightarrow{\bar{\partial}_z+d_{z'}} \mathcal{C}^\infty\tilde{\Lambda}^{m,1} \xrightarrow{\bar{\partial}_z+d_{z'}} \cdots \xrightarrow{\bar{\partial}_z+d_{z'}} \mathcal{C}^\infty\tilde{\Lambda}^{m,m+2n} \to 0.$$

We are using the notation $\mathcal{C}^\infty\tilde{\Lambda}^{m,q}$ to mean the sheaf of germs of smooth sections of $\tilde{\Lambda}^{m,q}$. This is a cohomological resolution since the Poincaré lemma is valid in the differential complex (III.2.3) (see [TREVES, 1992], Section 6.7).

However, it will often be more convenient to make use of the flabby resolution

$$\text{(III.2.5)}\quad 0 \to {}_m\mathcal{O}^{(m)} \xrightarrow{i} \mathcal{B}\tilde{\Lambda}^{m,0} \xrightarrow{\bar{\partial}_z+d_{z'}} \mathcal{B}\tilde{\Lambda}^{m,1} \xrightarrow{\bar{\partial}_z+d_{z'}} \cdots \xrightarrow{\bar{\partial}_z+d_{z'}} \mathcal{B}\tilde{\Lambda}^{m,m+2n} \to 0.$$

We are using the notation $\mathcal{B}\tilde{\Lambda}^{m,q}$ to mean the sheaf of germs of "currents"

$$\text{(III.2.6)}\quad f = \sum_{|I|+|J|+|K|=q} f_{I,J,K}\, dz_1\wedge\cdots\wedge dz_m\wedge d\bar{z}_I\wedge dz'_J\wedge d\bar{z}'_K$$

with coefficients $f_{I,J,K}$ that are hyperfunctions. Let us stress the fact that the term "hyperfunction" is used here in the sense of the standard C^ω structure of $\mathbb{R}^{2m+2n}$ (the C^ω structure of $\mathbb{R}^{2m+2n}$ is a maximal hypo—analytic structure; all the definitions and constructions of Chapters I and II, substantially simplified, are valid). The Cauchy—Riemann operators $\partial/\partial\bar{z}_j$ ($1 \leq j \leq m$) and the partial differentiations $\partial/\partial z'_k$ and $\partial/\partial\bar{z}'_l$ ($1 \leq k, l \leq n$) act on each coefficient of the current (III.2.6). Notice that the combined differential operator $\bar{\partial}_z + d_{z'}$ acts on f as the exterior derivative.

We are going to make use of the convolution of an analytic functional μ carried by real space $\mathbb{R}^N$ with a fundamental solution E of a linear partial differential operator $P(\partial/\partial x)$ with constant coefficients in $\mathbb{R}^N$. This can be viewed as the convolution in $\mathbb{R}^N$ of a hyperfunction with a compactly supported one, identified to μ. Let $K = \operatorname{supp} \mu \subset\subset \mathbb{R}^N$ and let $g \in \mathcal{B}(\mathbb{R}^N)$. If ν represents g in a ball $B_r = \{ x \in \mathbb{R}^N;\ |x| < r \}$ $(r > 0)$ the convolution $\mu\star\nu$ (see section I.1) is carried by $K+(\mathscr{C}\ell B_r)$; and if ν' represents g in $B_{r'}$ with $r' \geq r$, $\mu\star(\nu-\nu')$ is carried by $K+[(\mathscr{C}\ell B_{r'})\backslash B_r]$. Given any bounded, open subset Ω of $\mathbb{R}^N$ we will have $K+[(\mathscr{C}\ell B_{r'})\backslash B_r] \subset \mathbb{R}^N\backslash\Omega$ as soon as r is sufficiently large. This means that the hyperfunction defined by $\mu\star\nu'$ is equal, in Ω, to that defined by $\mu\star\nu$. By letting Ω expand to fill the whole space $\mathbb{R}^N$ this defines a hyperfunction in $\mathbb{R}^N$, which it is natural to denote by $\mu\star g$ (or $g\star\mu$) and to call the *convolution* of μ and g. We have, just like for distributions,

$$P(\partial/\partial x)(\mu\star g) = [P(\partial/\partial z)\mu]\star g = \mu\star[P(\partial/\partial x)g].$$

Apply this to $g = E$, a (distribution) fundamental solution E of $P(\partial/\partial x)$: $E \in \mathcal{D}'(\mathbb{R}^N)$ and in $\mathbb{R}^N$, $P(\partial/\partial x)E = \delta$, the Dirac distribution:

$$P(\partial/\partial x)(\mu\star E) = [P(\partial/\partial z)\mu]\star E = \mu. \tag{III.2.7}$$

In particular we see that $u = \mu \star E$ is a hyperfunction solution of the equation $P(\partial/\partial x)u = \mu$ in $\mathbb{R}^N$ (as announced above we are identifying μ to the compactly supported hyperfunction represented by μ in the whole of $\mathbb{R}^N$). Now suppose that μ represents a hyperfunction f in some open subset U of $\mathbb{R}^N$; then in U we have solved $P(\partial/\partial x)u = f$. Note, however, that if we replace μ by a different representative $\mu' \in \mathcal{O}'(\mathscr{C}\ell U)$ of f, in general $(\mu-\mu')\star E$ will not be carried by ∂U and therefore the hyperfunction $\mu' \star E$ will not be equal to $\mu \star E$, not even in U. But this will not matter to us since all we shall need (in the proof of Theorem III.2.1 below) is *some* solution u of the equation $P(\partial/\partial x)u = f$.

Our first application of (III.2.7) will be to prove Weyl's lemma for hyperfunctions:

LEMMA III.2.1.— *Let Ω be an open subset of $\mathbb{R}^N$. If a hyperfunction h in Ω is a solution of the Laplace equation in Ω, $\sum_{j=1}^{N} \partial^2 h/\partial x_j^2 = 0$, then $h \in C^\omega(\Omega)$.*

Proof: Let $U \subset\subset \Omega$ be an open ball centered at a point x_0. Let μ be an analytic functional in $\mathbb{C}^N$ carried by $\mathscr{C}\ell U$, ie., $\mu \in \mathcal{O}'(\mathscr{C}\ell U)$, and representing h in U. Thus $\Delta_z \mu = \sum_{j=1}^{N} \partial^2\mu/\partial z_j^2$ is carried by ∂U, the boundary of U in $\mathbb{R}^N$. By (III.2.7) we have $\mu = \Delta_z \mu \star E$, with $E = C_N/|x|^{N-2}$ if $N \geq 3$, $E(x) = -C_2 \log|x|$ if $N = 2$ (ie. E is the standard fundamental solution of the Laplacian in $\mathbb{R}^N$). We take advantage of the fact that $|x|$ extends as a holomorphic function in the cone $|\mathcal{Im}\, z| < |\mathcal{Re}\, z|$, and that the same is therefore true of $E(x)$. In the duality bracket

$$\mu_z = \langle \Delta\mu_w, E(z-w) \rangle$$

the variation of w can be restricted to any open neighborhood $\mathcal{N}$ of ∂U in

$\mathbb{C}^N$. If $\mathcal{N}$ is sufficiently contracted about ∂U and if $\epsilon > 0$ is sufficiently small, the function $(w,z) \to E(z-w)$ is holomorphic in $\mathcal{N}\times\{ z \in \mathbb{C}^m;\ |z-x_0| < \epsilon \}$. This implies that the function $z \to \langle \Delta\mu_w, E(z-w) \rangle$ is holomorphic in the open ball $|z-x_0| < \epsilon$. □

COROLLARY III.2.1.— *Let Ω be an open subset of $\mathbb{R}^N$. If a hyperfunction u in Ω is such that* $\sum_{j=1}^{N} \partial^2 u/\partial x_j^2 = f \in \mathcal{D}'(\Omega)$, *then* $u \in \mathcal{D}'(\Omega)$.

Proof: Let $\chi \in C_c^\infty(\Omega)$, $\chi \equiv 1$ in a neighborhood U of an arbitrary point x_0 of Ω. If E is any fundamental solution of the Laplacian and if $v = E\star(\chi f)$ then $\sum_{j=1}^{N} \partial^2 v/\partial x_j^2 = \chi f$ in $\mathbb{R}^N$; $u-v$ satisfies the Laplace equation in U, hence is real—analytic there. □

COROLLARY III.2.2.— *Let Ω be an open subset of $\mathbb{R}^N$. If a hyperfunction u in Ω is such that* $\sum_{j=1}^{N} \partial^2 u/\partial x_j^2 = f \in C^\infty(\Omega)$, *then* $u \in C^\infty(\Omega)$.

Proof: By Corollary III.2.1 we know that $u \in \mathcal{D}'(\Omega)$; the hypo—ellip—ticity of the Laplace operator demands $u \in C^\infty(\Omega)$. □

COROLLARY III.2.3.— *Let Ω be an open subset of $\mathbb{C}^{m+n}$ and h a hyperfunc— tion in Ω solution of* $(\bar\partial_z + \mathrm{d}_{z'})h = 0$. *Then h is a holomorphic function of z, independent of z'.*

Proof: The hyperfunction h is also a solution of the Laplace equation in Ω and therefore $h \in C^\omega(\Omega)$, by Lemma III.2.1. But then the result is obvious. □

We derive from Corollary III.2.3 the exactness of the short sequence

$$0 \to {}_m\mathcal{O}^{(m)} \xrightarrow{i} \mathcal{B}\tilde{\Lambda}^{m,0} \xrightarrow{\bar{\partial}_z + d_{z'}} \mathcal{B}\tilde{\Lambda}^{m,1}.$$

In order to prove the exactness of the remainder of the sequence (III.2.5) we need the analogue of the Poincaré lemma for the differential complex

$$\text{(III.2.8)} \qquad \bar{\partial}_z + d_{z'} : \mathcal{B}(\mathcal{U};\tilde{\Lambda}^{m,q}) \to \mathcal{B}(\mathcal{U};\tilde{\Lambda}^{m,q+1}),\ q = 0,1,\ldots,m+2n,$$

where $\mathcal{B}(\mathcal{U};\tilde{\Lambda}^{m,q})$ is the space of currents (III.2.6) whose coefficients are hyperfunctions in the open subset $\mathcal{U}$ of $\mathbb{R}^{2m+2n}$.

Actually it is to our advantage to prove a version of the Poincaré lemma *with parameters*: $\mathcal{U}$ will be an open subset of real space $\mathbb{R}^{2m+2n+d}$. But we must emphasize that the differential operator remains $\bar{\partial}_z + d_{z'}$ acting on currents (III.2.6) whose coefficients $f_{I,J,K}$ "depend" on $s \in \mathbb{R}^d$, ie., are hyperfunctions in $\mathcal{U}$. We continue to write $f \in \mathcal{B}(\mathcal{U};\tilde{\Lambda}^{m,q})$. It is convenient to take the set $\mathcal{U}$ to be an open polytope $\Pi\times\Pi'\times\Pi''$, with

$$\Pi = \{\, z \in \mathbb{C}^m;\ |x_i| < a_i,\ |y_j| < b_j,\ i, j = 1,\ldots,m \,\},$$

$$\Pi' = \{\, (x',y') \in \mathbb{R}^{2n};\ |x'_i| < a'_i,\ |y'_j| < b'_j,\ i, j = 1,\ldots,n \,\},$$

$$\Pi'' = \{\, s \in \mathbb{R}^d;\ |s_k| < c_k,\ k = 1,\ldots, d\}$$

$(a_i,\ b_j,\ a'_i,\ b'_j,\ c_k > 0)$.

Keep in mind that dealing with hyperfunctions in $\mathbb{R}^{2m+2n+d}$ for the standard C^ω structure of $\mathbb{R}^{2m+2n+d}$, hence with analytic functionals in $\mathbb{C}^{2m+2n+d}$ carried by $\mathbb{R}^{2m+2n+d}$, forces one to complexify both the real and the imaginary parts of the variables.

We shall deal with the following complex vector fields in $\mathbb{R}^{2m+2n+d}$:

$L_i = \partial/\partial\bar{z}_i$ if $1 \leq i \leq m$, $L_{m+j} = \partial/\partial x_j'$, $L_{m+n+k} = \partial/\partial y_k'$ if $1 \leq j, k \leq n$.

Here also we are going to apply (III.2.7) but now with E a fundamental solution of each one of those vector fields:

$$E_i = (1/\pi z_i)\otimes\left[\bigotimes_{\substack{j=1\\ j\neq i}}^{m} \delta(x_j)\otimes\delta(y_j)\right]\otimes\left[\bigotimes_{k=1}^{n} \delta(x_k')\otimes\delta(y_k')\right]\otimes\left[\bigotimes_{\ell=1}^{d} \delta(s_\ell)\right],$$

$$E_{m+\nu} = \mathfrak{H}(x_\nu')\otimes\delta(y_\nu')\otimes\left[\bigotimes_{j=1}^{m} \delta(x_j)\otimes\delta(y_j)\right]\otimes\left[\bigotimes_{\substack{k=1\\ k\neq\nu}}^{n} \delta(x_k')\otimes\delta(y_k')\right]\otimes\left[\bigotimes_{\ell=1}^{d} \delta(s_\ell)\right],$$

$$E_{m+n+\nu} = \delta(x_\nu')\otimes\mathfrak{H}(y_\nu')\otimes\left[\bigotimes_{j=1}^{m} \delta(x_j)\otimes\delta(y_j)\right]\otimes\left[\bigotimes_{\substack{k=1\\ k\neq\nu}}^{n} \delta(x_k')\otimes\delta(y_k')\right]\otimes\left[\bigotimes_{\ell=1}^{d} \delta(s_\ell)\right],$$

where $\mathfrak{H}(t)$ is the Heaviside function, $\mathfrak{H}(t) = 1$ for $t > 0$, $\mathfrak{H}(t) = 0$ for $t < 0$, and $i = 1,\dots,m$, $\nu = 1,\dots,n$.

THEOREM III.2.1.— *Let* $f \in \mathcal{B}(\Pi\times\Pi'\times\Pi'';\tilde{\Lambda}^{p,q})$ *satisfy* $(\bar{\partial}_z+d_{z'})f \equiv 0$. *If* $1 \leq q \leq m+2n$ *there is* $u \in \mathcal{B}(\Pi\times\Pi'\times\Pi'';\tilde{\Lambda}^{p,q-1})$ *such that* $(\bar{\partial}_z+d_{z'})u = f$.

Proof: We change notation slightly and write t_i instead of $\bar{z}_i$ if $1 \leq i \leq m$, t_{m+j} instead of x_j', and t_{m+n+k} instead of y_k' if $1 \leq j, k \leq n$. Notice that, given any C^∞ function (or any hyperfunction) g in $\mathbb{R}^{2m+2n+d}$ we have, in this notation,

$$(\bar{\partial}_z+d_{z'})g = \sum_{j=1}^{m+2n} L_j g\, dt_j.$$

It suffices to deal with a single value of the exponent p, e. g., p = m. In the new notation we rewrite the current (III.2.6) as follows:

$$f = dz_1\wedge\cdots\wedge dz_m\wedge\left[\sum_{|I|=q} g_I dt_I\right],$$

where the elements of the q–tuple I are integers i_α such that $1 \leq i_0 < \cdots < i_q \leq m+2n$ and $g_I \in \mathcal{B}(\Pi\times\Pi'\times\Pi'')$. We may as well disregard the exterior product $dz_1\wedge\cdots\wedge dz_m$ and assume that

$$f = \sum_{|I|=q} g_I dt_I,$$

which amounts to dealing with the case $p = 0$.

Let j_0 be any integer such that $1 \leq j_0 \leq m+2n$, and let $g \in \mathcal{B}(\Pi\times\Pi'\times\Pi'')$ satisfy the conditions

$$\text{(III.2.9)} \qquad L_j g = 0 \quad \textit{in } \Pi\times\Pi'\times\Pi'',\ j = j_0+1,\dots,m+2n.$$

[There are no equations (III.2.9) when $j_0 = m+2n$.] We seek a solution $u \in \mathcal{B}(\Pi\times\Pi'\times\Pi'')$ of the system of equations in $\Pi\times\Pi'\times\Pi''$,

$$\text{(III.2.10)} \qquad L_{j_0} u = g,\ L_j u = 0 \ \textit{if } j = j_0+1,\dots,m+2n.$$

If $\mu \in \mathcal{O}'(\mathcal{Cl}(\Pi\times\Pi'\times\Pi''))$ represents g in $\Pi\times\Pi'\times\Pi''$ the hyperfunction $u_0 = E_{j_0}\star\mu$ satisfies $L_{j_0} u_0 = g$ in $\Pi\times\Pi'\times\Pi''$. When $j_0 = m+2n$ $u = u_0$ solves (III.2.10). In passing note that this is automatically the case when the number $m+2n$ of vector fields L_j is equal to one. When $1 \leq j_0 < m+2n$, we have $L_j u_0 = E_{j_0}\star L_j\mu$ (with the understanding that, in order for L_j to act on the analytic functional μ it must be extended holomorphically). Now take $j > j_0$; (III.2.9) entails that $L_j\mu$ is carried by $\partial(\Pi\times\Pi'\times\Pi'')$. With a slight abuse of language let us write $\Pi\times\Pi'\times\Pi'' = \Pi_{j_0}\times\Pi'''$, where Π_{j_0} is a rectangle in z_0–space if $1 \leq j_0 \leq m$, an interval in the x'_{j_0}–line (resp., y'_{j_0}–line) if $m+1 \leq j_0 \leq m+n$ (resp., $m+n+1 \leq j_0 \leq m+2n$). We use a decomposition (see Proposition I.4.2 and Remark I.4.2)

$L_j\mu = \alpha_j+\beta_j$, where the analytic functional in $\mathbb{C}^m$, α_j (resp., β_j) is carried by $(\partial\Pi_{j_0})\times(\mathcal{C}\ell\Pi''')$ [resp., $(\mathcal{C}\ell\Pi_{j_0})\times(\partial\Pi''')$]. If we recall the definition of E_{j_0} we see right—away that the hyperfunctions $E_{j_0}\star\beta_j$ vanish identically in $\Pi\times\Pi'\times\Pi''$. Note also that, if $j_0 < j < k \leq m+2n$,

$$L_k\alpha_j - L_j\alpha_k = L_j\beta_k - L_k\beta_j$$

is carried by $(\partial\Pi_{j_0})\times(\partial\Pi''')$. Again thanks to the definition of E_{j_0} we conclude that

$$L_k(E_{j_0}\star\alpha_j) = L_j(E_{j_0}\star\alpha_k) \text{ in } \Pi\times\Pi'\times\Pi''.$$

Reasoning by induction on the number of vector fields L_j we can assume that the Poincaré lemma has been proved when this number is $< m + 2n$. As a consequence we can postulate the existence of a hyperfunction v in $\Pi\times\Pi'\times\Pi''$ such that $L_j v = E_{j_0}\star\alpha_j$, $j = j_0+1,\ldots,m+2n$. Notice that $L_{j_0}v$ is represented by α_j in $\Pi\times\Pi'\times\Pi''$, hence vanishes there. We conclude that $u = u_0 - v$ verifies (III.2.10).

Next we return to the current f and we reason by induction on the smallest number j_0 such that $g_I \equiv 0$ if the q—tuple I contains an integer j $> j_0$. The case $j_0 = m+2n$ is not precluded, nor the case $j_0 = 0$ (in which case $g_I \equiv 0$). Thanks to the hypothesis that $(\bar\partial_z + d_{z'})f \equiv 0$ we see that each coefficient g_I is a solution of (III.2.9) in $\Pi\times\Pi'\times\Pi''$ if $j_0 \in I$, and we can find a solution $u_I \in \mathcal{B}(\Pi\times\Pi'\times\Pi'')$ of (III.2.10) in which $g = g_I$, which allows us to form

$$u = (-1)^{m-q+1} \sum_{\substack{|I|=q \\ j_0\in I}} u_I \wedge dt_{I\setminus\{j_0\}}.$$

We have

$$(\bar\partial_z+d_{z'})u = (-1)^{m-q+1} \sum_{\substack{|I|=q \\ j_0\in I}} \sum_{j=1}^{j_0} L_j u_I \, dt_j \wedge dt_{I\setminus\{j_0\}} =$$

$$(-1)^{m-q+1} \sum_{\substack{|I|=q \\ j_0\in I}} \sum_{j=1}^{j_0-1} L_j u_I \, dt_j \wedge dt_{I\setminus\{j_0\}} + \sum_{\substack{|I|=q \\ j_0\in I}} g_I dt_I .$$

It follows that the form $g - (\bar\partial_z+d_{z'})u$ does not involve any differential dt_j for $j \geq j_0$. We may therefore proceed with the induction until we reach $j_0 = 0$. □

Theorem III.2.1 entails that the flabby resolution (III.2.5) is cohomological. It follows that the cohomology with coefficients in the sheaf ${}_m\mathcal{O}^{(m)}$ can be equated to the cohomology of the differential complex (III.2.8). We return to an open subset $\mathcal{U}$ of $\mathbb{C}^m\times\mathbb{C}^n$ (no parameters s_k are present). We define, for any closed subset $\mathcal{S}$ of $\mathcal{U}$,

(III.2.11)$_0$
$$H^0_{\mathcal{S}}(\mathcal{U};{}_m\mathcal{O}^{(m)}) = H^{m,0}_{\mathcal{S}}(\mathcal{U}) =$$
$$\{ h \in \mathcal{B}(\mathcal{U};\tilde\Lambda^{m,0});\ (\bar\partial_z+d_{z'})h = 0,\ \text{supp}\ h \subset \mathcal{S} \};$$

and for $1 \leq q \leq m+2n$,

(III.2.11)$_q$
$$H^q_{\mathcal{S}}(\mathcal{U};{}_m\mathcal{O}^{(m)}) = H^{m,q}_{\mathcal{S}}(\mathcal{U}) =$$
$$\{ f \in \mathcal{B}(\mathcal{U};\tilde\Lambda^{m,q});\ (\bar\partial_z+d_{z'})f = 0,\ \text{supp}\ f \subset \mathcal{S} \}/$$
$$\{ (\bar\partial_z+d_{z'})u;\ u \in \mathcal{B}(\mathcal{U};\tilde\Lambda^{m,q-1}),\ \text{supp}\ u \subset \mathcal{S} \}.$$

Now let U, V be two open subsets of Σ such that $V \subset U \subset\subset \Sigma$. Note that U is a closed subset of $\mathbb{C}^{m+n}\setminus\partial U$ (keep in mind that ∂U is the boundary of U in Σ) and that, likewise, V is closed in $\mathbb{C}^{m+n}\backslash\partial V$. This allows us to introduce the relative cohomology spaces $H_U^{m,q}(\mathbb{C}^{m+n}\backslash\partial U)$ ($0 \leq q \leq m+2n$) and their analogues with V in the place of U, defined in accordance to (III.2.11). The statement of the main theorem, Theorem III.2.2 below, will rely on the following

LEMMA III.2.2.— *The inclusion* $V \subset U$ *defines a natural restriction mapping*

$$r_V^U: H_U^{m,q}(\mathbb{C}^{m+n}\backslash\partial U) \to H_V^{m,q}(\mathbb{C}^{m+n}\backslash\partial V).$$

Proof: There is a natural restriction map

$$H_U^{m,q}(\mathbb{C}^{m+n}\backslash\partial U) \to H_V^{m,q}(\mathbb{C}^{m+n}\backslash[(\mathscr{C}\ell U)\backslash V]).$$

Indeed, if $f \in \mathcal{B}(\mathbb{C}^{m+n}\backslash\partial U;\tilde{\Lambda}^{m,q})$ vanishes identically off U, then the restriction of f to

$$\mathbb{C}^{m+n}\backslash[(\mathscr{C}\ell U)\backslash V] = [\mathbb{C}^{m+n}\backslash(\mathscr{C}\ell U)]\cup V$$

vanishes identically off V. Obviously

$$\partial V \subset (\mathscr{C}\ell U)\backslash V \Rightarrow \mathbb{C}^{m+n}\backslash[(\mathscr{C}\ell U)\backslash V] \subset \mathbb{C}^{m+n}\backslash\partial V;$$

and V is closed in both sets $\mathbb{C}^{m+n}\backslash[(\mathscr{C}\ell U)\backslash V]$ and $\mathbb{C}^{m+n}\backslash\partial V$. Identifying each element of $\mathcal{B}(\mathbb{C}^{m+n}\backslash[(\mathscr{C}\ell U)\backslash V];\tilde{\Lambda}^{m,q})$ supported in V to an element of $\mathcal{B}(\mathbb{C}^{m+n}\backslash\partial V;\tilde{\Lambda}^{m,q})$ defines a linear bijection

$$H_V^{p,q}(\mathbb{C}^{m+n}\backslash[(\mathscr{C}\ell U)\backslash V]) \to H_V^{p,q}(\mathbb{C}^{m+n}\backslash\partial V)$$

whose inverse is induced by the restriction map

$$\mathbb{C}^{m+n}\backslash\partial V \to \mathbb{C}^{m+n}\backslash[(\mathscr{C}\ell U)\backslash V].$$

{This is the natural isomorphism posited by the *excision axiom* for cohomology ([EILENBERG–STEENROD, 1952], p. 14).} □

THEOREM III.2.2.— *Let* q *be an integer*, $0 \leq q \leq n$. *There are natural isomorphisms* i_U^q *and* i_V^q *such that the diagram*

$$\text{(III.2.12)}\qquad \begin{array}{ccc} \mathfrak{Sol}^{(q)}(U) & \xrightarrow{\ i_U^q\ } & H_U^{m,m+n+q}(\mathbb{C}^{m+n}\backslash \partial U) \\ {\scriptstyle r_V^U}\downarrow & & \downarrow {\scriptstyle r_V^U} \\ \mathfrak{Sol}^{(q)}(V) & \xrightarrow[\ i_V^q\ ]{} & H_V^{m,m+n+q}(\mathbb{C}^{m+n}\backslash \partial V) \end{array}$$

is commutative.

In (III.2.12) r_V^U is the restriction map defined in (III.1.21).

REMARK III.2.1.— For technical reasons it will be convenient to replace, in the proof of Theorem III.2.2, the complements of ∂U and ∂V in the whole space $\mathbb{C}^{m+n}$ by their complements with respect to a "box" $\Pi\times\Pi'$, with Π and Π' polyhedrons like those in Theorem III.2.1, now selected to be suitably small, and such that $\mathcal{C}\ell U \subset\subset \Sigma\cap\Pi\times\Pi'$. This will turn out to be also useful when applying Theorem III.2.2 in Chapter IV. It has no effect on its statement since, according to the excision axiom, the spaces $H_U^{m,m+n+q}(\mathbb{C}^{m+n}\backslash\partial U)$ and $H_U^{m,m+n+q}(\Pi\times\Pi'\backslash\partial U)$ are naturally isomorphic.

□

Before embarking on the proof of Theorem III.2.2 we explain how it entails exactly what is needed, in order to define *invariantly* the sheaf $\mathcal{S}ol_{\mathcal{M}}$ of (germs of) hyperfunction solutions on $\mathcal{M}$, as well as the sheaves $\mathcal{S}ol_{\mathcal{M}}^{(q)}$ $(0 \le q \le n)$.

Assume the embedding (III.2.1) maps a point $p \in \mathcal{M}'$ to the origin of $\mathbb{C}^{m+n}$. Theorem III.2.2 implies that the stalk of the sheaf $\mathcal{S}ol_{\Sigma}^{(q)}$ at 0, $\mathcal{S}ol_{\Sigma}^{(q)}\big|_0$, is naturally isomorphic to the inductive limit (in the sense of the mappings r_V^U) of the relative cohomology spaces $H_U^{m,m+n+q}(\mathbb{C}^{m+n}\backslash \partial U)$. The latter inductive limit is associated with the differential operator $\bar{\partial}_z+d_{z'}$, ie., with the resolutions (III.2.4) or (III.2.5), and the differential complex (III.2.3), and therefore is an invariant of the hypo—analytic structure of Σ transferred from $\mathcal{M}' \subset \mathcal{M}$. As a consequence the same is true of $\mathcal{S}ol_\Sigma^{(q)}\big|_0$. The pullback of $\mathcal{S}ol_\Sigma^{(q)}\big|_0$ to $\mathcal{M}$ via the map (III.2.1) will be, by definition, the stalk at $\wp$ of the sheaf $\mathcal{S}ol_{\mathcal{M}}^{(q)}$.

REMARK III.2.2.— Theorem III.2.2, and the construction on which its statement and its proof are based, is valid when n = 0, ie., when there are no variables z'. In this case the hypo—analytic structure of $\mathcal{M}$ is maximal, and $\mathfrak{Sol}(U) = \mathcal{B}(U)$ (Definition I.5.1). According to the preceding remarks, the stalk at $\wp$ of the sheaf of hyperfunctions $\mathcal{B}(\mathcal{M})$ is naturally isomorphic to the stalk at the origin of the sheaf defined by the relative cohomology presheaf, $(H_U^{m,m}(\mathbb{C}^m\backslash \partial U), r_V^U)$. This is precisely Sato's definition of the sheaf of hyperfunctions when $\Sigma = \mathbb{R}^m$ ([SATO, 1959, 1960]; see [HARVEY, 1969] for the case of a totally real submanifold of $\mathbb{C}^m$). □

REMARK III.2.3.— When m = 0, n $\geq$ 1, the sheaf ${}_m\mathcal{O}^{(m)}$ is simply the constant sheaf with stalk $\cong \mathbb{C}$; its pullback to Σ is equal to the sheaf $\mathcal{S}ol_\Sigma$ (Σ is an open subset of $\mathbb{R}^n$). If U is an open subset of $\mathbb{R}^n$ Theorem III.2.2 states that the space $\mathfrak{Sol}^{(q)}(U)$ is isomorphic to $H_U^{n+q}(\mathbb{C}^n\backslash \partial U)$. The latter is naturally isomorphic to the qth De Rham cohomology space $H^q(U)$. □

REMARK III.2.4.— Suppose $\mathcal{M}$ carries two hypo—analytic structures $\mathcal{A}_1$ and $\mathcal{A}_2$ such that the following is true:

(III.2.13) *any function defined in an arbitrary open subset Ω of $\mathcal{M}$ that is hypo-analytic in Ω in the sense of $\mathcal{A}_1$ is also hypo-analytic in Ω in the sense of $\mathcal{A}_2$.*

Let $(\mathcal{M}',Z)$, $Z = (Z_1,...,Z_{m_2})$, be a hypo—analytic chart in $\mathcal{M}$ for the structure $\mathcal{A}_2$ such that $Z_1,...,Z_{m_1}$ $(m_1 \leq m_2)$ are hypo—analytic functions for $\mathcal{A}_1$. Assume there is a hypo—analytic embedding $\mathcal{M}' \ni p \to (Z(p),Z'(p)) \in \Sigma \subset \mathbb{C}^{m_2}\times\mathbb{C}^n$. The pullbacks to $\mathcal{M}'$ of the hyperfunctions f on Σ such that $\bar\partial_{z'} f \equiv 0$ are the solutions for $\mathcal{A}_2$; the pullbacks of those such that $\bar\partial_{z'} f \equiv 0$ and also $\partial f/\partial \bar z_j \equiv 0$ for every $j = m_1+1,...,m_2$ are the solutions for $\mathcal{A}_1$. It follows that every hyperfunction solution for $\mathcal{A}_1$ is a solution for $\mathcal{A}_2$ (cf. [TREVES, 1992], Proposition III.1.2).

This applies in particular to the case in which $\mathcal{A}_2$ is maximal (for instance $\mathcal{A}_2$ is a real—analytic structure) in which case the notion of hyperfunction on $\mathcal{M}$ in the sense of $\mathcal{A}_2$ is defined (in Chapter I). Assume that (III.2.13) holds; when $\mathcal{A}_2$ is real—analytic this simply means that every hypo—analytic function for $\mathcal{A}_1$ is of class C^ω. Then every hyper—function solution for $\mathcal{A}_1$ is a hyperfunction in the sense of $\mathcal{A}_2$. □

Remark III.2.4 may serve as an introduction to the following sub—section:

Hyperfunctions solutions in real analytic hypo—analytic structures

In this subsection we assume that $\mathcal{M}$ is a C^ω manifold equipped with a hypoanalytic structure of class C^ω, ie., $\mathcal{M}$ can be covered by the domains U^α of hypo–analytic charts $(U^\alpha,Z^\alpha)_{\alpha\in A}$ satisfying (I.5.7) and (I.5.8) and such that $Z^\alpha \in \mathcal{C}^\omega(U^\alpha)$ for every $\alpha \in A$. Since the tangent structure bundle $\mathcal{V}$ of $\mathcal{M}$ will then be locally spanned by real analytic vector fields and since such vector fields act on the standard hyper–functions in $\mathcal{M}$, ie., on the sections of the sheaf $\mathcal{B}_{\mathcal{M}}$ defined within the real analytic structure of $\mathcal{M}$ (see end of section I.5), it is natural to introduce the subsheaf $\mathcal{F}_{\mathcal{M}}^{(0)}$ of $\mathcal{B}_{\mathcal{M}}$ consisting of the germs of hyperfunctions annihilated by the germs of $\mathcal{C}^\omega$ sections of $\mathcal{V}$.

THEOREM III.2.3.— *There is a natural isomorphism* $\mathcal{F}_{\mathcal{M}}^{(0)} \cong \mathcal{Sol}_{\mathcal{M}}^{(0)}$.

Proof: Let $(\mathcal{M}',x,t,Z)$ be a "chart" like the one introduced at the start of section III.1. But here x_j $(1 \leq j \leq m)$ and t_k $(1 \leq k \leq n)$ are real analytic coordinates and $Z(x,t) = x+\imath\Phi(x,t)$ with $\Phi : \mathcal{M}' \to \mathbb{C}^m$ of class $\mathcal{C}^\omega$, $\Phi(0,0) = 0$, $\Phi_x(0,0) = 0$. The map $(x,t) \to (Z(x,t),t)$ is an analytic diffeomorphism of $\mathcal{M}'$ onto the submanifold Σ of $\mathbb{C}^m\times\mathbb{R}^n$. We complexify x and t as $z = x+\imath y$, $z' = t+\imath y'$. Set $U_0 = B_0\times\Theta_0$ with B_0 (resp., Θ_0) be an open ball cen–tered at 0 in x–space $\mathbb{R}^m$ (resp., t–space $\mathbb{R}^n$); and call U the image of U_0 under the map $(x,t) \to (Z(x,t),t)$. We assume $U \subset\subset \Sigma$. Actually we take B_0 and Θ_0 sufficiently small that this map extends as a biholomorphism

$$\chi : (z,z') \to (z+\imath\Phi(z,z'),z')$$

from an open neighborhood in $\mathbb{C}^m\times\mathbb{C}^n$ of $\mathcal{C}\ell U_0$ onto one of $\mathcal{C}\ell U$. Con–sider the natural isomorphism χ_*: $\mathcal{O}'(\mathcal{C}\ell U_0) \to \mathcal{O}'(\mathcal{C}\ell U)$ defined by

$$< \chi_*\mu,h > = < \mu,(h\circ\chi)\det D\chi >, \quad \mu \in \mathcal{O}'(\mathcal{C}\ell U_0),\ h \in \mathcal{O}(\mathbb{C}^{m+n})$$

($D\chi$ is the Jacobian matrix of the map χ). Clearly χ_* induces an isomor–

phism $\mathcal{O}'(\partial U_0) \cong \mathcal{O}'(\partial U)$ where ∂U is the boundary of U relative to Σ.

If $L_1,\dots,L_n$ are the vector fields (III.1.10) we have, for any distribution u in U_0,

$$^tL_j[u \det Z_x] = -(L_j u) \det Z_x,\ j = 1,\dots,n$$

(^tL_j is the formal transpose of L_j; see Lemma II.1.1, [TREVES, 1992]). Denote by $\tilde{L}_j$ the holomorphic extension of L_j. Since

$$\det D\chi = \det(I_m + \imath\Phi_z)\ (I_m\text{: m×m identity matrix})$$

unique continuation implies

$$^t\tilde{L}_j[(h\circ\chi) \det D\chi] = -[\tilde{L}_j(h\circ\chi)] \det D\chi,\ j = 1,\dots,n.$$

But

$$\tilde{L}_j(h\circ\chi) = (\partial h/\partial z_j')\circ\chi,$$

which shows that

$$L_j\mu = \chi_*^{-1}(\partial/\partial z_j')\chi_*\mu,\ j = 1,\dots,n.$$

This shows that $L_j\mu \in \mathcal{O}'(\partial U_0) \Leftrightarrow (\partial/\partial z_j')\chi_*\mu \in \mathcal{O}'(\partial U)$ and that χ_* induces an isomorphism of the space of continuous sections of $\mathcal{F}_{\mathcal{M}}^{(0)}$ over U_0 onto $\mathfrak{Sol}^{(0)}(U)$. □

More generally we may consider the cohomology sheaves $\mathcal{F}_{\mathcal{M}}^{(q)}$ (q = 0,1,...,n) associated to the differential complex analogous to (III.1.6) acting on spaces of currents with (standard) hyperfunction coefficients. The previous argument can be repeated with minor modifications and yields

THEOREM III.2.4.— *There are natural isomorphisms*

$$\mathcal{F}_{\mathcal{M}}^{(q)} \cong \mathcal{S}ol_{\mathcal{M}}^{(q)},\ q = 0,1,\dots,n.$$

I. *Preliminaries*

We shall of course assume m+n $\geq$ 1, otherwise there is nothing to prove. The case m = 0, ie., the case in which there are no variables z, is not precluded and thus the argument will prove the claim in Remark III.2.3. We are going to construct the linear map i_U^q and prove that it is an isomorphism. The analogous reasoning will be valid with V in the place of U.

A generic element of $\mathfrak{Sol}^{(q)}(U)$ is the cohomology class $[f]$ of a $\partial_{z'}$–closed current (III.1.19), f, that has a representative

(III.3.1)
$$\mu = \sum_{|J|=q} \mu_J dz'_J$$

with $\mu_J \in \mathcal{O}'(\mathscr{C}\ell U)$. [Henceforth we shall express such a property by wri– ting $\mu \in \mathcal{O}'(\mathscr{C}\ell U;\Lambda^q)$, $\partial_{z'}\mu \in \mathcal{O}'(\partial U;\Lambda^{q+1})$.]

We shall use systematically the notation $\Omega = \Pi\times\Pi'\backslash\partial U$ (see Remark III.2.1); note that U is closed in Ω and $\Omega\backslash U = \Pi\times\Pi'\backslash(\mathscr{C}\ell U)$. We are going to apply Theorems I.2.3 and I.4.2 in (z,z')–space $\mathbb{C}^{m+n}$. We avail ourselves of the natural isomorphism $f \to f\,dz$ from $\mathcal{O}$ to ${}_1\mathcal{O}^{(1)}$ when m = 1, n = 0; and of the map

$$dz_1\wedge\cdots\wedge dz_m\wedge dz'_1\wedge\cdots\wedge dz'_n\wedge f \to dz_1\wedge\cdots\wedge dz_m\wedge f$$

from $\mathscr{C}^\infty\Lambda^{m+n,q}$ to $\mathscr{C}^\infty\tilde{\Lambda}^{m,q}$ when n $\geq$ 1. Thus, applying Theorem I.2.3 when m = 1 and n = 0, we can assign to μ a one–form $F\,dz$, with $F \in \mathcal{O}(\mathbb{C}\backslash(\mathscr{C}\ell U))$, $F(z) \to 0$ as $z \to \infty$. Below we deal with the restriction of this form to $\Omega\backslash U$. And applying Theorem I.2.3 when m = 0, n = 1, and

Theorem I.4.2 (with the polyhedron $\Pi\times\Pi'$ substituted for the "biball" B_{01}) when $m+n \geq 2$, we can assign to μ a differential form

$$F = \sum_{|J|=q} F_J \wedge dz'_J \in C^{\infty}(\Omega\backslash U;\tilde{\Lambda}^{m,m+n-1,q}), \tag{III.3.2}$$

with $F_J \in C^{\infty}(\Omega\backslash U;\tilde{\Lambda}^{m,m+n-1})$, $(\bar{\partial}_z+\bar{\partial}_{z'})F_J = 0$. We are using the notation $\tilde{\Lambda}^{m,p,q}$ to mean the vector subbundle of $\Lambda^{m+p+q}\mathbb{C}T^*(\mathbb{C}^{m+n})$ spanned by the exterior products

$$dz_1\wedge\cdots\wedge dz_m\wedge d\bar{z}_I\wedge d\bar{z}'_K\wedge dz'_J,\ |I|+|K| = p,\ |J| = q.$$

When $n \geq 1$ we must also have $\partial_{z'}\mu \in \mathcal{O}'(\partial U;\Lambda^{q+1})$. This is automatic if $q = n$. When $0 \leq q < n$ Theorem I.3.1 tells us that there are differential forms

$$A = \sum_{|J|=q+1} A_J\wedge dz'_J \in C^{\infty}(\Omega;\tilde{\Lambda}^{m,m+n-1,q+1}),$$

$$G = \sum_{|J|=q+1} G_J\wedge dz'_J \in C^{\infty}(\Omega\backslash U;\tilde{\Lambda}^{m,m+n-2,q+1}),$$

such that

$$\partial_{z'}F = A - (\bar{\partial}_z+\bar{\partial}_{z'})G \ \ in\ \Omega\backslash U, \tag{III.3.3}$$

$$(\bar{\partial}_z+\bar{\partial}_{z'})A = 0 \ \ in\ \Omega. \tag{III.3.4}$$

We are going to make use of the long exact sequence for relative cohomology, specifically of the following segment of that sequence:

$$H^{m,m+n+q-1}(\Omega) \xrightarrow{\rho^{q-1}} H^{m,m+n+q-1}(\Omega\backslash U) \xrightarrow{\delta^q} H_U^{m,m+n+q}(\Omega) \xrightarrow{\iota^q}$$

$$H^{m,m+n+q}(\Omega) \xrightarrow{\rho^q} H^{m,m+n+q}(\Omega\backslash U).$$

It must be emphasized that $H^{m,q}$ is the qth cohomology space of the differential complex (III.2.7). We recall the meaning of the various maps in the sequence: ρ^q is defined by the restriction from Ω to $\Omega\backslash U$; to any form $f \in \mathcal{B}(\Omega;\tilde{\Lambda}^{m,m+n+q})$ such that $(\bar{\partial}_z+d_{z'})f = 0$ and supp $f \subset U$, ι^q assigns its class mod $(\bar{\partial}_z+d_{z'})\mathcal{B}(\Omega;\tilde{\Lambda}^{m,m+n+q-1})$. The map δ^q is defined as follows: if $u \in \mathcal{B}(\Omega\backslash U;\tilde{\Lambda}^{m,m+n+q-1})$, $(\bar{\partial}_z+d_{z'})u = 0$, and if $\tilde{u}$ is an arbitrary extension of u to Ω, then $v = (\bar{\partial}_z+d_{z'})\tilde{u} \in \mathcal{B}(\Omega;\tilde{\Lambda}^{m,m+n+q})$, $(\bar{\partial}_z+d_{z'})v = 0$ and supp $v \subset U$; and $\delta^q([u]) = [v]$. From the (obvious) exactness of the above sequence we derive the following short exact sequence

$$0 \to H^{m,m+n+q-1}(\Omega\backslash U)/\mathrm{Im}\,\rho^{q-1} \xrightarrow{\delta^q} H_U^{m,m+n+q}(\Omega) \xrightarrow{\iota^q} \mathrm{Ker}\,\rho^q \to 0.$$

This is equivalent to the direct sum decomposition

$$\text{(III.3.5)} \qquad H_U^{m,m+n+q}(\Omega) \cong (\mathrm{Im}\,\delta^q) \oplus (\mathrm{Ker}\,\rho^q).$$

The subspace Im δ^q will turn out to be the image under the isomorphism i_U^q of the subspace $\mathfrak{Sol}_0^{(q)}(U)$ of $\mathfrak{Sol}^{(q)}(U)$ consisting of the cohomology classes of currents (III.1.19) that have a representative μ of the type (III.3.1) such that $\partial_{z'}\mu = 0$.

Note that

$$\mathfrak{Sol}_0^{(n)}(U) = \mathfrak{Sol}^{(n)}(U), \; \mathfrak{Sol}_0^{(0)}(U) = 0.$$

The latter follows from the fact that each partial differentiation $\partial/\partial z_j'$ (j = 1,...,n) defines a surjection of $\mathcal{O}(\mathbb{C}^{m+n})$ onto itself.

Let μ be a current (III.3.1) representing a class in $\mathfrak{Sol}_0^{(q)}(U)$ (thus $\partial_{z'}\mu = 0$) and let F be the corresponding differential form, given by (III.3.2). When $n \geq 1$, $\partial_{z'} F \equiv 0$ if $m = 0$ or, if $m \geq 1$, $\partial_{z'} F$ is $(\bar\partial_z + \bar\partial_{z'})$—exact in $\Omega \backslash U$. In both cases this is equivalent to saying that, in (III.3.3), we may take $A \equiv 0$.

When $q = n$ (in particular, when $n = 0$) the situation just described always obtains. On the other hand, we have $H^{m,m+2n}(\mathcal{U}) = 0$ whatever the open subset $\mathcal{U}$ of $\mathbb{C}^{m+n}$ (Theorem I.4.3). From the exactness of the sequence for relative cohomology we derive that Ker $\rho^q = 0$ and $H_U^{m,m+2n}(\Omega) \cong \operatorname{Im} \delta^q$.

If $\mathcal{U}$ is an arbitrary open subset of $\mathbb{C}^{m+n}$ and q an integer, $0 \leq q \leq n$, we shall denote by $H^{m,p,q}(\mathcal{U})$ the cohomology spaces of the differential complex

(III.3.6) $\bar\partial_z + \bar\partial_{z'}: \mathcal{C}^\infty(\mathcal{U};\tilde\Lambda^{m,p,q}) \to \mathcal{C}^\infty(\mathcal{U};\tilde\Lambda^{m,p+1,q})$, $p = 0,1,\ldots,m+n$.

Below we are going to take advantage of the following two lemmas.

LEMMA III.3.1.– *If* $m+n \geq 2$ *and* $0 \leq q \leq n$, *the restriction mapping*

$$H^{m,m+n-1,q}(\Omega) \to H^{m,m+n-1,q}(\Omega \backslash U)$$

is injective.

Proof of Lemma III.3.1: By Theorem I.4.2 the injectivity of the restriction map $H^{m,m+n-1,n}(\Omega) \to H^{m,m+n-1,n}(\Omega \backslash U)$ is equivalent to the injectivity of the "inclusion" map $\mathcal{O}'(\partial U) \to \mathcal{O}'(\mathscr{C}\ell U)$. Division by dz'_J for each multi—index J such that $|J| = n-q$ shows that the restriction map $H^{m,m+n-1,q}(\Omega) \to H^{m,m+n-1,q}(\Omega \backslash U)$ is also injective. □

LEMMA III.3.2.— *Let* K *be a compact subset of* $\Sigma \cap (\Pi \times \Pi')$.

If m+n $\geq$ 2 *the restriction mapping defines an isomorphism* $\mathcal{O}(\Pi\times\Pi') \cong \mathcal{O}(\Pi\times\Pi'\backslash K)$.

If m+n $\geq$ 3, $H^{m,p,q}(\Pi\times\Pi'\backslash K) = 0$ *whatever the integers* p, q, $1 \leq p \leq m+n-2$, $0 \leq q \leq n$.

Proof of Lemma III.3.2: The first assertion is a direct consequence of Hartogs' theorem. The proof of the second assertion duplicates that of Theorem I.4.1, but in (z,z')—space and after substitution of $\Pi\times\Pi'$ for B_{01}. Given any differential form $g \in \mathcal{C}^{\infty}(\Pi\times\Pi'\backslash K;\Lambda^{m+n,p})$ such that $(\bar\partial_z+\bar\partial_{z'})g = 0$, one uses the operators $\mathcal{T}^p_\nu$ and $\mathcal{T}^{p+1}_\nu$ defined in (I.3.22) with $\mathcal{U} = \mathcal{U}_\nu$ in a suitably chosen basis of open neighborhoods of K in $\Pi\times\Pi'$ to construct a solution $u \in \mathcal{C}^{\infty}(\Pi\times\Pi'\backslash K;\Lambda^{m,p-1})$ of $(\bar\partial_z+\bar\partial_{z'})u = g$. Since $p \leq m+n-2$, the fact that $\mathcal{T}^{p+1}_\nu[(\bar\partial_z+\bar\partial_{z'})\psi_\nu\wedge g] = 0$ in $\Pi\times\Pi'\backslash\mathcal{U}_\nu$ is a direct consequence of (I.3.24), without the need for a hypothesis of the kind (I.3.25). We leave the details to the reader. This shows that $H^{m,p,n}(\Pi\times\Pi'\backslash K) = 0$ for all p, $1 \leq p \leq m+n-2$. From there we proceed as in the proof of Lemma III.2.1: division by dz'_J for each multi—index J such that $|J| = n-q$ shows that $H^{m,p,q}(\Pi\times\Pi'\backslash K) = 0$ for all p, q, $1 \leq p \leq m+n-2$, $0 \leq q \leq n$. □

II. *Construction of the classes* $i^q_U([f])$

II*a*. *Case* q = n

The form F in (III.3.2) is an element of $\mathcal{C}^{\infty}(\Omega\backslash U;\tilde\Lambda^{m,m+n-1,n})$; $(\bar\partial_z+d_{z'})F \equiv 0$. Let E be an arbitrary hyperfunction extension of F to Ω, ie., each coefficient of E is an extension of the corresponding coefficient of F; $(\bar\partial_z+d_{z'})E$ is $(\bar\partial_z+d_{z'})$—closed and supported in U. We define $i^n_U([f])$

to be the class of $(\overline{\partial}_z+d_{z'})E$ in $H_U^{m,m+2n}(\Omega)$. [The case q = n does not preclude n = 0.]

IIb. *Case* q = n—1

In this case, if A and G are the forms in (III.3.3)–(III.3.4) then perforce $\partial_{z'}A = 0$, $\partial_{z'}G = 0$, and Equation (III.3.3) reads

$$A = (\overline{\partial}_z+d_{z'})(F + G) \textit{ in } \Omega\backslash U.$$

Let E be a hyperfunction extension of $F + G$ to Ω; then $A-(\overline{\partial}_z+d_{z'})E$ is $(\overline{\partial}_z+d_{z'})$— closed in Ω and supported in U. We define the cohomology class of $A-(\overline{\partial}_z+d_{z'})E$ to be $i_U^{n-1}([f])$.

IIc. *Case* q $\leq$ n−2

Now suppose n $\geq$ q+2. We can find a solution $A^{(2)} \in \mathcal{C}^\infty(\Omega;\tilde{\Lambda}^{m,m+n-2,q+2})$ of the following equation in Ω,

$$\text{(III.3.7)} \qquad (\overline{\partial}_z+\overline{\partial}_{z'})A^{(2)} = \partial_{z'}A.$$

Indeed, by (III.3.3) we know that $\partial_{z'}A = -(\overline{\partial}_z+\overline{\partial}_{z'})\partial_{z'}G$ in $\Omega\backslash U$ and it suffices to apply Lemma III.3.1.

If n = q+2 we stop there. If q $\leq$ n—3 we can solve by induction the string of equations

$$\text{(III.3.8)} \qquad (\overline{\partial}_z+\overline{\partial}_{z'})A^{(k)} = \partial_{z'}A^{(k-1)} \quad (k = 3,\ldots,n-q)$$

with $A^{(k)} \in \mathcal{C}^\infty(\Omega;\tilde{\Lambda}^{m,m+n-k,q+k})$. Indeed, if we assume that (III.3.8) is valid for $3 \leq k < k_0$, we get $(\overline{\partial}_z+\overline{\partial}_{z'})\partial_{z'}A^{(k)} = 0$ for those indices k.

Then we can apply Lemma III.3.2 to find $A^{(k)}$ for $k = k_0$.

In the sequel $A^{(1)}$ will stand for A and we shall use the notation

(III.3.9) $$A' = \sum_{k=1}^{n-q} (-1)^{k-1} A^{(k)}.$$

Since $A^{(n-q)} \in C^{\infty}(\Omega;\tilde{\Lambda}^{m,m+q,n})$ we have $\partial_{z'} A^{(n-q)} = 0$ and therefore, thanks also to (III.3.7) and (III.3.8):

(III.3.10) $$(\bar{\partial}_z + d_{z'})A' = 0 \quad \textit{in } \Omega.$$

Recall that $\partial_{z'} G + A^{(2)} \in C^{\infty}(\Omega\backslash U;\tilde{\Lambda}^{m,m+n-2,q+2})$. By virtue of (III.3.3) and (III. 3.7) we have $(\bar{\partial}_z + \bar{\partial}_{z'})(\partial_{z'} G + A^{(2)}) = 0$ in $\Omega\backslash U$. When $m+n = 2$ (and $n \geq q+2$) necesarily $m = 0$, $n = 2$, $q = 0$ and $r = 1 < n-q$. We stop here this part of the argument. Assume $m+n \geq 3$. By Lemma III.3.2 there is a solution $G^{(2)} \in C^{\infty}(\Omega\backslash U;\tilde{\Lambda}^{m,m+n-3,q+2})$ of the equation in $\Omega\backslash U$,

(III.3.11) $$(\bar{\partial}_z + \bar{\partial}_{z'})G^{(2)} = A^{(2)} + \partial_{z'} G.$$

If $n = q+2$ the argument stops here.

If $n \geq q+3$ we can find solutions $G^{(k)} \in C^{\infty}(\Omega\backslash U;\tilde{\Lambda}^{m,m+n-k-1,q+k})$ of the equations in $\Omega\backslash U$,

(III.3.12) $$(\bar{\partial}_z + \bar{\partial}_{z'})G^{(k)} = A^{(k)} + \partial_{z'} G^{(k-1)} \quad (k = 3,\dots,r).$$

where $r = \mathrm{Min}(n-q, m+n-1)$. If $k \leq r$,

$$(\bar\partial_z+\bar\partial_{z'})(A^{(k)}+\partial_{z'}G^{(k-1)}) = \partial_{z'}[A^{(k-1)}-(\bar\partial_z+\bar\partial_{z'})G^{(k-1)}] = 0$$

by induction on $k \geq 3$, and by (III.3.7)–(III.3.8). It suffices to apply Lemma III.3.2.

We are going to use the notation $G^{(1)} = G$ and

$$G' = \sum_{k=1}^{r} (-1)^{k-1}G^{(k)}. \tag{III.3.13}$$

If $r = n-q$ we have $G^{(n-q)} \in C^{\infty}(\Omega\backslash U;\tilde\Lambda^{m,m+q-1,n})$ hence $\partial_{z'}G^{(n-q)} = 0$. In this case (III.3.11)–(III.3.12) entail

$$(\bar\partial_z+d_{z'})G' = (\bar\partial_z+\bar\partial_{z'})G - A + A' = -(\bar\partial_z+d_{z'})F + A',$$

that is,

$$A' = (\bar\partial_z+d_{z'})(F + G') \quad in\ \Omega\backslash U. \tag{III.3.14}$$

The case $r = m+n-1 < n-q$ requires that we modify slightly the form $F + G'$. Notice that necessarily $m = q = 0$ (hence $r = n-1$) and $G^{(r)} \in \in C^{\infty}(\Omega\backslash U;\tilde\Lambda^{0,0,n-1})$. As we have seen earlier

$$(\bar\partial_z+\bar\partial_{z'})(A^{(n)}+\partial_{z'}G^{(n-1)}) = 0 \quad in\ \Omega\backslash U.$$

This means that $A^{(n)} + \partial_{z'}G^{(n-1)} = \psi dz'_1\wedge\cdots dz'_n$ with $\psi \in \mathcal{O}(\Omega\backslash U)$. It follows from Hartogs' theorem that ψ extends holomorphically to Π'. We can find holomorphic functions h_j in Π' such that if

$$H = \sum_{j=1}^{n} (-1)^{j-1}h_j dz'_1\wedge\cdots\wedge dz'_{j-1}\wedge dz'_{j+1}\wedge\cdots\wedge dz'_n$$

then $\partial_{z'}H = (-1)^n[A^{(n)}+\partial_{z'}G^{(n-1)}]$. In this case (III.3.11) and (III.3.12) entail

$$(\bar\partial_z+d_{z'})(G' - H) = (\bar\partial_z+\bar\partial_{z'})G - A + A' = -(\bar\partial_z+d_{z'})F + A',$$

that is,

$$A' = (\bar{\partial}_z + d_{z'})(F + G' - H) \quad in\ \Omega\backslash U. \tag{III.3.15}$$

Let then E be a hyperfunction extension of $F + G' - H$ to Ω (with $H \equiv 0$ if m+n–1 $\geq$ n–q). By (III.3.10) we know that the current $A' - (\bar{\partial}_z + d_{z'})E$ is $(\bar{\partial}_z + d_{z'})$–closed in Ω; and by (III.3.14)–(III.3.15) that its support is contained in U. Thus that current defines a class in the relative cohomology space $H_U^{m,m+n+q}(\Omega)$ which we set to be $i_U^q([f])$.

We do not know yet that $[f] \to i_U^q([f])$ is a well–defined linear map from $\mathfrak{Sol}^{(q)}(U)$ into $H^{m,m+n+q}(\Omega)$. That is what we must now check.

III. ***The map*** i_U^q ***is well–defined***

Assume that $f \in \mathcal{B}(U;\Lambda^q)$ "cobounds" in the differential complex (III.1.20). This means that there are differential forms

$$A_0 \in \mathcal{C}^\infty(\Omega;\tilde{\Lambda}^{m,m+n-1,q}),\ G_0 \in \mathcal{C}^\infty(\Omega\backslash U;\tilde{\Lambda}^{m,m+n-2,q}),$$
$$\Phi \in \mathcal{C}^\infty(\Omega\backslash U;\tilde{\Lambda}^{m,m+n-1,q-1}),$$

such that $(\bar{\partial}_z + \bar{\partial}_{z'})A_0 = 0$ in Ω, $(\bar{\partial}_z + \bar{\partial}_{z'})\Phi = 0$ in $\Omega\backslash U$, and that

$$F - \partial_{z'}\Phi = A_0 - (\bar{\partial}_z + \bar{\partial}_{z'})G_0, \tag{III.3.16}$$

with the understanding that $G_0 \equiv 0$ if m+n = 1, and $\Phi \equiv 0$ if q = 0. The form Φ defines an element $\phi \in \mathcal{B}(U;\Lambda^{q-1})$ such that $\partial_{z'}\phi = f$.

If n = q, both A and G vanish identically and $\partial_{z'}A_0 \equiv \partial_{z'}G_0 \equiv 0$, hence $(\bar{\partial}_z + d_{z'})A_0 \equiv 0$ in Ω. In this case A_0 extends F –

$(\bar\partial_z + d_{z'})(\Phi - G_0)$ to Ω. If S extends $\Phi - G_0$ to Ω, $E = A_0 + (\bar\partial_z + d_{z'})S$ is an extension of F to Ω (cf. Part II*a*) and $(\bar\partial_z + d_{z'})E = 0$. We conclude that $i_U^n([f])$ vanishes.

Assume $n \geq q+1$. We derive from (III.3.3) and (III.3.16):

$$\text{(III.3.17)} \qquad A - \partial_{z'} A_0 = (\bar\partial_z + \bar\partial_{z'})(G + \partial_{z'} G_0) \quad \textit{in } \Omega \backslash U.$$

We may avail ourselves of Lemma III.3.1 to get a solution $B^{(1)} \in C^\infty(\Omega; \tilde\Lambda^{m,m+n-2,q+1})$ of the equation in Ω,

$$\text{(III.3.18)} \qquad (\bar\partial_z + \bar\partial_{z'})B^{(1)} = A - \partial_{z'} A_0.$$

If $n = q+1$ the argument stops here. If $n \geq q+2$ we solve recursively the equations in Ω,

$$\text{(III.3.19)} \qquad (\bar\partial_z + \bar\partial_{z'})B^{(k)} = A^{(k)} + \partial_{z'} B^{(k-1)} \quad (2 \leq k \leq r),$$

where $B^{(k)} \in C^\infty(\Omega; \tilde\Lambda^{m,m+n-k-1,q+k})$ and $r = \mathrm{Min}(m+n-1, n-q)$ as before. To solve the equations (III.3.19) we observe that

$$(\bar\partial_z + \bar\partial_{z'})(A^{(2)} + \partial_{z'} B^{(1)}) = (\bar\partial_z + \bar\partial_{z'})A^{(2)} - \partial_{z'}(A - \partial_{z'} A_0) = 0$$

by (III.3.7), whereas, if $k \geq 3$,

$$(\bar\partial_z + \bar\partial_{z'})(A^{(k)} + \partial_{z'} B^{(k-1)}) = (\bar\partial_z + \bar\partial_{z'})A^{(k)} - \partial_{z'} A^{(k-1)} = 0$$

by (III.3.8). It suffices then to apply the second part of Lemma III.5.3.

We are going to use the notation

$$\text{(III.3.20)} \qquad B' = \sum_{k=1}^{r} (-1)^{k-1} B^{(k)}.$$

Again we must distinguish between the cases $r = n-q$, and $r = m+n-1 < n-q$. First suppose $r = n-q$. Since $\partial_{z'}B^{(n-q)} \equiv 0$ it follows from (III.3.18) and (III.3.19) that

$$(\bar{\partial}_z + d_{z'})B' = A' - \partial_{z'}A_0.$$

Recalling that $(\bar{\partial}_z + \bar{\partial}_{z'})A_0 \equiv 0$ we obtain, in Ω,

$$A' = (\bar{\partial}_z + d_{z'})(A_0 + B') \quad in\ \Omega. \tag{III.3.21}$$

Now suppose $m = q = 0$, $r = n-1$. Earlier we have seen that

$$(\bar{\partial}_z + \bar{\partial}_{z'})(A^{(n)} + \partial_{z'}B^{(n-1)}) = 0 \ in\ \Omega.$$

This means that $A^{(n)} + \partial_{z'}B^{(n-1)} = h dz'_1 \wedge \cdots dz'_n$ with $h \in \mathcal{O}(\Omega)$. It follows from Hartogs' theorem that h extends holomorphically to Π'. We can find holomorphic functions p_j in Π' such that if

$$P = \sum_{j=1}^{n} (-1)^{j-1} p_j dz'_1 \wedge \cdots \wedge dz'_{j-1} \wedge dz'_{j+1} \wedge \cdots \wedge dz'_n$$

then $\partial_{z'}P = (-1)^n (A^{(n)} + \partial_{z'}B^{(n-1)})$. In this case (III.3.12) and (III.3.13) entail

$$(\bar{\partial}_z + d_{z'})(B' - P) = A' - \partial_{z'}A_0,$$

whence

$$A' = (\bar{\partial}_z + d_{z'})(A_0 + B' - P) \quad in\ \Omega. \tag{III.3.22}$$

Continue to assume $n \geq q+1$. We compare the pairs of equations (III.3.17)–(III.3.18) and (III.3.11)–(III.3.12). We solve the following string of equations, in $\Omega \backslash U$:

$$(\bar\partial_z+\bar\partial_{z'})v^{(1)} = B^{(1)} - G^{(1)} - \partial_{z'}G_0, \tag{III.3.23}$$

and if $n \geq q+2$,

$$(\bar\partial_z+\bar\partial_{z'})v^{(k)} = B^{(k)} - G^{(k)} + \partial_{z'}v^{(k-1)} \quad (k = 2,\dots,s), \tag{III.3.24}$$

where $v^{(k)} \in C^\infty(\Omega\backslash U;\tilde\Lambda^{m,m+n-k-2,q+k})$ and $s = \mathrm{Min}(m+n-2,n-q)$.

First of all, (III.3.17) and (III.3.18) entail, in $\Omega\backslash U$,

$$(\bar\partial_z+\bar\partial_{z'})(B^{(1)} - G^{(1)} - \partial_{z'}G_0) =$$

$$A - \partial_{z'}A_0 - (\bar\partial_z+\bar\partial_{z'})(G+\partial_{z'}G_0) = 0.$$

It suffices therefore to apply the second part of Lemma III.3.2.

Next, suppose $2 \leq k \leq r = \mathrm{Min}(m+n-1,n-q)$. From (III.3.11), (III.3.12) and (III.3. 19) we derive, in $\Omega\backslash U$,

$$(\bar\partial_z+\bar\partial_{z'})[B^{(k)} - G^{(k)} + \partial_{z'}v^{(k-1)}] =$$

$$(\bar\partial_z+\bar\partial_{z'})[B^{(k)} - G^{(k)}] - \partial_{z'}[B^{(k-1)} - G^{(k-1)}] = 0.$$

Again we apply the second part of Lemma III.3.2, until we reach the value $k = s$.

When $s = n-q$ we avail ourselves of the fact that $v^{(n-q)} \in C^\infty(\Omega\backslash U;\tilde\Lambda^{m,m+q-2,n})$ and therefore $\partial_{z'}v \equiv 0$. Using the notation

$$v' = \sum_{k=1}^{s} (-1)^{k-1}v^{(k)} \tag{III.3.25}$$

we obtain

$$(\bar\partial_z+d_{z'})v' = B' - G' - \partial_{z'}G_0 \quad in\ \Omega\backslash U. \tag{III.3.26}$$

Suppose now s = m+n−2 < n−q, hence m+q $\leq$ 1. In this case r = s+1 = m + n −1 $\leq$ n−q, and we have just seen that

$$(\bar\partial_z+\bar\partial_{z'})[B^{(r)} - G^{(r)} + \partial_{z'} v^{(s)}] = 0.$$

If m+q $\leq$ 1 we have $v^{(s)} \in C^\infty(\Omega\backslash U;\tilde\Lambda^{m,o,m+n+q-2})$. The coefficient of $B^{(r)} - G^{(r)} + \partial_{z'} v^{(s)} \in C^\infty(\Omega\backslash U;\tilde\Lambda^{m,o,m+n+q-1})$ extends holomorphically to $\Pi\times\Pi'$ and there is a form $Q \in C^\infty(\Omega\backslash U;\tilde\Lambda^{m,o,m+n+q-2})$ with holomorphic coefficients in $\Pi\times\Pi'$ such that

$$\partial_{z'} Q = (-1)^r(B^{(r)} - G^{(r)} + \partial_{z'} v^{(s)}).$$

In this case we have

$$(\bar\partial_z+d_{z'})(v' - Q) = B' - G' - \partial_{z'} G_0 \quad in\ \Omega\backslash U. \tag{III.3.27}$$

With the understanding that $Q \equiv 0$ when s = n−q we derive from (III.3.16), (III.3.26) and (III.3.27):

$$F + G' = A_0 + B' - (\bar\partial_z+d_{z'})\psi \quad in\ \Omega\backslash U. \tag{III.3.28}$$

where $\psi = G_0 + v' - Q - \Phi \in C^\infty(\Omega\backslash U;\tilde\Lambda^{m,m+n+q-1})$ [recall that $(\bar\partial_z + d_{z'})\Phi \equiv 0$]. Actually we shall extend ψ as an element of $\mathcal{B}(\Omega;\tilde\Lambda^{m,m+n+q-1})$. Combining (III.3.21) and (III.3.22) with (III.3.28) shows that

$$A' = (\bar\partial_z+d_{z'})(F+G' - P) \quad in\ \Omega\backslash U,$$

with the understanding that $P \equiv 0$ when r = n−q. Comparing the last equation to (III.3.15) shows that $(\bar\partial_z+d_{z'})(H-P) = \partial_{z'}(H-P) = 0$ in $\Omega\backslash U$, hence throughout $\Pi\times\Pi'$ since the coefficients of H and of P are holomorphic in $\Pi\times\Pi'$. But then we can find a form $\chi = \sum_{|I|=n-2} \chi_I dz_I$ with holomorphic coefficients in $\Pi\times\Pi'$, such that $\partial_{z'}\chi = H - P$.

We are now in a position to conclude the argument. Set

$$E_0 = A_0 + B' - H - (\bar\partial_z + d_{z'})\psi \in \mathcal{B}(\Omega;\tilde\Lambda^{m,m+n+q-1})$$

and let E be the extension of $F + G' - H$ introduced at the end of Part IIc. By (III.3.28) we have $\mathrm{supp}(E_0 - E) \subset U$. On the other hand, (III.3.22) can be rewritten as

$$A' = (\bar\partial_z + d_{z'})[A_0 + B' - H + (\bar\partial_z + d_{z'})\chi] =$$
$$(\bar\partial_z + d_{z'})[E_0 + (\bar\partial_z + d_{z'})(\chi + \psi)],$$

whence

$$A' - (\bar\partial_z + d_{z'})E = (\bar\partial_z + d_{z'})(E_0 - E)$$

which proves that the cohomology class of $A' - (\bar\partial_z + d_{z'})E$ in $H_U^{m,m+n+q}(\Omega)$, $i_U^q([f])$, vanishes.

IV. *The map i_U^q is injective*

We shall avail ourselves of the fact that the cohomology (in $\mathbb{C}^{m+n}$) with coefficients in the sheaf ${}_m\mathcal{O}^{(m)}$ is the same whether we use the "smooth resolution" (III.2.4) or the "hyperfunction resolution" (III.2.5). We shall repeatedly use the following observation:

LEMMA III.3.3.— *Let* F *be a closed subset of* $\Pi\times\Pi'$. *Given any form*

$$g = \sum_{k=0}^{n-q} (-1)^{k-1} g^{(k)} \in C^\infty(\Pi\times\Pi'\backslash F;\tilde\Lambda^{m,m+n+q}),$$

with $g^{(k)} \in C^\infty(\Pi\times\Pi'\backslash F;\tilde\Lambda^{m,m+n-k,q+k})$, *there is*

$$v \in C^\infty(\Pi\times\Pi'\backslash F;\tilde\Lambda^{m,m+n+q-1})$$

such that

$$g - (\bar\partial_z + d_{z'})v = \sum_{k=1}^{n-q} (-1)^{k-1}\tilde g^{(k)}$$

with $\tilde{g}^{(k)} \in C^\infty(\Pi\times\Pi'\backslash F;\tilde{\Lambda}^{m,m+n-k,q+k})$.

Proof: If $g^{(0)}$ does not vanish identically we write

$$g^{(0)} = \sum_{|J|=q} g_J^{(0)}\wedge dz'_J,\ g_J^{(0)} \in C^\infty(\Pi\times\Pi'\backslash F;\tilde{\Lambda}^{m,m+n}).$$

For each J, $|J| = q$, we have $(\bar{\partial}_z+\bar{\partial}_{z'})g_J^{(0)} \equiv 0$. We apply Theorem I.4.3: there is $v_J \in C^\infty(\Pi\times\Pi'\backslash F;\tilde{\Lambda}^{m,m+n-1})$ such that $(\bar{\partial}_z+\bar{\partial}_{z'})v_J = g_J^{(0)}$ in $\Pi\times\Pi'\backslash F$. Then $v = \sum_{|J|=q} v_J\wedge dz'_J$ belongs to $C^\infty(\Pi\times\Pi'\backslash F;\tilde{\Lambda}^{m,m+n-1,q})$ and solves $(\bar{\partial}_z+\bar{\partial}_{z'})v = g^{(0)}$ in $\Pi\times\Pi'\backslash F$; $g - (\bar{\partial}_z+d_{z'})v$ has the desired property. □

Let us prove that i_U^q is injective. Using the notation of Parts II and III suppose

$$A' - (\bar{\partial}_z+d_{z'})E = (\bar{\partial}_z+d_{z'})v,\ v \in \mathcal{B}(\Omega;\tilde{\Lambda}^{m,m+n+q-1}),\ \text{supp}\ v \subset U.$$

By the uniqueness of the cohomology with coefficients in the sheaf ${}_m\mathcal{O}^{(m)}$ and by virtue of (III.3.15) there is $u \in C^\infty(\Omega;\tilde{\Lambda}^{m,m+n+q-1})$ such that

$$A' = (\bar{\partial}_z+d_{z'})u \quad in\ \Omega, \tag{III.3.29}$$

$$u = F + G' - H + (\bar{\partial}_z+d_{z'})\chi \quad in\ \Omega\backslash U, \tag{III.3.30}$$

with $\chi \in C^\infty(\Omega\backslash U;\tilde{\Lambda}^{m,m+n+q-2})$. Using Lemma III.3.3 we can assume that

$$u = \sum_{k=1}^{n-q-1} (-1)^{k-1}u^{(k)},\ u^{(k)} \in C^\infty(\Omega;\tilde{\Lambda}^{m,m+n-k,q+k-1}).$$

Recalling that $A = A^{(1)} \in C^\infty(\Omega;\tilde{\Lambda}^{m,m+n-1,q+1})$ and using the notation (III.3.9) we obtain

(III.3.31) $$(\bar\partial_z+\bar\partial_{z'})u^{(1)} \equiv 0 \quad in\ \Omega.$$

Now recall that $F \in C^\infty(\Omega\backslash U;\tilde\Lambda^{m,m+n-1,q})$ and let G' be defined by (III.3.13). If we write

$$\chi = \sum_{k=1}^{m}(-1)^{k-1}\chi^{(k)},\ \chi^{(k)} \in C^\infty(\Omega\backslash U;\tilde\Lambda^{m,m+n-k,q+k-2}),$$

we derive from (III.3.30) that $u^{(1)} = F - (\bar\partial_z+\bar\partial_{z'})\chi^{(2)}$. In view of (III.3.29) this is precisely a decomposition of the kind (III.3.16), which proves that $[f] = 0$.

V. ***The map*** i^q_U ***is surjective***

Consider a form $\alpha = \sum_{k=1}^{n-q}(-1)^{k-1}\alpha^{(k)} \in C^\infty(\Omega;\tilde\Lambda^{m,m+n+q})$ such that $(\bar\partial_z+d_{z'})\alpha = 0$. The latter means that, in Ω,

(III.3.32) $$(\bar\partial_z+\bar\partial_{z'})\alpha^{(1)} = 0,$$

(III.3.33) $$(\bar\partial_z+\bar\partial_{z'})\alpha^{(k)} = \partial_{z'}\alpha^{(k-1)} \quad (k = 2,\ldots,n-q).$$

Now assume that the class $[\alpha] \in H^{m,m+n+q}(\Omega)$ belongs to Ker ρ^q. This is the same as saying that there is $\beta \in C^\infty(\Omega\backslash U;\tilde\Lambda^{m,m+n+q-1})$ satisfying $\alpha = (\bar\partial_z+d_{z'})\beta$ in $\Omega\backslash U$. Writing

$$\beta = \sum_{k=0}^{n-q}(-1)^{k-1}\beta^{(k)},\ \beta^{(k)} \in C^\infty(\Omega\backslash U;\tilde\Lambda^{m,m+n-k-1,q+k}),$$

we have in $\Omega\backslash U$,

(III.3.34) $$(\overline{\partial}_z+\overline{\partial}_{z'})\beta^{(0)} = 0,$$

(III.3.35) $$\alpha^{(k)} = (\overline{\partial}_z+\overline{\partial}_{z'})\beta^{(k)} - \partial_{z'}\beta^{(k-1)} \quad (k = 1,\dots,r),$$

where $r = \mathrm{Min}(m+n-1,n-q)$. The construction in Part II applies if we select

$$F = -\beta^{(0)} \in C^{\infty}(\Omega\backslash U;\tilde{\Lambda}^{m,m+n-1,q}),\ A = \alpha^{(1)} \in C^{\infty}(\Omega;\tilde{\Lambda}^{m,m+n-1,q+1}),$$
$$G = \beta^{(1)},\ A^{(k)} = \alpha^{(k)}\ (k = 2,\dots,n-k),\ G^{(l)} = \beta^{(l)}\ (l = 2,\dots,r).$$

Notice that $A' = \alpha$, $F + G' = \beta$ and (III.3.14) holds. If $f \in \mathcal{B}(U;\Lambda^q)$ corresponds to F we will have $\partial_{z'}f = 0$, and the construction in Part II shows that $i^q_U([f])$ is equal to the cohomology class in $H^{m,m+n+q}_U(\Omega)$ of the form $\alpha - (\overline{\partial}_z+d_{z'})E$, where E is a hyperfunction extension of β. This means that the natural projection of $i^q_U([f])$ into Ker ρ^q is equal to $[\alpha]$.

Note that if the differential form β above extends as a form which belongs to $C^{\infty}(\Omega;\tilde{\Lambda}^{m,m+n+q-1})$ and if $\alpha = (\overline{\partial}_z+d_{z'})\beta$ in Ω then F has the $(\overline{\partial}_z+\overline{\partial}_{z'})$—closed extension $\beta^{(0)} \in C^{\infty}(\Omega;\tilde{\Lambda}^{m,m+n-1,q})$, which means that $f \equiv 0$. Thus we have constructed a linear map

$$\mathrm{Ker}\ \rho^q \xrightarrow{\kappa^q} \mathfrak{Sol}^{(q)}(U)$$

such that $i^q_U \circ \kappa^q$ is a right inverse of the natural map

$$H^{m,m+n+q}_U(\Omega) \to \mathrm{Ker}\ \rho^q.$$

Let us now restrict the map i^q_U to $\mathfrak{Sol}^{(q)}_0(U)$. In the construction of Part II this amounts to assuming $A \equiv 0$. It means that $i^q_U([f])$ is the class of $(\overline{\partial}_z+d_{z'})E$ and therefore belongs to Im δ^q. We shall construct an inverse of the map

$$i_U^q : \mathfrak{Sol}_0^{(q)}(U) \to \operatorname{Im} \delta^q.$$

We begin by noting that there is a natural map

$$\tilde{\lambda}^q\!: H^{m,m+n+q-1}(\Omega\backslash U) \to \mathfrak{Sol}_0^{(q)}(U).$$

Indeed, consider

$$\beta = \sum_{k=1}^{r} (-1)^{k-1}\beta^{(k)} \in C^\infty(\Omega\backslash U; \tilde{\Lambda}^{m,m+n+q-1}),$$

with $\beta^{(k)} \in C^\infty(\Omega\backslash U;\tilde{\Lambda}^{m,m+n-k,q+k-1})$ and $r = \mathrm{Min}(m+n, n-q+1)$, such that $(\bar{\partial}_z+d_{z'})\beta = 0$ [we have availed ourselves of Lemma III.3.1 to eliminate $\beta^{(0)}$]. Then $(\bar{\partial}_z+\bar{\partial}_{z'})\beta^{(1)} = 0$ in $\Omega\backslash U$; thus $\beta^{(1)}$ can be identified to a form (III.3.2), $F \in C^\infty(\Omega\backslash U;\tilde{\Lambda}^{m,m+n-1,q})$.

If $r = 1$ we must also have $\partial_{z'}F \equiv 0$ in $\Omega\backslash U$. This means that, if μ is the "current" (III.3.1) corresponding to F, then $\partial_{z'}\mu \equiv 0$. In turn this implies that the element $f \in \mathcal{B}(U;\Lambda^q)$ represented by μ belongs to $\mathfrak{Sol}_0^{(q)}(U)$.

If $r \geq 2$ we also have, in $\Omega\backslash U$,

$$(\bar{\partial}_z+\bar{\partial}_{z'})\beta^{(k)} = \partial_{z'}\beta^{(k-1)} \quad (k = 2,\ldots,r), \tag{III.3.36}$$

$$\partial_{z'}\beta^{(r)} = 0. \tag{III.3.37}$$

Equation (III.3.36) with $k = 2$ shows that $\partial_{z'}f \equiv 0$ in U. It also shows that if we put $F = \beta^{(1)}$ and $G = \beta^{(2)}$ in (III.3.3) we must also put $A = 0$. Here also we conclude that $\partial_{z'}\mu \equiv 0$ if μ is the "current" (III.3.1) corresponding to F.

In all these cases we define $\tilde{\lambda}^q([\beta]) = [f]$.

Now suppose $\beta = (\bar{\partial}_z+d_{z'})\omega$, with

$$\omega = \sum_{k=1}^{s} (-1)^{k-1}\omega^{(k)} \in \mathcal{C}^{\infty}(\Omega\backslash U;\tilde{\Lambda}^{m,m+n+q-2}),$$

$$\omega^{(k)} \in \mathcal{C}^{\infty}(\Omega\backslash U;\tilde{\Lambda}^{m,m+n-k,q+k-2})$$

and s = Min(m+n,n−q+2). This implies, in $\Omega\backslash U$,

(III.3.38) $$(\bar{\partial}_z+\bar{\partial}_{z'})\omega^{(1)} = 0,$$

(III.3.39) $$\beta^{(1)} = -\,\partial_{z'}\omega^{(1)} + (\bar{\partial}_z+\bar{\partial}_{z'})\omega^{(2)}.$$

By virtue of (III.3.38) $-\omega^{(1)}$ represents a current $\nu \in \mathcal{O}'(\mathcal{C}\ell U;\Lambda^{q-1})$; (III.3.39) means that the current $\mu \in \mathcal{O}'(\mathcal{C}\ell U;\Lambda^{q})$ corresponding to $F = \beta^{(1)}$ satisfies the equation $\mu = \partial_{z'}\nu$. Consequently the currents $f \in \mathcal{B}(U;\Lambda^{q})$ represented by μ and $v \in \mathcal{B}(U;\Lambda^{q+1})$ represented by ν are related by the equation $\partial_{z'}v = f$; it shows that $\tilde{\lambda}^q$ is truly a linear map.

Without assuming that β cobounds let us now repeat the con— struction of Part II with $F = \beta^{(1)}$ and $G = \beta^{(2)}$. If r = 1, $\beta^{(k)} \equiv 0$ for all k $\geq$ 2 and $i_U^q([f])$ is the class of $(\bar{\partial}_z+d_{z'})\tilde{\beta}$, with $\tilde{\beta}$ an extension of β, ie., $i_U^q([f]) = \delta^q([\beta])$. Suppose r $\geq$ 2. Comparing the equations (III.3.11) and (III.3.12) where we put $A = A^{(k)} = 0$, with (III.3.36), we see that we can take $G^{(k)} = \beta^{(k+1)}$. [We are availing ourselves of the results of Part III, which allow us to vary our choices of the $G^{(k)}$.] Since $\partial_{z'}G^{(r-1)} = 0$ by (III.3.37), we obtain $F + G' = \beta$. This means that $i_U^q([f])$ is equal to $\delta^q([\beta])$. In other words, $i_U^q\circ\tilde{\lambda}^q([\beta]) = \delta^q([\beta])$. Since i_U^q is injective (by Part IV) we conclude that

$$\text{Ker } \delta^q = \text{Im } \rho^{q-1} = \text{Ker } \tilde{\lambda}^q.$$

By going to the quotient mod Im ρ^{q-1} we complete the definition of the map

$$\lambda^q\colon \operatorname{Im} \delta^q \ (\cong H^{m,m+n+q-1}(\Omega\backslash U)/\operatorname{Im} \rho^{q-1}) \to \mathfrak{Sol}_0^{(q)}(U).$$

Thus the map λ^q is injective and $\operatorname{Im} \delta^q = i_U^q(\mathfrak{Sol}_0^{(q)}(U))$. By virtue of the direct sum decomposition (III.3.5) this completes the proof of the surjectivity of i_U^q.

VI. ***Commutativity of the diagram*** (III.2.12)

Let $V \subset U$ be as in Theorem III.2.2. We go back to $f \in \mathcal{B}(U;\Lambda^q)$, given by (III. 1.19), such that $\partial_{z'} f \equiv 0$. Restricting each coefficient f_J of f to V yields an element $\mathfrak{r}_V^U f \in \mathcal{B}(V;\Lambda^q)$. We can represent $\mathfrak{r}_V^U f$ by a current μ_V analogous to (III.3.1) but whose coefficients are elements of $\mathcal{O}'(\mathscr{Cl}V)$; to any such current will correspond a differential form analogous to (III.3.2), F_V. By following the recipe in Part II we assign to it the class $i_V^q([\mathfrak{r}_V^U f]) \in H_V^{m,m+n+q}(\Omega_1)$, where $\Omega_1 = \Pi\times\Pi'\backslash\partial V$.

Let μ be a current (III.3.1) representing f, F the form (III.3.2) associated to μ. We recall that, by the results in Part III, we are free to select the representative μ_V. In particular, we can decompose each coefficient of μ [see (III.3.1)] as a sum of an element in $\mathcal{O}'((\mathscr{Cl}U)\backslash V)$ and an element in $\mathcal{O}'(\mathscr{Cl}V)$. This yields two currents μ_1 and μ_2. Note that $\partial_{z'}\mu = \partial_{z'}\mu_1 + \partial_{z'}\mu_2$ is carried by ∂U and therefore $\partial_{z'}\mu_1$ is carried by $(\mathscr{Cl}V)\cap[(\mathscr{Cl}U)\backslash V] = \partial V$. Thus μ_1 represents $\mathfrak{r}_V^U f$ and we are free to take $\mu_V = \mu_1$.

Below we write $\Omega_2 = \Pi\times\Pi'\backslash[(\mathscr{Cl}U)\backslash V]$. Let then

$$F_1 \in C^\infty(\Omega_1\backslash V;\tilde{\Lambda}^{m,m+n-1,q}),\ F_2 \in C^\infty(\Omega_2;\tilde{\Lambda}^{m,m+n-1,q})$$

correspond to μ_1 and μ_2 respectively; $(\bar\partial_z+\bar\partial_{z'})F_1 = 0$ in $\Omega_1\backslash V$,

$(\bar{\partial}_z+\bar{\partial}_{z'})F_2 = 0$ in Ω_2. In the construction of $i^q_U([f])$ we are at liberty to take F equal to the sum of the restrictions of F_1 and F_2 to $\Omega\backslash U = \Omega_2\backslash V \subset \Omega_1\backslash V$. The analogue of $A' - (\bar{\partial}_z+d_{z'})E$, $A'_1-(\bar{\partial}_z+d_{z'})E_1$, has the following properties:

$$A'_1 - (\bar{\partial}_z+d_{z'})E_1 \in C^\infty(\Omega_1;\tilde{\Lambda}^{m,m+n+q}),\ (\bar{\partial}_z+d_{z'})[A'_1 - (\bar{\partial}_z+d_{z'})E_1] = 0,$$

$$\mathrm{supp}[A'_1 - (\bar{\partial}_z+d_{z'})E_1] \subset V.$$

The class of $A'_1 - (\bar{\partial}_z+d_{z'})E_1$ in $H^{m,m+n+q}_V(\Omega_1)$ is $i^q_V([r^U_V f])$. We must show that the "restriction" of $i^q_V([r^U_V f])$ to Ω_2 is equal to that of $i^q_U([f])$; that is to say, we must show that the restrictions to Ω_2 of $A' - (\bar{\partial}_z+d_{z'})E$ and of $A'_1 - (\bar{\partial}_z+d_{z'})E_1$ differ by a current $(\bar{\partial}_z+d_{z'})v$ with $v \in \mathcal{B}(\Omega_2;\tilde{\Lambda}^{m,m+n+q-1})$, supp $v \subset V$.

Consider

$$A = \partial_{z'}F_1 + \partial_{z'}F_2 + (\bar{\partial}_z+\bar{\partial}_{z'})G,\ A_1 = \partial_{z'}F_1 + (\bar{\partial}_z+\bar{\partial}_{z'})G_1.$$

We recall that

$$A \in C^\infty(\Omega;\tilde{\Lambda}^{m,m+n-1,q+1}),\ A_1 \in C^\infty(\Omega_1;\tilde{\Lambda}^{m,m+n-1,q+1}),$$
$$(\bar{\partial}_z+\bar{\partial}_{z'})A = 0 \textit{ in } \Omega,\ (\bar{\partial}_z+\bar{\partial}_{z'})A_1 = 0 \textit{ in } \Omega_1,$$
$$G \in C^\infty(\Omega\backslash U;\tilde{\Lambda}^{m,m+n-2,q+1}),\ G_1 \in C^\infty(\Omega_1\backslash V;\tilde{\Lambda}^{m,m+n-2,q+1}).$$

We may restrict both A and A_1 to $\Omega_2 \subset \Omega\cap\Omega_1$; and since $\Omega_2\backslash V = \Omega\backslash U \subset \Omega_1\backslash V$ we may restrict F, G, F_1 and G_1 to $\Omega_2\backslash V$. We get

$$\partial_{z'}F_2 = A_2 - (\bar{\partial}_z+\bar{\partial}_{z'})G_2 \text{ in } \Omega_2\backslash V,$$

where

$$A_2 = A - A_1 \in C^\infty(\Omega_2;\tilde{\Lambda}^{m,m+n-1,q+1}),$$
$$G_2 = G - G_1 \in C^\infty(\Omega_2\backslash V;\tilde{\Lambda}^{m,m+n-2,q+1}).$$

Since $F_2 \in C^\infty(\Omega_2;\tilde{\Lambda}^{m,m+n-2,q+1})$ we are exactly in the situation studied in Part III, if we replace F by F_2, A by A_2, G by G_2, Ω by Ω_2 and U by V. The starting point in Part III, Formula (III.3.16), is valid with $\Phi \equiv 0$, A_0

$= F_2$ and $G_0 \equiv 0$. The subsequent argument shows that the corresponding current $A_2' - (\bar{\partial}_z + d_{z'})E_2$ is equal to $(\bar{\partial}_z + d_{z'})v$ with

$$v \in \mathcal{B}(\Omega_2; \tilde{\Lambda}^{m,m+n+q-1}), \text{ supp } v \subset V.$$

But the very conclusion of that argument implies that we have the right to take $A_2' = A' - A_1'$, $E_2 = E - E_1$. This proves the desired result and completes the proof of Theorem III.2.2.

CHAPTER IV

TRANSVERSAL SMOOTHNESS OF HYPERFUNCTION SOLUTIONS

INTRODUCTION

Most of the analysis in Chapter IV is carried out within the framework of the so—called *coarse local embedding* of an open subset $\mathcal{M}'$ of the hypo—analytic manifold $\mathcal{M}$ (*fine local embeddings* are considered in sections IV.6 and IV.7): in $\mathcal{M}'$ m hypoanalytic functions $Z_1,\dots,Z_m$ are given, as well local coordinates $x_1,\dots,x_m,t_1,\dots,t_n$ (they all vanish at a central point O). Things can be arranged in such a way that $Z_j = x_j + i\Phi_j(x,t)$, j = 1,...,m. The analysis is transferred to the image U under the map $(x,t) \to (Z(x,t),t)$ of a product set $U_0 \times \Pi_0'$, with $U_0 \subset\subset \Pi_0$, and Π_0 and Π_0' poly—hedrons in x—space $\mathbb{R}^m$ and t—space $\mathbb{R}^n$ respectively. The hypo—analytic structure on the image $\Sigma \subset \mathbb{C}^m \times \mathbb{R}^n$ of $\mathcal{M}'$ transferred from $\mathcal{M}'$ is induced by the elliptic structure on $\mathbb{C}^m \times \mathbb{R}^n$ associated to the differential operator $\bar{\partial}_z + d_t$: the hypo—analytic functions on Σ are the restrictions of functions that are holomorphic with respect to z and independent of t.

The distribution solutions in U, ie., the distributions annihilated by the same vector fields $L_1,\dots,L_n$ that annihilate $Z_1,\dots,Z_m$, are C^∞ func—tions of $t \in \Pi_0'$ valued in the space of distributions of $x \in \Pi_0$. The central result of Chapter IV addresses the question of whether the same is true of the hyperfunction solutions. To answer it we begin by defining what we mean by a hyperfunction in U that depends smoothly on t (in section IV.2, after the analytic functionals that depend smoothly on t have been defined in section IV.1). Those hyperfunctions make up a space $\mathcal{B}(U,C_t^\infty)$ which is definitely *not* invariant under hypo—analytic automorphisms. We focus on its subspace $\mathfrak{Sol}(U,C_t^\infty)$ consisting of those hyperfunctions u such that $d_t u \equiv 0$. We also introduce the cohomology spaces $\mathfrak{Sol}^{(q)}(U,C_t^\infty)$ of the differential complex d_t: $\mathcal{B}(U,C_t^\infty;\Lambda^{q-1}) \to \mathcal{B}(U,C_t^\infty;\Lambda^q)$ for any q = 1,...,n, in analogy with what was done in Chapter III in connection with the local embedding into $\mathbb{C}^{m+n}$.

This leads to a version of Theorem III.2.2 suited to the coarse embedding. In section IV.3 we describe the changes in the proof of Theorem III.2.2 needed to prove this variant. The basic isomorphism is now

$$\mathfrak{Sol}^{(q)}(U,\mathcal{C}^{\infty}_t) \cong H_U^{m,m+q}(\mathbb{C}^m\times\mathbb{R}^n\backslash\partial U).$$

The right hand side is the relative cohomogy space in the sense of the differential complex $\overline{\partial}_z+d_t$.

In section IV.4 we show that all hyperfunction solutions (according to Chapter III) are smooth with respect to t (in the sense of section IV.3) and more generally, for $q \geq 1$, that every cohomology class in $\mathfrak{Sol}^{(q)}(U)$ has a (d_t–closed) representative

$$f = \sum_{|J|=q} f_J dt_J$$

whose coefficients belong to $\mathcal{B}(U,\mathcal{C}^{\infty}_t)$. This is done by showing that integration along the fibres, in the fibration $\mathbb{C}^m\times\mathbb{C}^n\backslash\partial U \to \mathbb{C}^m\times\mathbb{R}^n\backslash\partial U$, leads to a natural isomorphism $H_U^{m,m+n+q}(\mathbb{C}^m\times\mathbb{C}^n\backslash\partial U) \cong H_U^{m,m+q}(\mathbb{C}^m\times\mathbb{R}^n\backslash\partial U)$.

Section IV.5 discusses some of the consequences of this result: The trace of a hyperfunction solution h on the submanifolds t = *const.*, and more generally on any maximally real submanifold $\mathcal{X}$ in the domain of definition of h, is well defined. The hyperfunction solution h is uniquely determined by its trace on $\mathcal{X}$: if the latter vanishes identically, $h \equiv 0$ in a full neighborhood of $\mathcal{X}$. A consequence of the latter is that supp h is equal to the union of a family of *orbits* of the involutive structure of $\mathcal{M}$ ([TREVES, 1992], Section I.11 and Theorem II.3.3). The (hyperfunction) boundary value of a regular [ie., $\mathcal{C}^{\infty}$ with respect to (z,t) and holomorphic with respect to z] function $\tilde{h}$ in wedges in $\mathbb{C}^m\times\mathbb{R}^n$ with edge $U \subset \Sigma$ is a solution if and only if $\tilde{h}$ is constant with respect to t.

Section IV.6 looks at the FBI transform of analytic functionals in z–space $\mathbb{C}^m$ depending smoothly on t and describes how it can be used to define the FBI transform of hyperfunctions on $\mathcal{M}$ (or on Σ) which depend smoothly on t. Theorem IV.6.4 extends to the latter the decomposition Theorem II.3.1, representing locally an arbitrary hyperfunction as the sum of boundary values of regular functions in wedges. For the germ of a hyperfunction u at $O \in \mathcal{M}$ to be hypo–analytic it is necessary and suffi– cient that its FBI transform (at O) be exponentially decaying (Theorem IV.6.5). And the hypo–analytic wave–front set at O of the trace on a maximally real submanifold $\mathcal{X}$ through O of every germ at O of a hyper– function solution is contained in the characteristic set of $\mathcal{M}$ (Theorem IV.6.6).

With the adaptations that are obviously required section IV.7 re– peats for a fine embedding $(z,s,t) \to (z,w,t)$ [z varies in a polyhedron in $\mathbb{C}^\nu$ and t in one in $\mathbb{R}^{n-\nu}$; $w = s+i\phi(z,s,t) \in \mathbb{C}^d$] the theory developed in sec– tions IV.1 to IV.4 for a coarse embedding. In that fine embedding, for a hyperfunction h which is smooth with respect to (z,t) to be a solution, it is necessary and sufficient that h be holomorphic with respect to z and locally constant with respect to t. The CR case is characterized by the propperty that there are no variables t_k: for h to be a CR hyperfunction it is necessary and sufficient that h be holomorphic with respect to z.

The FBI minitransform of analytic functionals in w–space $\mathbb{C}^d$ depending smoothly on (z,t) is introduced in section IV.8 and used to define the FBI minitransform of hyperfunctions on Σ which depend smoothly on (z,t). When $d = 1$, ie., when Σ is a hypersurface in $\mathbb{C}^\nu\times\mathbb{C}\times\mathbb{R}^{n-\nu}$, in a sufficiently small neighborhood of 0 every hyperfunction solution is equal to the sum of the boundary values of a pair of functions defined on the opposite sides of Σ, holomorphic with respect to (z,w) and constant with respect to t (Theorem IV.8.2).

Section IV.9 is devoted to the proof of a necessary condition for local solvability, ie., for the vanishing of the sheaves associated to the presheaves

$$U \to \mathfrak{Sol}^{(q)}(U) \ (q \geq 1),$$

when the hypo–analytic manifold $\mathcal{M}$ is of the hypersurface type (ie., d = 1). The condition is the vanishing of the *intersection number* of the level sets of a hypoanalytic function w whose differential spans the characteris– tic set at O. This is the same result as for distributions and shows that most of the "classical" examples of nonsolvability, such as the Lewy or the Mizohata equations, and many others, are also nonsolvable in hyperfunc– tions.

IV.1 ANALYTIC FUNCTIONALS DEPENDING SMOOTHLY ON t

In this section we associate to the coarse local embedding

$$(x,t) \to (Z(x,t),t)$$

the analytic functionals in z–space $\mathbb{C}^m$ that depend smoothly on t.

If $\mathcal{U}$ is an open subset of $\mathbb{C}^m \times \mathbb{R}^n$ by a ***regular function*** in $\mathcal{U}$ we shall mean a function $f \in C^\infty(\mathcal{U})$ which is holomorphic with respect to z. We denote by $\mathfrak{F}(\mathcal{U})$ the linear space of all regular functions in $\mathcal{U}$; it is a closed subspace of $C^\infty(\mathcal{U})$ (it is the kernel of the differential operator $\bar{\partial}_z$). We equip $\mathfrak{F}(\mathcal{U})$ with the topology induced by $C^\infty(\mathcal{U})$; it is a Fréchet space. Thanks to the Cauchy inequalities (in z–space) we see that its topology is the same as the topology of uniform convergence of each derivative $\partial_t^\alpha f$ ($\alpha \in \mathbb{Z}_+^n$) on every compact subset of $\mathcal{U}$.

REMARK IV.1.1.— In the terminology of [TREVES, 1992] (Definition I.1.3), the elements of $\mathfrak{F}(\mathcal{U})$ are the C^∞ *solutions* in $\mathcal{U}$ when $\mathbb{C}^m \times \mathbb{R}^n$ is equipped with the involutive structure defined by the vector fields $\partial/\partial \bar{z}_i$ (i = 1,...,m). This is the *standard flat CR structure* on $\mathbb{C}^m \times \mathbb{R}^n$ (Example I.2.4, *loc. cit.*), not to be confused with its standard elliptic structure, defined by the vector fields $\partial/\partial \bar{z}_i$ (i = 1,...,m) *and* $\partial/\partial t_j$ (j = 1,...,n). The solutions in the elliptic structure are the functions $h(z)$ holomorphic with respect to z and locally independent of t ([TREVES, 1992], p. 77). □

We set $Z_j(x,t) = x_j + i\Phi_j(x,t)$ $(1 \leq j \leq m)$, $Z = (Z_1,...,Z_m) \in \mathbb{C}^m$ (cf., Section II.1, [TREVES, 1992]). As in Chapter III we call Σ the image of $\mathcal{U}'$ under the map $(x,t) \to (Z(x,t),t)$. In the sequel we limit the variation of x to the polyhedron

$$\Pi_0 = \{\, x \in \mathbb{R}^m;\ |x_i| < a_i,\ \ i = 1,...,m \,\}$$

and that of t to the polyhedron

$$\Pi_0' = \{ t \in \mathbb{R}^n;\ |t_k| < a_k',\ k = 1,...,n \}.$$

We identify $\Pi_0 \times \Pi_0'$ to a subset of $\mathcal{M}'$ and, as a matter of fact, to a relatively compact open subset of $\mathcal{M}'$. We shall assume that $\Phi = (\Phi_1,...,\Phi_m)$ maps $\Pi_0 \times \Pi_0'$ into

$$\Pi_1 = \{ y \in \mathbb{R}^m;\ |y_j| < b_j,\ j = 1,...,m \}.$$

Thus $(x,t) \to (Z(x,t),t)$ induces a diffeomorphism of $\Pi_0 \times \Pi_0'$ onto a relatively compact open subset Σ_0 of $\Sigma \cap (\Pi \times \Pi_0')$ ($\Pi = \Pi_0 \times \Pi_1$, cf. Theorem III.2.1). It is convenient to assume

(IV.1.1) $$\Phi(0,0) = 0,\ d_x\Phi(0,0) = 0.$$

We take the "radii" a_i, a_k' of Π_0 and of Π_0' sufficiently small that the following holds:

(IV.1.2) $$|\Phi(x,t) - \Phi(x_*,t)| \leq \tfrac{1}{2}|x - x_*|,\ \forall\ x,\ x_* \in \Pi_0,\ t \in \Pi_0'.$$

If A is a subset of $\mathbb{C}^m \times \mathbb{R}^n$ and if $t_0 \in \mathbb{R}^n$ we write $A_{t_0} = \{ z \in \mathbb{C}^m;\ (z,t_0) \in A\}$. We are going to deal with subsets S of $\Pi \times \Pi_0'$ ($\subset \mathbb{C}^m \times \mathbb{R}^n$) endowed with the following two properties:

(IV.1.3) $$\forall\ t \in \Pi_0',\ S_t \neq \emptyset;$$

(IV.1.4) *To each compact subset* K' *of* Π_0' *there is a compact subset* $\mathcal{K}$ *of* Π *such that* $(z,t) \in S,\ t \in K' \Rightarrow z \in \mathcal{K}$.

Applying (IV.1.3) with $K' = \{t\}$ shows that $\mathscr{C}\ell S_t$ is a compact subset of the polyhedron Π.

In most of the sequel S will be a *closed* subset of Σ_0 satisfying (IV.1.3) and (IV.1.4). The image of the compact subset $\mathcal{K}$ in (IV.1.4) under the map $x+iy \to x$ is a compact subset K of Π_0. It contains the image $\tilde{S}_t$ of S_t under the map $(Z(x,t),t) \to x$ provided $t \in K'$. Thus (IV.1.4) entails that the function $t \to \mathrm{dist}(\tilde{S}_t,\partial\Pi)$ is uniformly bounded from below (away from zero) on every compact subset of Π_0'. We can therefore select a C^∞ function $d(t)$ in Π_0' such that

$$0 < d(t) < \frac{1}{4}\mathrm{dist}(\tilde{S}_t,\partial\Pi),\ \forall\ t \in \Pi_0'. \tag{IV.1.5}$$

We are going to need a fairly precise version with parameters (here $t \in \Pi_0'$), valid for regular functions, of the approximation Lemma II.2.1 in [TREVES, 1992]. We introduce a pair of special neighborhoods of S in $\mathbb{C}^m \times \Pi_0'$ (for k = 1,–1):

$$\mathcal{N}_k = \{ (z,t) \in \mathbb{C}^m \times \Pi_0';$$
$$z = Z(x,t)+i\gamma,\ \gamma \in \mathbb{R}^m,\ \mathrm{dist}(x,S_t) < 2^k d(t),\ |\gamma| < 2^{k-1}d(t) \}.$$

LEMMA IV.1.1.— *Every function $F \in \mathfrak{F}(\mathcal{N}_1)$ is the limit in $\mathfrak{F}(\mathcal{N}_{-1})$ of a sequence of functions $F_\nu \in \mathfrak{F}(\mathbb{C}^m \times \Pi_0')$ $(\nu = 1,2,\dots)$.*

Proof: Select $\chi \in C^\infty(\Pi \times \Pi_0')$, $\chi(x,t) = 1$ if $\mathrm{dist}(x,S_t) \le d(t)$, $\chi(x,t) = 0$ if $\mathrm{dist}(x,S_t) \ge \frac{3}{2}d(t)$, and define (cf. [TREVES, 1992], p. 85, also (I.5.4))

$$F_\nu(z,t) = (\nu/\pi)^{m/2}\int_{x'\in\mathbb{R}^m} e^{-\nu<z-Z(x',t)>^2} F(Z(x',t),t)\chi(x',t)\mathrm{d}Z(x',t),$$

where $\mathrm{d}Z(x',t) = \det[Z_x(x',t)]\mathrm{d}x'$. It is obvious that $F_\nu \in \mathfrak{F}(\mathbb{C}^m\times\Pi_0')$.

Let us first show that $F_\nu \to F$ uniformly on compact subsets of $\mathcal{N}_{-1}$. We take $(z,t) \in \mathcal{U}_{-1}$ and write $z = Z(x,t)+\imath\gamma$. To visualize better what we are doing let us set $z' = Z(x',t)$, $\tilde{\chi}(z',t) = \chi(\mathcal{R}e\, z',t) = \chi(x,t)$. The slice Σ_t of Σ can be regarded as an m–dimensional chain; we may write

$$F_\nu(z,t) = (\nu/\pi)^{m/2}\int_{\Sigma_t} e^{-\nu<Z(x,t)-z'+\imath\gamma>^2} F(z',t)\tilde{\chi}(z',t)\mathrm{d}z'.$$

We are going to move the integration from Σ_t to $\Sigma_t+\imath\gamma$. Since supp $\tilde{\chi}(\cdot,t) \subset\subset \Sigma_t$ and since $\bar{\partial}_z F \equiv 0$ Stokes' theorem yields

(IV.1.6) $$F_\nu(z,t) =$$

$$(\nu/\pi)^{m/2}\int_{\Sigma_t} e^{-\nu<Z(x,t)-z'>^2} F(z'+\imath\gamma,t)\tilde{\chi}(z',t)\mathrm{d}z' \; -$$

$$(\nu/\pi)^{m/2}\int_{\mathfrak{C}_t} e^{-\nu<z-z'>^2} F(z',t)(\bar{\partial}_z\tilde{\chi})(z',t)\wedge\mathrm{d}z',$$

where $\mathfrak{C}_t$ is the (m+1)–chain $\{\, z' \in \mathbb{C}^m;\ \exists\ s \in [0,1],\ z'-\imath s\gamma \in \Sigma_t \,\}$. According to Lemma II.2.1, [TREVES, 1992], the first integral at the right in (IV.1.6) converges uniformly to the function $F(z,t)\tilde{\chi}(z,t)$ in every compact subset of $\mathcal{N}_{-1}$. That function is equal to F on $\mathcal{N}_{-1}$.

On the other hand, if $\bar{\partial}_z\chi(z',t) \neq 0$ we have $d(t) < \mathrm{dist}(x',S_t) < \frac{3}{2}d(t)$. Let us write, as we may, $\mathfrak{C}_t \ni z' = Z(x',t) + \imath s\gamma$, $0 \leq s \leq 1$. In the second integral at the right in (IV.1.6), we have

$$\mathcal{R}e<z-z'>^2 = \mathcal{R}e<Z(x,t)-Z(x',t)+\imath(1-s)\gamma>^2 =$$

$$|x-x'|^2 - |\Phi(x,t)-\Phi(x',t)+(1-s)\gamma|^2 \geq$$
$$|x-x'|^2 - 2(|\Phi(x,t)-\Phi(x',t)|^2+|\gamma|^2) \geq \tfrac{1}{2}|x-x'|^2 - 2|\gamma|^2$$

thanks to (IV.1.2). Now suppose $z \in \mathcal{N}_{-1}$: $\mathrm{dist}(x,S_t) < \frac{1}{2}d(t)$ hence $|x-x'| > \frac{1}{2}d(t)$; $|\gamma| < \frac{1}{4}d(t)$. We conclude that $\mathcal{Re}<z-z'>^2/d(t)^2$ is bounded away from zero, uniformly if (z,t) stays in a compact subset of $\mathcal{N}_{-1}$; and therefore that the second integral at the right in (III.6.1) converges to zero in $C^0(\mathcal{N}_{-1})$ as $\nu \to +\infty$.

We must now look at the derivatives of $F_\nu(z,t)$ with respect to t. We make use of Lemma II.2.2, in particular of (II.2.9), [TREVES, 1992]. We obtain, for any $\alpha \in \mathbb{Z}^m_+$, $\alpha \neq 0$,

$$\partial^\alpha_t F_\nu(z,t) =$$
$$(\nu/\pi)^{m/2}\int_{x'\in\mathbb{R}^m} e^{-\nu<z-Z(x',t)>^2} L^\alpha[F(Z(x',t),t)\chi(x',t)]\, dZ(x',t),$$

where the differential operator $L^\alpha = L_1^{\alpha_1}\cdots L_n^{\alpha_n}$ [see (III.1.9), (III.1.10) in present text] acts in the variables (x',t). Since $F(z,t)$ is holomorphic with respect to z,

$$L^\alpha[F(Z(x',t),t)\chi(x',t)] = \sum_{\beta\leq\alpha} \binom{\alpha}{\beta} (\partial^\beta_t F)(Z(x',t),t)L^{\alpha-\beta}\chi(x',t).$$

By our choice of χ, if $\beta \neq \alpha$, $\mathrm{dist}(x',\tilde{S}_t) \leq d(t) \Rightarrow L^{\alpha-\beta}\chi(x',t) = 0$. The same reasoning used in dealing with the second integral in the right—hand side of (IV.1.6) shows that, for $\beta \neq \alpha$ and as $\nu \to +\infty$,

$$\int_{x'\in\mathbb{R}^m} e^{-\nu<z-Z(x',t)>^2}(\partial^\beta_t F)(Z(x',t),t)L^{\alpha-\beta}\chi(x',t)\, dZ(x',t) \to 0$$

uniformly on the compact subsets of $\mathcal{N}_{-1}$. On the other hand, by using Lemma II.2.1, *loc. cit.*, as we have done in dealing with the first integral at the right in (IV.1.6), we see that

$$(\nu/\pi)^{m/2}\int_{x'\in\mathbb{R}^m} e^{-\nu<z-Z(x',t)>^2}(\partial^\alpha_t F)(Z(x',t),t)\chi(x',t)\, dZ(x',t)$$

converges uniformly to $\partial_t^\alpha F(z,t)$ in every compact subset of $\mathcal{N}_{-1}$. □

If S is a closed subset of Σ_0 satisfying (IV.1.3) & (IV.1.4) we shall denote by $\mathfrak{F}(S)$ the locally convex inductive limit of the spaces $\mathfrak{F}(\mathcal{U})$ as $\mathcal{U}$ ranges over the family of all the open subsets of $\mathbb{C}^m \times \Pi_0'$ that contain S. By letting $d \to 0$ in $\mathcal{C}^\infty(\Pi_0')$ we derive from Lemma IV.1.1:

COROLLARY IV.1.1.— *The restriction map* $\mathfrak{F}(\mathbb{C}^m \times \Pi_0') \to \mathfrak{F}(S)$ *has a dense image.*

We must now define analytic functionals in $\mathbb{C}^m$ that depend smoothly on $t \in \Pi_0'$. As in Chapters I and II, let $\mathcal{O}'(\mathbb{C}^m)$ denote the space of analytic functionals in $\mathbb{C}^m$. For any $k \in \mathbb{Z}_+$ or $k = +\infty$, consider the space $\mathcal{C}^k(\Pi_0';\mathcal{O}'(\mathbb{C}^m))$: its elements are the $\mathcal{C}^k$ functions $\Pi_0' \ni t \to \mu(t) \in \mathcal{O}'(\mathbb{C}^m)$. That $\mu(t)$ is a continuous function of $t \in \Pi_0'$ at t_0 means the following: given any bounded subset $\mathcal{B}$ of $\mathcal{O}(\mathbb{C}^m)$,

(IV.1.7)
$$\forall\, \epsilon > 0,\ \exists\, \delta > 0,$$
$$|t - t_0| \leq \delta \Rightarrow \forall\, h \in \mathcal{B},\ |\langle \mu(t) - \mu(t_0), h\rangle| \leq \epsilon.$$

But suppose (IV.1.7) holds for individual entire functions h, ie., let $\delta = \delta(h)$ depend on h. The image of a compact set $K' \subset \Pi_0'$ under the map $t \to \mu(t)$ is a weakly compact, thus strongly bounded (actually, strongly compact) subset $\mathcal{K}'$ of $\mathcal{O}'(\mathbb{C}^m)$. By reflexivity the topology of $\mathcal{O}(\mathbb{C}^m)$ is the same as that of uniform convergence on the bounded subsets of $\mathcal{O}'(\mathbb{C}^m)$, hence there is a neighborhood $\mathcal{N}_h$ of h in $\mathcal{O}(\mathbb{C}^m)$ such that

$$\forall\, t \in K',\ g \in \mathcal{N}_{h},\ |<\mu(t)-\mu(t_0),g-h>| \leq \epsilon.$$

The closure of $\mathcal{B}$ is compact and can be covered by a finite number of sets $\mathcal{N}_{h_1},\ldots,\mathcal{N}_{h_r}$. Setting $\delta = \underset{1\leq i\leq r}{\text{Min}}\ \delta(h_i)$ we see that (IV.1.7) will be true for all $h \in \mathcal{B}$ — provided 2ϵ is substituted for ϵ. In other words, to say that $\mu(t)$ is continuous is the same as saying that $\mu(t)$ is weakly continuous, ie., that the complex—valued function $\Pi_0' \ni t \to <\mu(t),h>$ is continuous for each $h \in \mathcal{O}(\mathbb{C}^m)$. The same argument applies to the existence and continuity of derivatives: $\mu \in \mathcal{C}^k(\Pi_0';\mathcal{O}'(\mathbb{C}^m))$ if and only if $\Pi_0' \ni t \to <\mu(t),h>$ is of class $\mathcal{C}^k$ for each $h \in \mathcal{O}(\mathbb{C}^m)$.

With $K' \subset\subset \Pi_0'$ and using the reflexivity of $\mathcal{O}(\mathbb{C}^m)$ as above we see that

$$h \to \underset{t\in K'}{\text{Max}}\ |<\mu(t),h>|$$

is a continuous seminorm on $\mathcal{O}(\mathbb{C}^m)$. As a consequence there are numbers $R,\ C > 0$ such that

$$\text{(IV.1.8)} \qquad \underset{t\in K'}{\text{Max}}\ |<\mu(t),h>| \leq C \underset{|z|\leq R}{\text{Max}}\ |h(z)|.$$

Thus, for all $t \in K'$, $\mu(t)$ is carried by the closed ball $\{\, z \in \mathbb{C}^m;\ |z| \leq R \,\}$. Henceforth we hypothesize that $\mu \in \mathcal{C}^\infty(\Pi_0';\mathcal{O}'(\mathbb{C}^m))$. An inequality similar to (IV.1.8), possibly with increased R and C, will be valid for each partial derivative $\partial_t^\beta \mu(t)$.

We can also make use of the natural Hamel basis $\{z^\alpha/\alpha!\}_{\alpha\in\mathbb{Z}_+^m}$ in $\mathcal{O}(\mathbb{C}^m)$ and of the dual basis in $\mathcal{O}'(\mathbb{C}^m)$, $\{(-1)^{|\alpha|}\,\delta^{(\alpha)}\}_{\alpha\in\mathbb{Z}_+^m}$. Indeed, we can write

$$\mu(t) = \sum_{\alpha \in \mathbb{Z}_+^m} c_\alpha(t)\delta^{(\alpha)}.$$

From (IV.1.8) we derive at once that, for some $C > 0$,

$$\text{(IV.1.9)} \qquad \operatorname*{Max}_{K'} |c_\alpha| \leq CR^{|\alpha|}/\alpha!, \ \forall\, \alpha \in \mathbb{Z}_+^m.$$

The analogue is valid for all the t—derivatives of the c_α.

Suppose $\mu \in C^\infty(\Pi_0';\mathcal{O}'(\mathbb{C}^m))$ and $f \in C^\infty(\Pi_0';\mathcal{O}(\mathbb{C}^m)) = \mathfrak{F}(\mathbb{C}^m \times \Pi_0')$; Taylor expansion

$$f(z,t) = \sum_{\alpha \in \mathbb{Z}_+^m} \partial_z^\alpha f(0,t) z^\alpha/\alpha!$$

allows us to define the product $f(z,t)\mu(t)$ in the obvious manner

$$<f(z,t)\mu_z(t), h(z)> = \sum_{\alpha \in \mathbb{Z}_+^m} \frac{1}{\alpha!}\partial_z^\alpha f(0,t) <\mu_z(t), z^\alpha h(z)>.$$

By using the Cauchy inequalities for f and the inequalities (IV.1.8) for μ and its partial derivatives with respect to t one can easily check that $f(z,t)\mu(t)$ defines an element of $C^\infty(\Pi_0';\mathcal{O}'(\mathbb{C}^m))$ carried, for each $t \in K'$, by the ball $\{ z \in \mathbb{C}^m;\ |z| \leq R \}$. More explicitly,

$$f(z,t)\mu(t) = \sum_{\alpha,\beta \in \mathbb{Z}_+^m} \frac{1}{\alpha!}\partial_z^\alpha f(0,t) c_\beta(t) z^\alpha \delta^{(\beta)} =$$

$$\sum_{\alpha,\beta \in \mathbb{Z}_+^m} (-1)^{|\alpha|} \binom{\alpha+\beta}{\alpha} \partial_z^\alpha f(0,t) c_{\alpha+\beta}(t)\delta^{(\beta)}.$$

The "scalar product" of $\mu(t)$ with $f(z,t)$ is defined in the obvious manner:

$$<\mu_z(t), f(z,t)> = <\mu_z(t) f(z,t), 1> = \sum_{\alpha \in \mathbb{Z}_+^m} (-1)^{|\alpha|} \partial_z^\alpha f(0,t) c_\alpha(t).$$

The Leibniz formula applies:

(IV.1.10) $$\partial_t^\alpha[f(z,t)\mu(t)] = \sum_{\beta\preceq\alpha} \binom{\alpha}{\beta} \partial_t^{\alpha-\beta} f(z,t)\, \partial_t^\beta \mu(t),$$

where α, $\beta \in \mathbb{Z}_+^n$ and $\beta \preceq \alpha$ means $\beta_j \leq \alpha_j$, $j = 1,\ldots,n$. As a consequence,

(IV.1.11) $$\partial_t^\alpha <\mu_z(t), f(z,t)> = \sum_{\beta\preceq\alpha} \binom{\alpha}{\beta} <\partial_t^\beta \mu_z(t), \partial_t^{\alpha-\beta} f(z,t)>.$$

REMARK IV.1.2.— For the readers familiar with topological tensor products and nuclear spaces, we point out that both spaces $C^\infty(\Pi_0')$ and $\mathcal{O}'(\mathbb{C}^m)$ are Fréchet nuclear, and that $\mathcal{O}'(\mathbb{C}^m)$ is dual of Fréchet nuclear. The maps $h(z)g(t) \to h(z)\otimes g(t)$ and $\mu_z g(t) \to \mu_z \otimes g(t)$ define (natural) isomorphisms

$$\mathfrak{F}(\mathbb{C}^m \times \Pi_0') = C^\infty(\Pi_0'; \mathcal{O}(\mathbb{C}^m)) \cong C^\infty(\Pi_0')\hat{\otimes}\mathcal{O}(\mathbb{C}^m),$$

$$C^\infty(\Pi_0'; \mathcal{O}'(\mathbb{C}^m)) \cong C^\infty(\Pi_0')\hat{\otimes}\mathcal{O}(\mathbb{C}^m).$$

The multiplication

$$C^\infty(\Pi_0'; \mathcal{O}(\mathbb{C}^m)) \times C^\infty(\Pi_0'; \mathcal{O}'(\mathbb{C}^m)) \ni (f, \mu) \to f\mu \in C^\infty(\Pi_0'; \mathcal{O}'(\mathbb{C}^m))$$

is then simply the extension of the multiplication

$$(C^\infty(\Pi_0')\otimes\mathcal{O}(\mathbb{C}^m)) \times (C^\infty(\Pi_0')\otimes\mathcal{O}'(\mathbb{C}^m)) \ni (g_1\otimes h, g_2\otimes\mu) \to g_1 g_2 \otimes h\mu.$$

The bracket $<\mu_z(t), f(z,t)>$ is simply the pairing of values between an element of $C^\infty(\Pi_0'; \mathbf{E})$ and one of $C^\infty(\Pi_0'; \mathbf{E}')$, when $\mathbf{E}$ is the Fréchet—Montel space $\mathcal{O}(\mathbb{C}^m)$ and $\mathbf{E}'$ is its dual. It is of course the natural exten—sion of the pairing

$$(C^\infty(\Pi_0')\otimes\mathcal{O}(\mathbb{C}^m)) \times (C^\infty(\Pi_0')\otimes\mathcal{O}'(\mathbb{C}^m)) \ni (g_1\otimes h, g_2\otimes\mu) \to g_1 g_2 <\mu, h>.$$

□

Let $\mathcal{U}$ be an open subset of $\mathbb{C}^m \times \Pi_0'$ such that $\mathcal{U}_t \neq \emptyset$ for every $t \in$

Π_0'. We shall denote by $\mathfrak{G}(\mathcal{U})$ the space of analytic functionals $\mu \in \mathcal{C}^\infty(\Pi_0';\mathcal{O}'(\mathbb{C}^m))$ that satisfy the following condition:

(IV.1.12) *To each* $K' \subset\subset \Pi_0'$ *and each* $\alpha \in \mathbb{Z}_+^n$ *there is* $C_\alpha(K') > 0$ *such that*

$$|<\partial_t^\alpha \mu(t),h>| \leq C_\alpha(K') \sup_{\mathcal{U}_t} |h|, \quad \forall\, t \in K',\ h \in \mathcal{O}(\mathbb{C}^m).$$

Notice that if $f \in \mathfrak{F}(\mathcal{U})$ and $\mu \in \mathfrak{G}(\mathcal{U})$ then by (IV.1.11), for all $t \in K'$,

$$\text{(IV.1.13)} \quad |\partial_t^\alpha <\mu_z(t), f(z,t)>| \leq C_\alpha'(K') \sum_{\beta \leq \alpha} \sup_{z \in \mathcal{U}_t} |\partial_t^\beta f(z,t)|.$$

We return to the submanifold Σ of $\mathbb{C}^m \times \mathbb{R}^n$ and to the closed subset S of Σ_0 endowed with Properties (IV.1.3) and (IV.1.4).

DEFINITION IV.1.1.— *We shall denote by* $\mathfrak{G}(S)$ *the intersection of the spaces* $\mathfrak{G}(\mathcal{U})$ *as* $\mathcal{U}$ *ranges over the family of all the open subsets of* $\mathbb{C}^m \times \Pi_0'$ *that contain* S.

The topology of $\mathfrak{G}(S)$ will be defined by the seminorms

$$\mu \to \mathfrak{p}_{\mathcal{U},K',N}(\mu) = \sup_h \{\sup_{t\in K'} \{ \sum_{|\alpha| \leq N} |<\partial_t^\alpha \mu(t),h>| / \sup_{\mathcal{U}_t} |h| \}\}$$

where $h \in \mathcal{O}(\mathbb{C}^m)$ does not vanish identically and $\alpha \in \mathbb{Z}_+^m$. By letting N vary in $\mathbb{Z}_+$, while $\mathcal{U} \supset\supset S$ ($\mathcal{U} \subset \mathbb{C}^m \times \Pi_0'$) and $K' \subset\subset \Pi_0'$ range over count—able "bases", one can easily check that $\mathfrak{G}(S)$ is metrizable; it is also complete, and thus it is a Fréchet space.

One may say that $\mathfrak{F}(S)$ is the space of germs of regular functions in $\mathbb{C}^m\times\Pi_0'$ at S, and that $\mathfrak{G}(S)$ is the space of analytic functionals depending smoothly on $t\in\Pi_0'$ carried by S. Indeed, note that whatever $t\in\Pi_0'$, $\partial_t^\alpha\mu(t)\in\mathcal{O}'(S_t)$.

The bilinear map

$$\text{(IV.1.14)}\quad \mathcal{C}^\infty(\Pi_0';\mathcal{O}(\mathbb{C}^m))\times\mathfrak{G}(S)\ni(f,\mu)\to\langle\mu_z(t),f(z,t)\rangle\in\mathcal{C}^\infty(\Pi_0'),$$

is well–defined since $\mathfrak{G}(\mathcal{U})$ a linear subspace of $\mathcal{C}^\infty(\Pi_0';\mathcal{O}'(\mathbb{C}^m))$. Keep in mind, however, that the topology of $\mathfrak{G}(S)$ is strictly finer than that induced by $\mathcal{C}^\infty(\Pi_0';\mathcal{O}'(\mathbb{C}^m))$. The coupling (IV.1.14) extends in a unique manner as a continuous bilinear map $\mathfrak{F}(S)\times\mathfrak{G}(S)\to\mathcal{C}^\infty(\Pi_0')$. Indeed it follows directly from (IV.1.13) that the bilinear map (IV.1.14) is continuous for the topology on $\mathcal{C}^\infty(\Pi_0';\mathcal{O}(\mathbb{C}^m))$ induced by $\mathfrak{F}(S)$; the uniqueness follows from Corollary IV.1.1.

Integral representations of elements of $\mathfrak{G}(S)$

For any pair of integers p, q, $0\le p, q\le m$, $\Lambda_*^{p,q}$ will denote the vector subbundle of $\mathbb{C}\Lambda^{p+q}T^*(\mathbb{C}^m\times\mathbb{R}^n)$ spanned by the exterior products $dz_I\wedge d\bar{z}_J$, $|I| = p$, $|J| = q$. If $\mathcal{U}$ is an open subset of $\mathbb{C}^m\times\mathbb{R}^n$, $\mathcal{C}^\infty(\mathcal{U};\Lambda_*^{p,q})$ is the space of $\mathcal{C}^\infty$ sections of the vector bundle $\Lambda_*^{p,q}$. The Cauchy–Riemann differential operator $\bar\partial$ $(=\bar\partial_z)$ with respect to the variables z (which defines the *Levi flat* CR structure on $\mathbb{C}^m\times\mathbb{R}^n$), defines the differential complexes (for p = 0,1,...,m)

(IV.1.15) $$\bar{\partial}_z : C^\infty(\mathcal{U};\Lambda_*^{p,q}) \to C^\infty(\mathcal{U};\Lambda_*^{p,q+1}),\ q = 0,1,\ldots.$$

Call $\mathfrak{F}(\mathcal{U},\Lambda_*^{p,o})$ the space of p—forms with coefficients that are regular functions in $\mathcal{U}$, $\sum_{|I|=p} f_I(z,t)dz_I$, $f_I \in \mathfrak{F}(\mathcal{U})$. By switching to the associated sheaves of germs, $\mathscr{C}^\infty\Lambda_*^{p,q}$ and $\mathscr{F}^{(p)}$, we obtain a cohomo—logical resolution of the latter:

(IV.1.16) $$0 \to \mathscr{F}^{(p)} \to \mathscr{C}^\infty\Lambda_*^{p,o} \xrightarrow{\bar{\partial}_z} \cdots \xrightarrow{\bar{\partial}_z} \mathscr{C}^\infty\Lambda_*^{p,q} \xrightarrow{\bar{\partial}_z} \mathscr{C}^\infty\Lambda_*^{p,q+1} \xrightarrow{\bar{\partial}_z} \cdots,$$

and this leads to the cohomology spaces with coefficients in the sheaf $\mathscr{F}^{(p)}$:

(IV.1.17) $$\mathfrak{H}^{p,o}(\mathcal{U}) = \Gamma(\mathcal{U};\mathscr{F}^{(p)}) \cong \mathfrak{F}(\mathcal{U};\Lambda_*^{p,o});$$

and if $q \geq 1$,

(IV.1.18) $$\mathfrak{H}^{p,q}(\mathcal{U}) = \frac{\{\ \sigma \in \Gamma(\mathcal{U};\mathscr{C}^\infty\Lambda_*^{p,q});\ \bar{\partial}_z\sigma = 0\ \}}{\{\ \sigma \in \Gamma(\mathcal{U};\mathscr{C}^\infty\Lambda_*^{p,q});\ \exists\ \tau \in \Gamma(\mathcal{U};\mathscr{C}^\infty\Lambda_*^{p,q-1}),\ \sigma = \bar{\partial}_z\tau\ \}} \cong \frac{\{\ f \in C^\infty(\mathcal{U};\Lambda_*^{p,q});\ \bar{\partial}_z f = 0\ \}}{\{\ f \in C^\infty(\mathcal{U};\Lambda_*^{p,q});\ \exists\ u \in C^\infty(\mathcal{U};\Lambda_*^{p,q-1}),\ f = \bar{\partial}_z u\ \}}.$$

Let S be a closed subset of Σ_0 satisfying (IV.1.3) and (IV.1.4). We select a function $\chi \in \mathcal{C}^{\infty}(\Pi\times\Pi_0')$ such that

(IV.1.19) $\chi \equiv 1$ *in a neighborhood of* S *in* $\Pi\times\Pi_0'$;

(IV.1.20) *to every compact set* $\mathrm{K}' \subset \Pi_0'$ *there is a compact set* $\mathrm{K} \subset \Pi$ *such that* $\chi(z,t) \neq 0,\ t \in \mathrm{K}' \Rightarrow z \in \mathrm{K}$.

In particular, (IV.1.20) implies that for each $t \in \Pi_0'$ the function $z \to \chi(z,t)$ has a compact support contained in Π.

By virtue of Stokes' theorem, whatever $g \in \mathcal{C}^{\infty}(\Pi\times\Pi_0'\backslash \mathrm{S};\Lambda_*^{\mathrm{m},\mathrm{m}-1})$, $\bar{\partial}_z g = 0$, and $h \in \mathcal{O}(\mathbb{C}^{\mathrm{m}})$, the integral in z–space, depending on $t \in \Pi_0'$,

$$I(t,g,h) = \int h(z)\bar{\partial}_z\chi(z,t)\wedge g,$$

does not depend on the choice of χ [submitted to the requirements (IV.1.19) and (IV.1.20)]: it only depends on the cohomology class of g, $[g] \in \mathfrak{H}^{\mathrm{m},\mathrm{m}-1}(\Pi\times\Pi_0'\backslash \mathrm{S})$. Differentiating with respect to t under the integral sign shows that $h \to I(t,g,h)$ defines an analytic functional $\mu_g(t)$ in Π which depends smoothly on $t \in \Pi_0'$. This allows us to verify (IV.1.12) with $\mu = \mu_g$ and an arbitrary open neighborhood $\mathcal{U}$ of S in $\Pi\times\Pi_0'$. In other words, we have defined a linear map

(IV.1.21) $$\mathfrak{H}^{\mathrm{m},\mathrm{m}-1}(\Pi\times\Pi_0'\backslash \mathrm{S}) \ni [g] \to \mu_g \in \mathfrak{G}(\mathrm{S}).$$

Case $\mathrm{m} = 1$

In the case $\mathrm{m} = 1$ we must consider a class $[g] \in \mathfrak{H}^{1,0}(\Pi\times\Pi_0'\backslash \mathrm{S})$. But by (IV.1. 17) this simply means a one–form $\mathrm{g} = g_0 dz$ with $g_0 \in$

$\mathfrak{F}(\Pi\times\Pi_0'\backslash S)$. [Here Π is the square $\{ (x,y) \in \mathbb{R}^2;\ |x| < a,\ |y| < b \}$.]

THEOREM IV.1.1.— *Suppose* m = 1. *The map* (IV.1.21) *induces naturally a linear bijection*

(IV.1.22) $$\mathfrak{H}^{1,0}(\Pi\times\Pi_0'\backslash S)/\mathfrak{H}^{1,0}(\Pi\times\Pi_0') \ni [\dot{g}] \to \mu_g \in \mathfrak{G}(S).$$

Proof: I. *Injectivity of* (IV.1.22). Define

$$G(z,t) = (2\imath\pi)^{-1}\int (z-z')^{-1}(\partial\chi/\partial\bar{z})(z',t)g(z',t)\mathrm{d}\bar{z}'\wedge\mathrm{d}z'.$$

We have $G \in C^\infty(\mathbb{C}\times\Pi_0')$ and $\partial G/\partial\bar{z} = g\partial\chi/\partial\bar{z}$. If $\mu_g(t) \equiv 0$ then, for any fixed $t \in \Pi_0'$, $G(z,t) \equiv 0$ outside some compact subset $K(t)$ of Π, and therefore the function

$$\tilde{g}(z,t) = [1-\chi(z,t)]g(z,t) + G(z,t)$$

is equal to $g(z,t)$ when $z \in \Pi\backslash K(t)$. But applying $\partial/\partial\bar{z}$ to both sides of the last equation shows that $\tilde{g}$ is a C^∞ function of t in Π_0' valued in $\mathcal{O}(\Pi)$.

II. *Surjectivity of* (IV.1.22). Let $\mu \in \mathfrak{G}(S)$ and for each $t \in \Pi_0'$ consider the Cauchy transform of $\mu(t) \in \mathcal{O}'(S_t)$ [see (I.2.12)]:

$$\Gamma\mu(z,t) = (2\imath\pi)^{-1}\langle\mu_w(t),(z-w)^{-1}\rangle,\ w \notin S_t.$$

By looking at $\Gamma\mu$ in product sets $A\times B \subset \mathbb{C}\times\Pi_0'\backslash S$, for any possible choice of the open sets $A \subset \mathbb{C}$, $B \subset \Pi_0'$, it is readily checked that $\Gamma\mu \in \mathfrak{F}(\mathbb{C}\times\Pi_0'\backslash S)$. Set $g = -\Gamma\mu(z,t)\mathrm{d}z \in \mathfrak{H}^{0,1}(\mathbb{C}\times\Pi_0'\backslash S)$. With χ satisfying (IV.1.19) and (IV.1.20) we consider, for $h \in \mathcal{O}(\mathbb{C})$ and $t \in \Pi_0'$, the integral

$$\int h(z)\bar{\partial}_z\chi(z,t)\wedge g = -\int h(z)\Gamma\mu(z,t)(\partial\chi/\partial\bar{z})(z,t)\mathrm{d}\bar{z}\wedge\mathrm{d}z =$$

$$\langle \mu_w(t),(2\imath\pi)^{-1}\int (w-z)^{-1}h(z)(\partial\chi/\partial\bar{z})(z,t)\mathrm{d}\bar{z}\wedge\mathrm{d}z \rangle.$$

If w varies in a neighborhood of S_t in which $\chi \equiv 1$ we have

$$h(w) = (2\imath\pi)^{-1}\int (w-z)^{-1}h(z)(\partial\chi/\partial\bar{z})(z,t)d\bar{z}\wedge dz$$

by the inhomogeneous Cauchy formula. We conclude that $\mu = \mu_g$. □

REMARK IV.1.3.— Given any open subset $\mathcal{U}$ of $\mathbb{C}$, the map $g \to g dz$ is a bijection of $\mathfrak{F}(\mathcal{U})$ onto $\mathfrak{F}(\mathcal{U};\Lambda_*^{1,0}) \cong \mathfrak{H}^{1,0}(\mathcal{U})$. We transfer onto $\mathfrak{H}^{1,0}(\Pi\times\Pi_0'\backslash S)$ the topology of $\mathfrak{F}(\Pi\times\Pi_0'\backslash S)$ and onto $\mathfrak{H}^{1,0}(\Pi\times\Pi_0')$ that of $\mathfrak{F}(\Pi\times\Pi_0')$; $\mathfrak{H}^{1,0}(\Pi\times\Pi_0'\backslash S)$ and $\mathfrak{H}^{1,0}(\Pi\times\Pi_0')$ become Frechet spaces; $\mathfrak{H}^{1,0}(\Pi\times\Pi_0')$ can be identified to a closed subspace of $\mathfrak{H}^{1,0}(\Pi\times\Pi_0'\backslash S)$. Thus the quotient space $\mathfrak{H}^{1,0}(\Pi\times\Pi_0'\backslash S)/\mathfrak{H}^{1,0}(\Pi\times\Pi_0')$ carries a natural Fréchet space topology. It is a direct consequence of the closed graph theorem that the bijection in Theorem IV.1.1 is a homeomorphism. □

Case m $\geq$ 2

The multidimensional version of Theorem IV.1.1 can be regarded as a version with parameters (namely, the "variables" $t \in \Pi_0'$) of Theorem I.4.2. It is convenient to assume that all derivatives of the map $\Phi\colon \Pi_0\times\Pi_0' \to \Pi_1$ are bounded. This can be achieved by decreasing the radii a_i and a_k'.

The argument will closely duplicate that in Section I.3 and be— cause of this, we shall skip large portions of it. The main difference is that the parameter t must be inserted everywhere. Thus we define the function

$$\rho(w,t) = |v-\Phi(u,t)|^2$$

where $w = u+\imath v \in \Pi$, $t \in \Pi_0'$. For each $(w,t) \in \Pi\times\Pi_0'$, we consider the quadratic form in z–space,

$$Q(z,w,t) = \sum_{j,k=1}^{m} \frac{\partial^2\rho}{\partial w_j\partial\bar{w}_k}(w,t)z_j\bar{z}_k.$$

Taylor's expansion yields

$$\rho(w,t) = \rho(z,t) + 2\,\mathcal{R}e\,\lambda(w-z,w,t) - \mathcal{Q}(w-z,w,t) + R(w-z,w,t),$$

where

$$\lambda(z,w,t) = \sum_{j=1}^{m} \lambda_j(z,w,t) z_j,$$

$$\lambda_j(z,w,t) = \frac{\partial \rho}{\partial w_j}(w,t) - \tfrac{1}{2} \sum_{j,k=1}^{m} \frac{\partial^2 \rho}{\partial w_j \partial w_k}(w,t) z_k \ (j = 1,\dots,m);$$

and for some $C > 0$, $|R(w-z,w,t)| \leq C|w-z|^3$, $\forall\ w,\ z \in \Pi$, $t \in \Pi'_0$. Note that $\lambda(z,w,t)$ is a holomorphic polynomial of degree two with respect to z whose coefficients are C^∞ functions of (w,t).

We have $\rho(w,0) = |v|^2 + O(|u|^3+|v|^3)$. It follows that the quadratic form $z \to \mathcal{Q}(z,0,0)$ is definite positive. After contracting Π_0, Π_1 and Π'_0 about 0 we may assume that, for some $c > 0$, and all w, $z \in \Pi$ ($= \Pi_0+\imath\Pi_1$), $t \in \Pi'_0$,

(IV.1.23) $$2\,\mathcal{R}e\,\lambda(w-z,w,t) \geq \rho(w,t) - \rho(z,t) + c|w-z|^2.$$

Writing $\lambda = \lambda(z,w,t)$, $\lambda_j = \lambda_j(z,w,t)$ we introduce the kernel

(IV.1.24) $$G_\lambda(z,w,t) =$$

$$(m-1)!(2\imath\pi\lambda)^{-m} \sum_{j=1}^{m} (-1)^{j-1} \lambda_j \bar{\partial}_w \lambda_1 \wedge \cdots \wedge \widehat{\bar{\partial}_w \lambda_j} \wedge \cdots \wedge \bar{\partial}_w \lambda_m \wedge dw_1 \wedge \cdots \wedge dw_m.$$

We are going to make use of the kernel $G_\lambda(w-z,w,t)$ obtained by substituting $w-z$ for z in the coefficients of $G_\lambda(z,w,t)$. Since the λ_j are affine functions of z, $G_\lambda(w-z,w,t)$ is the pullback of $G_\lambda(z,w,t)$ under the map $(z,w) \to (w-z,w)$ (of $\mathbb{C}^m \times \Pi$ into itself). Note that $G_\lambda(w-z,w,t)$ is a current of bidegree (m,m—1) in the open biball Π in w—space, depending on $z \in \mathbb{C}^m$ and on $t \in \Pi_0'$. By virtue of (IV.1.23) the coefficients of $G_\lambda(w-z,w,t)$ are C^∞ functions of (z,w,t) when $z \in \Sigma_0$, $w \in \Pi$, $t \in \Pi_0'$, $w \neq z$, holomorphic with respect to z.

Let S be the closed subset of Σ_0 under consideration [S satisfies (IV.1.3) and (IV.1.4)]. Consider two open subsets $\mathcal{U}$ and $\mathcal{U}'$ of $\Pi \times \Pi_0'$ such that $S_t \subset \mathcal{U}_t' \subset\subset \mathcal{U}_t \subset\subset \Pi$ for each $t \in \Pi_0'$. Select a cutoff function $\psi \in C^\infty(\Pi \times \Pi_0')$, $0 \leq \psi \leq 1$, $\psi = 0$ in $\mathcal{U}'$, $\psi = 1$ in $(\Pi \times \Pi_0') \backslash \mathcal{U}$. We define

$$k_j(z,w,t) = \psi(w,t)\lambda_j(w-z,w,t) + [1-\psi(w,t)](\bar{w}_j - \bar{z}_j),\ j = 1,\dots,m.$$

Then

$$k(z,w,t) = \sum_{j=1}^{m} k_j(z,w,t)(w_j - z_j) = \psi(w,t)\lambda(w-z,w,t) + [1-\psi(w,t)]\,|w-z|^2.$$

If we take into account (IV.1.23) we see that $k(z,w,t) \neq 0$ when $z \in \Sigma_0$, $w \in \Pi$, $t \in \Pi_0'$ and $z \neq w$. With this agreed upon we define the kernel N_k according to (I.3.9) or (I.3.10). Keep in mind, however, that in all this d continuous to denote the exterior derivative in (z,w)—space and does not include any differentiation with respect to t.

PROPOSITION IV.1.1.— *Let* S *be a closed subset of* Σ_0 *that satisfies* (IV.1.3) *and* (IV.1.4), *and let* $\mathcal{U}$ *and* $\mathcal{U}'$ *be open sets in* $\mathbb{C}^m \times \mathbb{R}^n$ *such that* $S \subset \mathcal{U}' \subset \mathcal{U} \subset \Pi \times \Pi_0'$, *and that* $S_t \neq \emptyset$ *for all* $t \in \Pi_0'$. *Suppose furthermore that* $S_t \subset\subset \mathcal{U}_t' \subset\subset \mathcal{U}_t \subset\subset \Pi$ *for each* $t \in \Pi_0'$. *Then there is an open subset* $\mathcal{U}''$ *of* $\Pi \times \Pi_0'$, *with* $S_t \subset\subset \mathcal{U}_t''$ *for each* $t \in \Pi_0'$, *such that the kernel* N_k *satisfies the following conditions:*

(IV.1.25) $|N_k(z,w,t)| \leq const.|w-z|^{1-2m}$, $\forall\ (z,t) \in \mathcal{U}''$, $w \in \Pi$;

(IV.1.26)
$$dN_k(z,w,t) = (2\imath)^{-m}\delta(w-z)d(\bar{w}-\bar{z})\wedge d(w-z) \quad \textit{if } (z,t) \in \mathcal{U}'',\ w \in \Pi;$$

(IV.1.27) $N_k(z,w,t)\wedge dz = G_\lambda(w-z,w,t)\wedge dz$ *if* $(z,t) \in \mathcal{U}''$, $(w,t) \notin \mathcal{U}$,

and furthermore, such that, for each $t \in \Pi_0'$, *the coefficients of* $G_\lambda(w-z,w,t)$ *are holomorphic functions of* z *in* $\mathcal{U}_t''$ *when* $t \notin \mathcal{U}_t$.

The proof of Proposition IV.1.1 is identical to that of Proposition I.3.1 except for the presence of the parameter t.

Let $S \subset \mathcal{U}'' \subset \mathcal{U} \subset \Pi\times\Pi_0'$ and the functions $k_j \in \mathcal{C}^\infty(\Pi\times\Pi\times\Pi_0')$ be as in Proposition IV.1.1. For $1 \leq q \leq m$ we define a bounded linear operator

$$\mathcal{T}^q\colon \mathcal{C}_c^\infty(\mathcal{U}'';\Lambda_*^{m,q}) \to \mathcal{C}^\infty(\Pi\times\Pi_0';\Lambda_*^{m,q-1})$$

by the formula

(IV.1.28) $$(\mathcal{T}^q\varphi)(w,t) = (-1)^{m-q}\int \varphi(z,t)\wedge N_k(z,w,t),$$

with the integration carried out in z-space (only the component of degree 2m with respect to z of the current $\varphi(z,t)\wedge N_k(z,w,t)$ contributes to the integral). Thanks to (IV.1.25) we know that $\varphi(z,t)\wedge N_k(z,w,t)$ is of class L^1 with respect to z in $\mathcal{U}''$.

PROPOSITION IV.1.2.— *Assume* $1 \leq q \leq m$ *and* $\varphi \in \mathcal{C}_c^\infty(\mathcal{U}'';\Lambda_*^{m,q})$; *then*

(IV.1.29) $$\bar{\partial}\varphi = 0 \Rightarrow \bar{\partial}\mathcal{T}^{q}\varphi = \varphi \quad in\ \Pi\times\Pi_0'.$$

Moreover

(IV.1.30) $$\mathcal{T}^{q}\varphi = 0 \quad in\ (\Pi\times\Pi_0')\backslash\mathcal{U}$$

if either $q \leq m-1$ *or, when* $q = m$, *if*

(IV.1.31) $$\int f(z)\varphi(z,t) = 0, \ \ \forall\, f \in \mathcal{O}(\mathcal{U}''),\ t \in \Pi_0'.$$

Same proof as Proposition I.3.2.

As before let S be a closed subset of Σ_0 that satisfies (IV.1.3) and (IV.1.4). We define the *Cauchy–Fantappiè transform relative to* $\mathcal{X}$ of an arbitrary analytic functional $\mu \in \mathfrak{G}(S)$:

(IV.1.32) $$\Gamma_{\mathcal{X}}\mu(w,t) = <\mu_z(t), G_\lambda(w-z,w,t)>.$$

When m = 1, the Cauchy—Fantappiè transform (relative to any C^∞ curve passing through the origin) coincides with $\Gamma\mu(w,t)dw$, where $\Gamma\mu$ is the Cauchy transform [see (I.2.12)]. We have $\Gamma_{\mathcal{X}}\mu \in C^\infty((\Pi\times\Pi_0')\backslash S;\Lambda_*^{m,m-1})$ and $\bar{\partial}_z\Gamma_{\mathcal{X}}\mu = 0$ in $(\Pi\times\Pi_0')\backslash S$. Thus $\Gamma_{\mathcal{X}}\mu$ defines a cohomology class $[\Gamma_{\mathcal{X}}\mu] \in \mathfrak{H}^{m,m-1}((\Pi\times\Pi_0')\backslash S)$.

Let us select a cutoff function $\chi \in C_c^\infty(\Pi\times\Pi_0')$ satisfying (IV.1.19) and (IV.1.20).

THEOREM IV.1.2.— *Let* S, μ, χ *be as above. Then, for all entire holomor– phic functions* h,

(IV.1.33) $$<\mu(t),h> = -\int h(w)\bar{\partial}\chi(w,t)\wedge\Gamma_{\chi}\mu(w,t).$$

Same proof as Theorem I.3.1. Theorems I.4.1 and I.4.2 also have their counterparts in the present set—up. It suffices to state

THEOREM IV.1.3.— *Suppose* m $\geq$ 2. *If the polyhedrons* Π_0, Π_1 *and* Π_0' *are sufficiently small the map* (IV.1.21) *is a linear bijection* $\mathfrak{H}^{m,m-1}(\Pi\times\Pi_0'\backslash S) \to \mathfrak{G}(S)$.

IV.2 HYPERFUNCTIONS DEPENDING SMOOTHLY ON t

We are going to adapt the constructions in sections I.4 and I.5 to the coarse embedding of $\mathcal{M}'$ into $\mathbb{C}^m \times \mathbb{R}^n$, allowing for dependence on the parameter $t \in \Pi_0'$. This leads to a definition of the hyperfunctions on the "generic" submanifold Σ of $\mathbb{C}^m \times \mathbb{R}^n$ that depend smoothly on $t \in \Pi_0'$. The class of these hyperfunctions is *not invariant*, in the sense that it is *not* preserved under changes of the hypo—analytic structure of Σ, ie., diffeo—morphisms of Σ onto itself (near one of its points) induced by automor—phisms of the *elliptic structure* of $\mathbb{C}^m \times \mathbb{R}^n$.

We recall that $\Sigma_0 \subset \Sigma$ is the image of $\Pi_0 \times \Pi_0'$ under the map $(x,t) \to (Z(x,t),t)$; $\Sigma_0 = \Sigma \cap (\Pi \times \Pi_0')$. In this section we shall deal with sets $F \subset \Sigma_0$ of a special type:

(IV.2.1) *There is a compact subset* K *of* Π_0 *such that* F *is the image of* $K \times \Pi_0'$ *under the map* $(x,t) \to (Z(x,t),t)$.

Clearly (IV.2.1) $\Rightarrow$ (IV.1.3) & (IV.1.4).

We need the analogue of Lemma I.4.1:

LEMMA IV.2.1.— *Let* F *be a closed subset of* Σ_0 *satisfying* (IV.2.1) *and let* $\Delta_{x,y} = \sum_{j=1}^{m} 4\partial^2/\partial z_j \partial \bar{z}_j$ *denote the Laplacian in* $\mathbb{C}^m \cong \mathbb{R}^{2m}$. *Then*

$$\Delta_{x,y} C^\infty(\Pi \times \Pi_0' \backslash F) = C^\infty(\Pi \times \Pi_0' \backslash F).$$

Proof: Let $\mathcal{K}$ be a compact subset of $\Pi \times \Pi_0' \backslash F$, and let $\varphi \in C_c^\infty(\Pi \times \Pi_0' \backslash F)$ be such that supp $\Delta_{x,y}\varphi \subset \mathcal{K}$. Then, for each $t \in \Pi_0'$, the function $(x,y) \to$

$\varphi(x,y,t)$ is harmonic in $\Pi\backslash\mathcal{K}_t$ and vanishes identically in each connected component of $\Pi\backslash\mathcal{K}_t$ whose closure is not a compact subset of $\Pi\backslash F_t$. Call $\hat{\mathcal{K}}_t$ the union of $\mathcal{K}_t$ and of all the connected components of $\Pi\backslash\mathcal{K}_t$ whose closures are compact subsets of $\Pi\backslash F_t$. We have just seen that supp φ is contained in the closure of the union of all those sets $\hat{\mathcal{K}}_t\times\{t\}$, $t \in \Pi_0'$. Call $\hat{\mathcal{K}}$ the union in question; it will suffice to prove that the closure of $\hat{\mathcal{K}}$ is a compact subset of $\Pi\times\Pi_0'\backslash F$. For then a theorem in [MALGRANGE, 1955] implies the sought conclusion. Indeed, $\hat{\mathcal{K}} \subset\subset \Pi\times\Pi_0'\backslash F$ means that $\Pi\times\Pi_0'\backslash F$ is $\Delta_{x,y}$—convex (see [HÖRMANDER, 1983], Section 10.6).

The coordinate projection $\wp$: $(z,t) \to t$ maps $\hat{\mathcal{K}}$ onto a compact set $K'' \subset K' = \wp(\mathcal{K}) \subset\subset \Pi_0'$ since $\hat{\mathcal{K}}_t = \emptyset$ if $\mathcal{K}_t = \emptyset$. For fixed $t \in K''$, any segment in $\mathbb{C}^m$ joining a point z_0 of $\hat{\mathcal{K}}_t$ to a point of the boundary of $\Pi\backslash F_t$ must contain a point of $\mathcal{K}_t$, hence $\text{dist}(\hat{\mathcal{K}}_t,\partial(\Pi\backslash F_t)) = \text{dist}(\mathcal{K}_t,\partial(\Pi\backslash F_t))$. Clearly the function $t \to 1/\text{dist}(\mathcal{K}_t,\partial(\Pi\backslash F_t))$ is locally, hence globally, bounded in K'. Let (z_ν,t_ν) be a sequence of points in $\hat{\mathcal{K}}$ converging to (z_0,t_0). Suppose $(z_0,t_0) \notin \Pi\times\Pi_0'\backslash F$. Of course $t_0 \in K'$ and since $\text{dist}(z_\nu,\partial\Pi) \geq c$, with $c > 0$ independent of ν, necessarily $z_0 = Z(x_0,t_0) \in F_{t_0}$, ie. $x_0 \in K$ [see (IV.2.1)]. But then $|z_\nu - Z(x_0,t_\nu)| \to 0$, which contradicts the fact that $\text{dist}(\hat{\mathcal{K}}_t,F_t)$ is bounded away from zero. □

We are now in a position to repeat the proof of Theorem I.4.3, to obtain:

THEOREM IV.2.1.— *Let* F *be a closed subset of* Σ_0 *satisfying* (IV.2.1). *Then*

$$\mathfrak{H}^{m,m}(\Pi\times\Pi'_0\backslash F) = 0.$$

Next we consider a pair of subsets F_1 and F_2 of Σ_0, both satisfying (IV.2.1). It is clear that both $F_1\cup F_2$ and $F_1\cap F_2$ satisfy (IV.2.1).

PROPOSITION IV.2.1.— *Let* F_1 *and* F_2 *be two closed subsets of* Σ_0 *satisfying* (IV.2.1).

(IV.2.2) *If* $\mu \in \mathfrak{G}(F_1)\cap\mathfrak{G}(F_2)$ *then* $\mu \in \mathfrak{G}(F_1\cap F_2)$;

(IV.2.3) *If* $\mu \in \mathfrak{G}(F_1\cup F_2)$ *then there exist* $\mu_j \in \mathfrak{G}(F_j)$, $j = 1,2$, *such that* $\mu = \mu_1+\mu_2$.

The proof of (IV.2.2) is essentially the same as that of Proposition I.4.1; that of (IV.2.3) is the same as that of Proposition I.4.2. It is based on Theorems IV.1.3 and IV.2.1. As before the only difference with the argument in section I.4 is the presence of the parameter t.

DEFINITION IV.2.1.— *Let* U *be an open subset of* Σ_0, $\overline{U}$ *its closure and* $\dot{U}$ *its boundary in* Σ_0. *Suppose that* $\overline{U}$ *satisfies* (IV.2.1). *We define* $\mathcal{B}(U,\mathcal{C}^\infty_t) = \mathfrak{G}(\overline{U})/\mathfrak{G}(\dot{U})$.

We shall loosely refer to any element of $\mathcal{B}(U,\mathcal{C}^\infty_t)$ as a hyperfunction in U which depends smoothly on $t \in \Pi'_0$.

Let $V \subset U$ be two open subsets of Σ_0 whose closures in Σ_0, $\overline{U}$ and $\overline{V}$, satisfy (IV.2.1); let $\dot{U}$ and $\dot{V}$ be their boundaries in Σ_0. There is a natural restriction map $\mathcal{B}(U,\mathcal{C}^\infty_t) \to \mathcal{B}(V,\mathcal{C}^\infty_t)$. By (IV.2.3) where $F_1 = \overline{U}\backslash V$ and $F_2 = \overline{V}$ any $\mu \in \mathfrak{G}(\overline{U})$ can be decomposed as a sum $\mu = \mu_1 + \mu_2$. We assign to the class of μ mod $\mathfrak{G}(\mathring{\mathrm{U}})$ the class of μ_2 mod $\mathfrak{G}(\mathring{\mathrm{V}})$. If $\mu \in \mathfrak{G}(\mathring{\mathrm{U}})$ then $\mu_2 = \mu - \mu_1 \in \mathfrak{G}(\overline{\mathrm{U}}\backslash \mathrm{V}) \cap \mathfrak{G}(\overline{\mathrm{V}}) \subset \mathfrak{G}(\mathring{\mathrm{V}})$ by (IV.2.2). Thus the map $\mathfrak{G}(\overline{\mathrm{U}}) \ni \mu \to \mu_2 \in \mathfrak{G}(\overline{\mathrm{V}})$ induces a map $\mathcal{B}(\mathrm{U},\mathcal{C}^{\infty}_t) \to \mathcal{B}(\mathrm{V},\mathcal{C}^{\infty}_t)$.

The preceding construction was carried out under the hypothesis that U is the image of a set $\mathrm{U}_0 \times \Pi'_0 \subset \Pi_0 \times \Pi'_0$ under the map $(x,t) \to (Z(x,t),t)$, with $\overline{\mathrm{U}}_0 \subset \Pi'_0$. Of course, exactly the same construction could have been carried out after replacement of Π'_0 by any open polyhedron $\Pi''_0 \subset \Pi'_0$. Call U' the corresponding subset of Σ. Restricting the variation of t from Π'_0 to Π''_0 defines the restriction mapping $\mathcal{B}(\mathrm{U},\mathcal{C}^{\infty}_t) \to \mathcal{B}(\mathrm{U}',\mathcal{C}^{\infty}_t)$. We may compose the two kinds of restriction mappings just introduced: let V_0 be an open subset of U_0 and Π''_0 one of Π'_0, and call now V the image of $\mathrm{V}_0 \times \Pi''_0$ under the map $(x,t) \to (Z(x,t),t)$. Following up the restriction in t–space $\mathcal{B}(\mathrm{U},\mathcal{C}^{\infty}_t) \to \mathcal{B}(\mathrm{U}',\mathcal{C}^{\infty}_t)$ with the restriction "in z–space" $\mathcal{B}(\mathrm{U}',\mathcal{C}^{\infty}_t) \to \mathcal{B}(\mathrm{V},\mathcal{C}^{\infty}_t)$ defines the natural restriction map $\mathcal{B}(\mathrm{U},\mathcal{C}^{\infty}_t) \to \mathcal{B}(\mathrm{V},\mathcal{C}^{\infty}_t)$.

In order to duplicate the argument in section I.5 we need the analogue of Lemma I.5.1. This requires that we take a closer look at the natural topology of $\mathfrak{G}(\mathrm{F})$. Throughout F satisfies (IV.2.1).

Let $\mathcal{U} \supset \mathrm{F}$ be an open subset of $\Pi \times \Pi'_0$, K' be a compact subset of Π'_0 and N an integer ≥ 0. Keep in mind that $\mathfrak{G}(\mathrm{F})$ is made into a Fréchet space by the seminorms $\mathfrak{p}_{\mathcal{U},\mathrm{K}',\mathrm{N}}$ introduced right after Definition IV.1.1. Let us go back to the map (IV.1.21) for $\mathrm{S} = \mathrm{F}$; recall that, if $g \in \mathcal{C}^{\infty}(\Pi \times \Pi'_0 \backslash \mathrm{F};\Lambda^{m,m-1}_{*})$, such that $\bar{\partial}_z g = 0$, is a representative of $[g] \in \mathfrak{H}^{m,m-1}(\Pi \times \Pi'_0 \backslash \mathrm{F})$, then

(IV.2.4) $$<\mu_g(t),h> = \int h(z)\bar{\partial}_z\chi(z,t)\wedge g(z,t)$$

where χ satisfies (IV.1.19) (with S = F) and (IV.1.20). We shall limit ourselves to the case m $\geq$ 2 and take advantage of Theorem IV.1.3; when m = 1 the argument is essentially the same, but based on Theorem IV.1.1.

The injectivity of the map (IV.1.21) tells us that, if $g \in \mathcal{C}^{\infty}(\Pi\times\Pi_0'\backslash F;\Lambda_*^{m,m-1})$, $\bar{\partial}_z g = 0$, in order that there be $u \in \mathcal{C}^{\infty}(\Pi\times\Pi_0'\backslash F;\Lambda_*^{m,m-2})$ such that $\bar{\partial}_z u = g$ in $\Pi\times\Pi_0'\backslash F$, it is necessary and sufficient that $\mu_g = 0$, ie., the integral at the right in (IV.2.4) vanish for all $h \in \mathcal{O}(\mathbb{C}^m)$. This proves that both the space of cocycles, and the subspace of coboundaries in $\mathcal{C}^{\infty}(\Pi\times\Pi_0'\backslash F;\Lambda_*^{m,m-1})$ are closed. It follows that the quotient topology on $\mathfrak{H}^{m,m-1}(\Pi\times\Pi_0'\backslash F)$ turns the latter into a Fréchet space. Now the map which to $g \in \mathcal{C}^{\infty}(\Pi\times\Pi_0'\backslash F;\Lambda_*^{m,m-1})$, $\bar{\partial}_z g = 0$, assigns $\mu_g \in \mathfrak{G}(F)$ is linear and continuous; its kernel is the space of coboundaries. It follows that the map (IV.1.21) is continuous. Theorem IV.1.2 provides us with an inverse: to $\mu \in \mathfrak{G}(F)$ it assigns the cohomology class of the Cau—chy—Fantappiè transform of μ, $[\Gamma_{\chi}\mu]$. The inverse of the map (IV.1.21) has perforce a closed graph, hence is continuous. We conclude that *the map* (IV.1.21) *is a Fréchet space isomorphism of* $\mathfrak{H}^{m,m-1}(\Pi\times\Pi_0'\backslash F)$ *onto* $\mathfrak{G}(F)$.

If $\mathcal{A}$ is an open subset of $\mathbb{C}^m\times\mathbb{R}^n$ let us denote by $\mathcal{E}'(\mathcal{A};\Lambda^{p,q,r})$ the space of currents in $\mathcal{A}$ of the form

$$u = \sum_{|I|=p}\ \sum_{|J|=q}\ \sum_{|K|=r} u_{IJK}dz_I\wedge d\bar{z}_J\wedge dt_K$$

whose coefficients $u_{IJK} \in \mathcal{E}'(\mathcal{A})$, ie., are compactly supported distributions in $\mathcal{A}$. There is a natural isomorphism of $\mathcal{E}'(\mathcal{A};\Lambda^{p,q,n})$ onto the dual of

$\mathcal{C}^\infty(\Pi\times\Pi'_0\backslash F;\Lambda^{m-p,m-q})$ (here "dual" always refers to the space of *continuous* linear functionals). We shall denote by $\mathcal{H}_c^{p,q,r}(\mathcal{A})$ the quotient space

$$\{\ u \in \mathcal{E}'(\mathcal{A};\Lambda^{p,q,r});\ \bar\partial_z u = 0\ \}/\{\ \bar\partial_z v\ ;\ v \in \mathcal{E}'(\mathcal{A};\Lambda^{p,q-1,r})\ \}.$$

PROPOSITION IV.2.2.— *The natural bracket pairing $\mathcal{C}^\infty$ test-forms and compactly supported currents,*

$$\mathcal{E}'(\Pi\times\Pi'_0\backslash F;\Lambda^{0,1,n}) \ni u \to \left\{ \mathcal{C}^\infty(\Pi\times\Pi'_0\backslash F;\Lambda_*^{m,m-1}) \ni g \to \int g\wedge u \in \mathbb{C} \right\},$$

induces a bijection of $\mathcal{H}_c^{0,1,n}(\Pi\times\Pi'_0\backslash F)$ onto the dual of $\mathcal{H}^{m,m-1}(\Pi\times\Pi'_0\backslash F)$.

Proof: Consider the sequence of linear maps,

$$\mathcal{C}^\infty(\Pi\times\Pi'_0\backslash F;\Lambda_*^{m,m-2}) \xrightarrow{\bar\partial_z} \mathcal{C}^\infty(\Pi\times\Pi'_0\backslash F;\Lambda_*^{m,m-1}) \xrightarrow{\bar\partial_z} \mathcal{C}^\infty(\Pi\times\Pi'_0\backslash F;\Lambda_*^{m,m}).$$

Theorem IV.2.1 states that the second map $\bar\partial_z$ is surjective. The range of the first map $\bar\partial_z$ is closed, as pointed out above. It suffices then to apply Serre's lemma (see Appendix to section I.6). □

The isomorphism (IV.1.21) allows us to identify $\mathcal{H}_c^{0,1,n}(\Pi\times\Pi'_0\backslash F)$ to the dual of $\mathfrak{G}(F)$. This identification can be extended to the strong dual topologies.

Now consider two closed subsets F_j ($j = 1,2$) of Σ_0, both satisfying (IV.2.1). If $F_1 \subset F_2$ we have a commutative diagram

$$\text{(IV.2.5)}\qquad \begin{array}{ccc} \mathfrak{H}^{m,m-1}((\Pi\times\Pi'_0)\backslash F_1) & \longrightarrow & \mathfrak{H}^{m,m-1}((\Pi\times\Pi'_0)\backslash F_2) \\ \downarrow & & \downarrow \\ \mathfrak{G}(F_1) & \longrightarrow & \mathfrak{G}(F_2)\,. \end{array}$$

The upper horizontal arrow is the restriction map; the lower one is the natural inclusion map. The vertical arrows stand for the map (IV.1.21) with $S = F_1$ and $S = F_2$ respectively. By transposing Diagram (IV.2.5) we can identify the transpose of the inclusion (lower horizontal arrow) to that of the upper one, namely to the "inclusion map"

$$\mathcal{H}_c^{0,1,n}(\Pi\times\Pi_0'\backslash F_2) \to \mathcal{H}_c^{0,1,n}(\Pi\times\Pi_0'\backslash F_1)$$

induced by the natural injection

$$\mathcal{E}'(\Pi\times\Pi_0'\backslash F_2;\Lambda^{0,1,n}) \to \mathcal{E}'(\Pi\times\Pi_0'\backslash F_1;\Lambda^{0,1,n}).$$

PROPOSITION IV.2.3.— *Suppose both* F_1 *and* F_2 *have Property* (IV.2.1) *and* $F_1 \subset F_2$. *Let* $K_j \subset \Pi_0$ *be the image of* F_j *under the coordinate projection* $(z,t) \to x$ $(j = 1,2)$. *If the following is true,*

(IV.2.6) *every connected component of* K_2 *intersects* K_1,

then the "inclusion" map $\mathfrak{G}(F_1) \to \mathfrak{G}(F_2)$ *has a dense image.*

Proof: By the preceding considerations it suffices to prove the injectivity of the "inclusion" map $\mathcal{H}_c^{0,1,n}(\Pi\times\Pi_0'\backslash F_2) \to \mathcal{H}_c^{0,1,n}(\Pi\times\Pi_0'\backslash F_1)$. Suppose $f \in \mathcal{E}'(\Pi\times\Pi_0'\backslash F_2;\Lambda^{0,1,n})$ is such that $f = \bar{\partial}_z u$ in $\Pi\times\Pi_0'\backslash F_1$, for some $u \in \mathcal{E}'(\Pi\times\Pi_0'\backslash F_1;\Lambda^{0,0,n})$. We shall make use of the standard mollifiers: select $\rho \in C_c^\infty(\mathbb{C}^m\times\mathbb{R}^n)$, $\rho \geq 0$ everywhere, $\int \rho(z,t)dxdydt = 1$, and set $\rho_\epsilon(z,t) = \epsilon^{-2m-n}\rho(z/\epsilon,t/\epsilon)$. Then define $f_\epsilon = \rho_\epsilon\star f$, $u_\epsilon = \rho_\epsilon\star u$, where the convolution $\rho_\epsilon\star$ acts coefficientwise. We shall take $0 < \epsilon < \epsilon_0$, with ϵ_0 small enough to ensure that the support of f_ϵ (resp., u_ϵ) is carried by a compact subset $\mathcal{K}_2$ (resp., $\mathcal{K}_1$) of $\Pi\times\Pi_0'\backslash F_2$ (resp., $\Pi\times\Pi_0'\backslash F_1$). We have $\bar{\partial}_z u_\epsilon = f_\epsilon$.

But now

$$f_\epsilon \in \mathcal{C}_c^\infty(\Pi\times\Pi_0'\backslash F_2;\Lambda^{0,1,n}),\ u_\epsilon \in \mathcal{C}_c^\infty(\Pi\times\Pi_0'\backslash F_1;\Lambda^{0,0,n}).$$

For fixed $\epsilon < \epsilon_0$ and $t \in \Pi_0'$, $u_\epsilon(\cdot,t)$ is holomorphic in $\mathbb{C}^m\backslash(\mathcal{K}_2)_t \supset (F_2)_t$ and $u_\epsilon(z,t) \equiv 0$ for all $z \in \mathbb{C}^m\backslash(\mathcal{K}_1)_t \supset (F_1)_t$. Hypotheses (IV.2.1) and (IV.2.6) imply that $\mathbb{C}^m\backslash(\mathcal{K}_2)_t$ contains an open neighborhood $\mathcal{U}_t$ of $(F_2)_t$ endowed with the following properties: $\mathcal{U} = \bigcup_{t\in\Pi_0'} (\mathcal{U}_t\times\{t\})$ is open in $\Pi\times\Pi_0'$ and every connected component of $\mathcal{U}_t$ intersects $(F_1)_t$. But then $u_\epsilon(z,t) \equiv 0$ for all $z \in \mathcal{U}_t$ and all ϵ, $0 < \epsilon < \epsilon_0$. One concludes easily from this that the supp $u_\epsilon \subset \mathcal{K}$, a compact subset of $\Pi\times\Pi_0'\backslash F_2$ independent of $\epsilon < \epsilon_0$; and by letting ϵ go to 0, that supp $u \subset \mathcal{K}$. This means that

$$f \in \bar\partial_z \mathcal{C}_c^\infty(\Pi\times\Pi_0'\backslash F_2;\Lambda^{0,0,n}),$$

ie., the cohomology class of f in $\mathfrak{H}_c^{0,0,n}(\Pi\times\Pi_0'\backslash F_2)$ vanishes. □

Proposition IV.2.3 is the analogue, when there are variables t_j, of Lemma I.5.1. It enables us to prove the analogue of Lemma I.5.2. Let U be an open subset of Σ whose closure satisfies (IV.2.1):

LEMMA IV.2.2.— *Let* U *be the union of an increasing sequence of open subsets* U_j *of* Σ_0, $j = 1,2,...$, *whose closures satisfy* (IV.2.1). *Let* K (*resp.,* K_j) *denote the image of* $\overline{U}$ (*resp.,* $\overline{U}_j$) *under the coordinate projection* $(z,t) \to x$. *Suppose that, whatever* $j = 1, 2,...$, *every connected component of* $K\backslash K_j$ *intersects the boundary of* K.

Let be given, for each $j = 1,2,...$, *a hyperfunction* $f_j \in \mathcal{B}(U_j,\mathcal{C}_t^\infty)$ *whose restriction to* U_k *is equal to* f_k *whatever* $k \le j$. *Then there is* $f \in \mathcal{B}(U,\mathcal{C}_t^\infty)$ *whose restriction to* U_j *is equal to* f_j *for every* $j = 1,2,....$

The proof is the same as that of Lemma I.5.2; here one exploits the fact that the space $\mathfrak{G}(\overline{U}\backslash U_j)$ are Fréchet spaces. In turn Lemma IV.2.2 leads to the generalization of Theorem I.5.1:

THEOREM IV.2.2.— *Let* U *be an open subset of* Σ_0 *whose closure satisfies* (IV.2.1) *and let* $\{U_\iota\}_{\iota\in I}$ *be a covering of* U *by open subsets of* Σ_0 *whose closures satisfy* (IV.2.1). *Suppose given, for each* $\iota \in I$, $f_\iota \in \mathcal{B}(U_\iota,\mathcal{C}^\infty_t)$ *such that, for each pair of indices* ι, $\kappa \in I$, *the restrictions to* $U_\iota \cap U_\kappa$ *of* f_ι *and* f_κ *are equal. Then there exists a unique element* f of $\mathcal{B}(U,\mathcal{C}^\infty_t)$ *whose restriction to* U_ι *is equal to* f_ι *for every* $\iota \in I$.

We are now in a position to define the sheaf of hyperfunctions depending smoothly on t, over Σ, $\mathscr{B}_\Sigma(\mathcal{C}^\infty_t)$. We do this by using special bases of neighborhoods U of W an arbitrary point (z_0,t_0) of Σ_0: each such neighborhood U will be the image of a set $(\mathscr{Int}\,K)\times\Pi''_0 \subset\subset \Pi_0\times\Pi'_0$ under the map $(x,t) \to (Z(x,t),t)$. Here K is a compact subset of Π_0 and

$$\Pi''_0 = \{\, t_0 \in \mathbb{R}^n;\ |t_j - t_{0j}| < a''_j,\ j = 1,\dots,n \,\} \subset \Pi'_0;$$

$x_0 \in \mathscr{Int}\,K$. The construction of $\mathcal{B}(U,\mathcal{C}^\infty_t)$ described above applies to U (applying it only requires moving the origin of t–space to t_0 and substituting Π''_0 for Π'_0).

What is *not* valid is the generalization of Lemma I.5.3. It is not true that, if U and V are two open subsets of Σ_0 whose closures satisfy (IV.2.1), and if $V \subset U$, the restriction map $\mathcal{B}(U,\mathcal{C}^\infty_t) \to \mathcal{B}(V,\mathcal{C}^\infty_t)$ is a surjection. This is due to the fact that the sheaf of germs of $\mathcal{C}^\infty$ functions in t–space is not flabby. It ensues that $\mathscr{B}_\Sigma(\mathcal{C}^\infty_t)$ is not flabby when the number n of t—variables is ≥ 1.

EXAMPLE IV.2.1.— *Boundary values of regular functions.*

To develop the counterpart of section II.1 when there are variables t_j is fairly routine. The generalization of the concept of wedge is obvious: select an open subset U of Σ, an open cone $\Gamma \subset \mathbb{R}^m\backslash\{0\}$, a number $\delta > 0$ and define [cf. (II.1.1)]

(IV.2.7)
$$\mathcal{W}_\delta(\mathrm{U},\Gamma) = \{ (z+\imath\gamma,t) \in \mathbb{C}^m\times\mathbb{R}^n;\ (z,t) \in \mathrm{U},\ \gamma \in \Gamma,\ |\gamma| < \delta \}.$$

It is convenient to limit one's attention to sets U of the following kind:

$$\text{(IV.2.8)} \qquad (z,t) \in \mathbb{C}^m\times\mathbb{R}^n;\ x \in \mathrm{U}_0,\ y = \Phi(x,t),\ t \in \Pi_0',$$

where U_0 is an open subset of Π_0.

Consider a regular function g in the wedge $\mathcal{W}_\delta(\mathrm{U},\Gamma)$. Let V_0 be an arbitrary open subset of Π_0 such that $\mathrm{V}_0 \subset\subset \mathrm{U}_0$ and whose boundary is smooth, and call V (resp. $\overline{\mathrm{V}}$) the image of $\mathrm{V}_0\times\Pi_0'$ [resp., $(\mathcal{C}\ell\mathrm{V}_0)\times\Pi_0'$] under the map $(x,t) \to (Z(x,t),t)$; $\overline{\mathrm{V}}$ is the closure of V in $\Pi\times\Pi_0'$. Define, for all $\gamma \in \Gamma$, $|\gamma| < \delta$, and all $t \in \Pi_0'$,

$$\text{(IV.2.9)} \qquad \langle\mu^\gamma_{g,\mathrm{V}}(t),h\rangle = \int_{\mathrm{V}_t+\imath\gamma} g(z,t)h(z)\mathrm{d}z,\quad h \in \mathcal{O}(\mathbb{C}^m).$$

[cf. (II.1.2)]. Obviously the difference between the situation here and in section II.1 is the presence of t; and the need to prove smoothness with respect to t. This is done by noting that

(IV.2.10)
$$\int_{V_t+\imath\gamma} g(z,t)h(z)\mathrm{d}z =$$
$$\int_{V_0} g(Z(x,t)+\imath\gamma,t)h(Z(x,t)+\imath\gamma)\det[Z_x(x,t)]\mathrm{d}x.$$

We derive at once from (IV.2.10), for any $\alpha \in \mathbb{Z}_+^n$:

$$\partial_t^\alpha\left\{\int_{V_t+\imath\gamma} g(z,t)h(z)\mathrm{d}z\right\} =$$
$$\sum_{\beta\leq\alpha}\binom{\alpha}{\beta}\int_{V_0}(\partial_t^\beta[h(Z(x,t)+\imath\gamma)])\,\partial_t^{\alpha-\beta}\left[g(Z(x,t)+\imath\gamma,t)\det[Z_x(x,t)]\right]\mathrm{d}x,$$

We get

$$\left|\partial_t^\alpha\left\{\int_{V_t+\imath\gamma} g(z,t)h(z)\mathrm{d}z\right\}\right| \leq C_\alpha(\gamma)\ \underset{V_t+\imath\gamma}{\mathrm{Max}}\ \sum_{|p|\leq|\alpha|}|\partial_z^p h|.$$

Let then $\mathcal{U}$ be an arbitrary open neighborhood of $\mathscr{C}\ell V$ in $\Pi\times\Pi_0'$ (recalling that $\Pi = \Pi_0+\imath\Pi_1$) and K' an arbitrary compact subset of Π_0'. If $\varepsilon > 0$ is sufficiently small and if $|\gamma| < \varepsilon$, then $V_t+\imath\gamma \subset \mathcal{U}_t$ for all $t \in K'$; and by the Cauchy inequalities, for a suitably large constant $C = C_\alpha(\gamma,\mathcal{U},K',N) > 0$ and all $t \in K'$,

(IV.2.11)
$$\sum_{|\alpha|\leq N}\left|\partial_t^\alpha\left\{\int_{V_t+\imath\gamma} g(z,t)h(z)\mathrm{d}z\right\}\right| \leq C\,\underset{\mathcal{U}_t}{\mathrm{Max}}\,|h|.$$

By combining the proof of Theorem II.1.1 with estimates of the kind (IV.2.11) we obtain the following result:

THEOREM IV.2.3.— *Let* U *be the set* (IV.2.8), Γ *be an open and convex cone in* $\mathbb{R}^m\backslash\{0\}$ *and* δ *a number* > 0. *Let* $V_0 \subset\subset U_0$ *be open and have a smooth boundary and set* $V = \{ (z,t) \in \Pi\times\Pi_0'; \ x \in V_0, \ y = \Phi(x,t) \}$; *denote by* $\overline{V}$ *and* $\dot{V}$ *the closure and the boundary, respectively, of* V *in* Σ_0. *If* $g \in \mathfrak{F}(\mathcal{W}_\delta(U,\Gamma))$ *and if* $\mu^\gamma_{g,V}(t)$ *is the analytic functional in* (IV.2.9) *there is* $\mu_{g,V}(t) \in \mathfrak{G}(\overline{V})$ *such that*

(IV.2.12) *to every open neighborhood* $\mathcal{N}$ *of* $\dot{V}$ *in* $\Pi\times\Pi_0'$ *and to every open set* $W_0 \subset\subset \Pi_0'$ *there is* ϵ, $0 < \epsilon < \delta$, *such that* $\mu_{g,V}(t) - \mu^\gamma_{g,V}(t) \in \mathfrak{G}(\mathcal{N}_0)$ *whatever* $\gamma \in \Gamma$, $|\gamma| < \epsilon$, *where* $\mathcal{N}_0 = \{ (\zeta,t) \in \mathcal{N}; \ t \in W_0 \}$.

If $V' \subset V$ *is the image of* $V_0'\times\Pi_0'$ *under the map* $(x,t) \to (Z(x,t),t)$, *with* V_0' *an open subset of* V_0, *and if* $\mu_{g,V'}(t)$ *is the analogue of* $\mu_{g,V}(t)$, *then* $\mu_{g,V}(t) - \mu_{g,V'}(t) \in \mathfrak{G}(\overline{V}\backslash V')$.

The space $\mathfrak{G}(\mathcal{N})$ of analytic functionals depending on parameters is defined in accordance with (IV.1.12).

The element of $\mathcal{B}(V,\mathcal{C}^\infty_t)$ defined by $\mu_{g,V}(t)$ is the boundary value $b_V g$ in V of $g \in \mathfrak{F}(\mathcal{W}_\delta(U,\Gamma))$. If $V' \subset V$ is the set in the last part of Theorem IV.2.3, $b_V g = b_{V'} g$ in V'. This allows us to let V expand and fill U. At the limit we will have defined a hyperfunction in U, smooth with respect to t, $b_U g$ — *the boundary value of* g *in* U.

THEOREM IV.2.4.— *Let* U *be the open subset* (IV.2.8) *of* Σ, Γ *be an open and convex cone in* $\mathbb{R}^m\backslash\{0\}$ *and* δ *a number* > 0. *The boundary value map* $g \to b_U g$ *is an injection of* $\mathfrak{F}(\mathcal{W}_\delta(U,\Gamma))$ *into* $\mathcal{B}(U,\mathcal{C}_t^\infty)$.

Indeed, the boundary value map is injective from $\mathcal{O}(\mathcal{W}_\delta(U_t,\Gamma))$ to $\mathcal{B}(U_t)$ for each $t \in \Pi_0'$ (Theorem II.1.2).

IV.3 HYPERFUNCTION SOLUTIONS DEPENDING SMOOTHLY ON t

As we have already stated, hyperfunctions (in contrast to distributions) that depend smoothly on t are not invariant under automorphisms of the elliptic structure of $\mathbb{C}^m \times \mathbb{R}^n$, ie., under changes of coordinates

$$z = H(z'),\ t = f(z',t') \tag{IV.3.1}$$

with H holomorphic, valued in $\mathbb{C}^m$, and f merely C^∞ (and valued in $\mathbb{R}^n$). The elliptic structure of $\mathbb{C}^m \times \mathbb{R}^n$ is defined by the sheaf of germs of functions that are holomorphic with respect to z and locally independent of t. These are the functions annihilated by the basic differential operator in the elliptic structure of $\mathbb{C}^m \times \mathbb{R}^n$, $\bar{\partial}_z + d_t$. The sheaf of germs of those functions is not affected by changes of variables (IV.3.1).

EXAMPLE IV.3.1.— Consider an element of $C^\infty(\mathbb{R}^n;\mathcal{O}'(\mathbb{C}^m))$,

$$\mu(t) = \sum_{\alpha \in \mathbb{Z}_+^m} c_\alpha(t)\delta^{(\alpha)} \tag{IV.3.2}$$

$[\delta^{(\alpha)} = (\partial/\partial z)^\alpha \delta;\ \delta = \delta^{(0)}$, the Dirac distribution in $\mathbb{C}^m]$. The coefficients c_α and their partial derivatives with respect to t must satisfy estimates of the kind (IV.1.9). If we carry out a change of variables (IV.3.1), say with $H(z') \equiv z'$ to simplify, each term $c_\alpha(t)\delta^{(\alpha)}$ transforms into

$$c_\alpha(f(z,t'))\delta^{(\alpha)} = \sum_{\beta \leq \alpha} (-1)^{|\beta|} \binom{\alpha}{\beta} (\partial/\partial z)^\beta [c_\alpha(f(z,t'))] \Big|_{z=0} \delta^{(\alpha-\beta)}.$$

This is certainly possible when the sum at the right in (IV.3.2) is finite, ie., when $\mu(t)$ is a distribution with support at the origin in $\mathbb{C}^m$. But, in general, when the sum is not finite the rearrangements of the series at the right in (IV.3.2) after the above computations will not converge; and even when they converge the resulting coefficients will not satisfy estimates of the kind (IV.1.9). □

The lack of invariance of $\mathscr{B}_\Sigma(C^\infty_t)$ prevents us from defining, on a hypo—analytic manifold, hyperfunctions that depend smoothly on t. How— ever we are going to show that it makes sense to speak of hyperfunction solutions in Σ that depend smoothly on t. The definition of the latter will be shown to be invariant, and can therefore be transferred to an arbitrary hypo—analytic manifold. As a matter of fact, we are going to show (in section IV.4) that all hyperfunction solutions are of this type: all hyper— function solutions are smooth with respect to "transversal" variables t. We shall also show that the invariance extends to the cohomology classes of currents in Σ that are closed in the sense of the differential complex on Σ induced by $\bar\partial_z + d_t$; and they also, therefore, can be defined in a general hypo—analytic structure.

For now our aim is to define the sheaf of (the germs of) the hyperfunction solutions depending smoothly on t, as well as the sheaves of the cohomology classes of d$'$— closed hyperfunction currents depending smoothly on t — in the hypo—analytic chart $(\mathcal{U}', x, t, Z)$. We must define the stalks of these sheaves at an arbitrary point $\wp$ of $\mathcal{U}'$. But then we shall transfer the analysis onto the image Σ of $\mathcal{U}'$ under the map $(x,t) \to (Z(x,t),t)$. We must define the corresponding sheaves on Σ, at least their stalks at $\wp$. And we must show that those stalks are invariant under coor— dinates changes of the type (IV.3.1). In other words we must show that the stalks are intrinsically tied to Σ and to the elliptic structure of the surrounding space $\mathbb{C}^m \times \mathbb{R}^n$. Of course, it suffices to do so for the stalk at the central point, the origin.

Below we denote by ${}_m\mathcal{O}$ the sheaf in $\mathbb{C}^m\times\mathbb{R}^n$ of the germ of functions that are holomorphic with respect to z and *constant* with respect to t; and by ${}_m\mathcal{O}^{(p)}$ the sheaf in $\mathbb{C}^m\times\mathbb{R}^n$ of the germs of of differential forms

$$h = \sum_{|I|=p} h_I(z)dz_I \tag{IV.3.3}$$

with holomorphic coefficients h_I. We are going to make use of the flabby resolution

$$0 \to {}_m\mathcal{O}^{(m)} \overset{i}{\to} \mathcal{B}\Lambda^{m,0} \overset{\bar\partial_z+d_t}{\to} \mathcal{B}\Lambda^{m,1} \overset{\bar\partial_z+d_t}{\to} \cdots \overset{\bar\partial_z+d_t}{\to} \mathcal{B}\Lambda^{m,m+n} \to 0. \tag{IV.3.4}$$

The continuous sections of the sheaf $\mathcal{B}\Lambda^{p,q}$ are the "currents"

$$f = \sum_{|I|=p}\ \sum_{|J|+|K|=q} f_{I,J,K}dz_I\wedge d\bar z_J\wedge dt_K \tag{IV.3.5}$$

with coefficients $f_{I,J,K}$ that are hyperfunctions (in the sense of the standard C^ω structure of $\mathbb{R}^{2m+n}$; cf. section III.2). The Cauchy–Riemann operators $\partial/\partial\bar z_j$ $(1 \le j \le m)$ and the partial differentiations $\partial/\partial t_k$ $(1 \le k \le n)$ act on each coefficient of the current (IV.3.5). The combined differential operator $\bar\partial_z+d_t$ acts on f itself according to the usual recipe:

$$(\bar\partial_z+d_t)f = \sum_{|I|=p}\ \sum_{|J|+|K|=q} \Big[\sum_{j'=1}^{m} (\partial f_{I,J,K}/\partial\bar z_{j'})d\bar z_{j'} + \tag{IV.3.6}$$

$$\sum_{k'=1}^{n} (\partial f_{I,J,K}/\partial t_{k'})dt_{k'}\Big] \wedge dz_I \wedge d\bar{z}_J \wedge dt_K.$$

When $p = m$, $(\bar{\partial}_z + d_t)f = df$, the exterior derivative of f.

The analogue of Corollary III.2.3 is valid:

Lemma IV.3.1.— *Let Ω be an open subset of $\mathbb{C}^m \times \mathbb{R}^n$ and h a hyperfunction in Ω solution of $(\bar{\partial}_z + d_t)h = 0$. Then h is a holomorphic function of z, locally independent of t.*

Lemma IV.3.1 implies the exactness of the short sequence

$$\text{(IV.3.7)} \qquad 0 \to {}_m\mathcal{O}^{(m)} \overset{i}{\to} \mathscr{B}\Lambda^{m,0} \overset{\bar{\partial}_z + d_t}{\to} \mathscr{B}\Lambda^{m,1}.$$

In order to prove the exactness of the remainder of the sequence (IV.3.4) we need the analogue of the Poincaré lemma for the differential complex

$$\text{(IV.3.8)} \qquad \bar{\partial}_z + d_t \colon \mathcal{B}(\mathcal{A};\Lambda^{m,q}) \to \mathcal{B}(\mathcal{A};\Lambda^{m,q+1}),\ q = 1,\dots,m+n,$$

where $\mathcal{B}(\mathcal{A};\Lambda^{m,q})$ is the space of currents (IV.3.5) whose coefficients are hyperfunctions in the open subset $\mathcal{A}$ of $\mathbb{C}^m \times \mathbb{R}^n$. This is proved exactly like Theorem III.2.1; the only difference is that there are no variables y' or, if one prefers, that the variation of z' is restricted to the real space $\mathbb{R}^n$. The analogue of Theorem III.2.1 reads:

THEOREM IV.3.1.— *Let $f \in \mathcal{B}(\Pi \times \Pi_0';\Lambda^{m,q})$ satisfy $(\bar{\partial}_z + d_t)f \equiv 0$. If $q \geq 1$ there is a solution $u \in \mathcal{B}(\Pi \times \Pi_0';\Lambda^{m,q-1})$ of $(\bar{\partial}_z + d_t)u = f$.*

Theorem IV.3.1 entails that the flabby resolution (IV.3.4) is co–homological. It follows that the cohomology with coefficients in ${}_{\mathrm{m}}\mathcal{O}^{(\mathrm{m})}$ can be equated to the cohomology of the differential complex (IV.3.8). If $\mathcal{A}$ is an open subset of $\mathbb{C}^{\mathrm{m}}\times\mathbb{R}^{\mathrm{n}}$ and $\mathcal{S}$ a closed subset of $\mathcal{A}$,

$$(\mathrm{IV.3.9})_0 \qquad \mathrm{H}^0_{\mathcal{S}}(\mathcal{A};{}_{\mathrm{m}}\mathcal{O}^{(\mathrm{m})}) = \mathrm{H}^{\mathrm{m},\mathrm{q}}_{\mathcal{S}}(\mathcal{A}) = \{\, h \in \mathcal{B}(\mathcal{A};\Lambda^{\mathrm{m},0});\ (\bar\partial_z + \mathrm{d}_t)h = 0,\ \operatorname{supp} h \subset \mathcal{S}\,\};$$

and for q $\geq$ 1,

$$(\mathrm{IV.3.9})_{\mathrm{q}} \qquad \mathrm{H}^{\mathrm{q}}_{\mathcal{S}}(\mathcal{A};{}_{\mathrm{m}}\mathcal{O}^{(\mathrm{m})}) = \mathrm{H}^{\mathrm{m},\mathrm{q}}_{\mathcal{S}}(\mathcal{A}) = \{\, f \in \mathcal{B}(\mathcal{A};\Lambda^{\mathrm{m},\mathrm{q}});\ (\bar\partial_z + \mathrm{d}_t)f = 0,\ \operatorname{supp} f \subset \mathcal{S}\,\}/ \{\, (\bar\partial_z + \mathrm{d}_t)u;\ u \in \mathcal{B}(\mathcal{A};\Lambda^{\mathrm{m},\mathrm{q}-1}),\ \operatorname{supp} u \subset \mathcal{S}\,\}.$$

The use of the cohomological resolutions based on the flabby sheaf $\mathscr{B}$ facilitates the definition of the relative cohomology spaces.

In what follows, U_0 will be an open subset of $\mathbb{R}^{\mathrm{m}}$, $\mathrm{U}_0 \subset\subset \Pi_0$. We shall denote by U the image of $\mathrm{U}_0\times\Pi_0'$ under the map $(x,t) \to (Z(x,t),t)$, by $\overline{\mathrm{U}}$ that of $(\mathscr{C}\ell\mathrm{U}_0)\times\Pi_0'$ and by $\dot{\mathrm{U}}$ that of $(\partial\mathrm{U}_0)\times\Pi_0'$; $\overline{\mathrm{U}} = (\mathscr{C}\ell\mathrm{U})\cap\Sigma_0$, $\dot{\mathrm{U}} = (\partial\mathrm{U})\cap\Sigma_0$. We shall deal with the linear space $\mathcal{B}(\mathrm{U},\mathcal{C}^\infty_t)$ of the hyperfunc–tions in U which depend smoothly on $t \in \Pi_0'$ (Definition IV.2.2).

DEFINITION IV.3.1.— *By a* ***hyperfunction solution in*** U ***depending smoothly on*** $\boldsymbol{t}$ *we shall mean an element u of* $\mathcal{B}(\mathrm{U},\mathcal{C}^\infty_t)$ *that has a representative* $\mu(t) \in \mathfrak{G}(\overline{\mathrm{U}})$ *whose partial derivatives* $\partial\mu/\partial t_{\mathrm{j}}$ (j = 1,...,n) *belong to* $\mathfrak{G}(\dot{\mathrm{U}})$. *The* *space of all hyperfunction solutions in* U *depending smoothly on t will be denoted by* $\mathfrak{Sol}(U,\mathcal{C}^{\infty}_t)$.

To say that $u \in \mathcal{B}(U,\mathcal{C}^{\infty}_t)$ is a hyperfunction solution depending smoothly on t is the same as saying that $d_t u \equiv 0$.

The discussion in Example III.1.2 can readily be generalized to show that all distribution solutions in an open subset (of the kind considered above) of Σ, U, are hyperfunction solutions that depend smoothly on t.

We shall also consider differential forms (or rather, currents) with hyperfunction coefficients. We denote by $\mathcal{B}(U,\mathcal{C}^{\infty}_t;\Lambda^q)$ (q = 1,...,n) the space of currents

(IV.3.10) $$f = \sum_{|J|=q} f_J dt_J$$

with coefficients $f_J \in \mathcal{B}(U,\mathcal{C}^{\infty}_t)$. We set $\mathcal{B}(U,\mathcal{C}^{\infty}_t;\Lambda^0) = \mathcal{B}(U,\mathcal{C}^{\infty}_t)$, $\mathcal{B}(U,\mathcal{C}^{\infty}_t;\Lambda^q) = \{0\}$ if $q > n$. Recalling that the partial differentiations ∂^{α}_t define endomorphisms of $\mathcal{B}(U,\mathcal{C}^{\infty}_t)$ the following is a differential complex

(IV.3.11) $$d_t\colon \mathcal{B}(U,\mathcal{C}^{\infty}_t;\Lambda^q) \to \mathcal{B}(U,\mathcal{C}^{\infty}_t;\Lambda^{q+1}),\quad q = 0,1,\dots,$$

with the standard definition of the exterior derivative:

$$d_t f = \sum_{|J|=q} \sum_{k=1}^{n} (\partial f_J/\partial t_k) dt_k \wedge dt_J$$

(clearly $d_t^2 = 0$).

DEFINITION IV.3.2.— *We shall denote by* $\mathfrak{Sol}^{(q)}(U,C^{\infty}_t)$ *the* q^{th} *cohomology space of the differential complex* (IV.3.11) $(0 \le q \le n)$.

The space $\mathfrak{Sol}^{(0)}(U,C^{\infty}_t)$ is naturally isomorphic to the space $\mathfrak{Sol}(U,C^{\infty}_t)$ of hyperfunction solutions in U that depend smoothly on t (Definition IV.3.1). Needless to say the definitions analogous to Definitions IV.3.1 & IV.3.2 are valid after replacing Π'_0 by a smaller polytope.

EXAMPLE IV.3.2.— Suppose $1 \le q \le n$ and let $f \in C^{\infty}(\mathcal{M}';\Lambda^{m,q}_{\mathcal{M}})$ be given by (III.1.11) (with $p = m$) and satisfy $Lf \equiv 0$ [see (III.1.12) and (III.1.13)]. Let $\tilde{f}$ denote the pushforward of f to Σ under the map $(x,t) \to (Z(x,t),t)$; $\tilde{f}$ is given by (III.1.14). Let U be an open subset of Σ_0 of the type considered above, with a smooth boundary $\dot{U}$ in Σ_0. As in Example III.1.3 the map

$$\mathcal{O}(\mathbb{C}^m) \ni h \to \int_{U_t} h\tilde{f},$$

(with the integration carried out in z–space) defines a current

$$\mu_f(t) = \sum_{|J|=q} \mu_J(z)dt_J$$

with $\mu_J \in \mathfrak{G}(\mathscr{C}\ell U)$ and such that $d_t\mu_f$ is carried by ∂U. Thus μ_f defines an element $\tilde{f}$ of $\mathcal{B}(U,C^{\infty}_t;\Lambda^q)$ such that $d_t\tilde{f} \equiv 0$ in U, ie., a class $[\tilde{f}] \in \mathfrak{Sol}^{(q)}(U,C^{\infty}_t)$. To assert that $[\tilde{f}] = 0$ is the same as saying that $d_t u = \tilde{f}$ in U for some $u \in \mathcal{B}(U,C^{\infty}_t;\Lambda^{q-1})$. This is the correct version of the statement that $Lu = f$ in U [see (III.1.12); in general, the action of the vector fields L_j on elements of $\mathcal{B}(U,C^{\infty}_t)$ is not defined]. □

Below U_0, V_0 will denote two open subsets of $\mathbb{R}^m$, $V_0 \subset U_0 \subset\subset \Pi_0$, $\Pi_0'' \subset \Pi_0'$ an open polytope; U and V will denote the image under the map $(x,t) \to (Z(x,t),t)$ of $U_0 \times \Pi_0'$ and $V_0 \times \Pi_0''$ respectively. Of course $V \subset U$. The restriction mapping $\mathcal{B}(U,\mathcal{C}_t^\infty) \to \mathcal{B}(V,\mathcal{C}_t^\infty)$ induces a restriction mapping

$$\text{(IV.3.12)} \qquad \mathfrak{r}_V^U : \mathfrak{Sol}^{(q)}(U,\mathcal{C}_t^\infty) \to \mathfrak{Sol}^{(q)}(V,\mathcal{C}_t^\infty).$$

We shall deal with the relative cohomology space $H_U^{p,q}(\Pi\times\Pi_0'\backslash\overset{\bullet}{U})$ and its analogue with V in the place of U. The reader will notice that the definition of these spaces agrees with (IV.3.9) because U is a closed subset of $\Pi\times\Pi_0'\backslash\overset{\bullet}{U}$. The statement of the main theorem of this section, Theorem IV.3.2, will rely on the following result, whose proof is completely analogous to that of Lemma III.2.2 (considering that the restriction in t–space is routine):

LEMMA IV.3.2.— *The inclusion* $V \subset U$ *defines a natural restriction map*

$$\text{(IV.3.13)} \qquad r_V^U : H_U^{p,q}(\Pi\times\Pi_0'\backslash\overset{\bullet}{U}) \to H_V^{p,q}(\Pi\times\Pi_0''\backslash\overset{\bullet}{V}).$$

THEOREM IV.3.2.— *Let* q *be an integer,* $0 \le q \le n$. *There are natural isomorphisms* i_U^q *and* i_V^q *such that the diagram*

$$\text{(IV.3.14)} \qquad \begin{array}{ccc} \mathfrak{Sol}^{(q)}(U,\mathcal{C}_t^\infty) & \xrightarrow{\;i_U^q\;} & H_U^{m,m+q}(\Pi\times\Pi_0'\backslash\overset{\bullet}{U}) \\ \mathfrak{r}_V^U \big\downarrow & & \big\downarrow r_V^U \\ \mathfrak{Sol}^{(q)}(V,\mathcal{C}_t^\infty) & \xrightarrow[\;i_V^q\;]{} & H_V^{m,m+q}(\Pi\times\Pi_0''\backslash\overset{\bullet}{V}) \end{array}$$

is commutative.

The proof of Theorem IV.3.2 will make frequent use of the following lemmas:

LEMMA IV.3.3.— *If* m $\geq$ 2 *the restriction map*

$$\mathfrak{H}^{m,m-1}(\Pi\times\Pi_0'\backslash\mathring{U}) \to \mathfrak{H}^{m,m-1}(\Pi\times\Pi_0'\backslash\overline{U})$$

is injective.

We recall that $\mathfrak{H}^{p,q}$ are the cohomology spaces of the $\overline{\partial}_z$—complex in $\mathbb{C}^m\times\mathbb{R}^n$ [see (IV.1.17), (IV.1.18)]. According to Theorem IV.1.3, Lemma IV.3.3 is a restatement of the fact that the inclusion map $\mathfrak{S}(\mathring{U}) \to \mathfrak{S}(\overline{U})$ is injective.

LEMMA IV.3.4.— *Let* K_0 *be a compact subset of* Π_0 *and let* K *denote the image of* $K_0\times\Pi_0'$ *under the map* $(x,t) \to (Z(x,t),t)$. *If* m $\geq$ 2 *the restriction mapping defines an isomorphism* $\mathcal{F}(\Pi\times\Pi_0') \cong \mathcal{F}(\Pi\times\Pi_0'\backslash K)$. *If* m $\geq$ 3,

(IV.3.15) $\mathfrak{H}^{m,q}(\Pi\times\Pi_0'\backslash K) = 0$ *for all integers* q, $1 \leq q \leq m-2$.

The proof of Lemma IV.3.4 mimicks that of Lemma III.3.2, where now we make use of the operators introduced in Proposition IV.1.2. These operators can also be used to establish the version with parameters of Hartog's theorem, needed to prove the first part of the claim.

Theorem IV.3.2 entails exactly what is needed, in order to define invariantly the sheaf of hyperfunction solutions on $\mathcal{M}$ depending smoothly on *t*, as well as the sheaves of the germs of the cohomology classes that belong to $\mathfrak{Sol}^{(q)}(U,\mathcal{C}^\infty_t)$ $(1 \le q \le n)$. We introduce the inductive limit,

$$\mathcal{S}ol^{(q)}_\Sigma(\mathscr{C}^\infty_t)\big|_0,$$

in the sense of the mappings r^U_V, of the spaces $\mathfrak{Sol}^{(q)}(U,\mathcal{C}^\infty_t)$ for all open subsets U of $\Sigma_0 \subset \Sigma \subset \mathbb{C}^m\times\mathbb{R}^n$ of the type considered above: U is the image of $U_0\times\Pi_0^*$ under the map $(x,t) \to (Z(x,t),t)$; U_0 is a relatively compact open subset of Π_0, Π_0^* one of Π_0', both containing the origin (we may as well take U_0 and Π_0^* to be open polytopes). Theorem IV.3.2 implies that $\mathcal{S}ol^{(q)}_\Sigma(\mathscr{C}^\infty_t)\big|_0$ is naturally isomorphic to the inductive limit (for the mappings r^U_V) of the relative cohomology spaces $H^{m,m+q}_U(\Pi\times\Pi_0^*\backslash\mathring{U})$. But the latter, associated as it is with the differential operator $\bar\partial_z + d_t$, ie., with the resolution (IV.3.4) and the differential complex (IV.3.8), is an invariant of the hypo—analytic structure of Σ induced by the elliptic structure of $\mathbb{C}^m\times\mathbb{R}^n$. The same is therefore true of $\mathcal{S}ol^{(q)}_\Sigma(\mathscr{C}^\infty_t)\big|_0$. The pullback of $\mathcal{S}ol^{(q)}_\Sigma(\mathscr{C}^\infty_t)\big|_0$ to $\mathcal{M}$ via the map $(x,t) \to (Z(x,t),t)$, will be, by definition, the stalk at the point $O \in \mathcal{M}'$ (whose coordinates x_i and t_j all vanish) of the sheaf $\mathcal{S}ol^{(q)}_{\mathcal{M}}(\mathscr{C}^\infty_t)$.

The proof of Theorem IV.3.2 to a large extent duplicates that of Theorem III. 2.2. We shall content ourselves with developing only those segments of the argument where it differs from that in section III.3.

Proof of Theorem IV.3.2

For the duration of this subsection we shall use the notation $\Omega_0 = \Pi\times\Pi_0'\backslash\mathring{U}$. Note that U is closed in Ω_0 and that $\Omega_0\backslash U = \Pi\times\Pi_0'\backslash\bar{U}$. We are going to construct the linear map i^q_U and prove that it is an isomorphism.

The analogous reasoning will be valid with V in the place of U (and Π_0'' in the place of Π_0').

An element of $\mathfrak{Sol}^{(q)}(U,\mathcal{C}_t^\infty)$ is a cohomology class of currents (IV.3.10) that have a representative

(IV.3.16) $$\mu = \sum_{|J|=q} \mu_J dt_J$$

with $\mu_J \in \mathfrak{G}(\overline{U})$, such that the coefficients of $d_t\mu$ belong to $\mathfrak{G}(\mathring{U})$. The isomorphisms in Theorem IV.1.1 when m = 1, and in Theorem IV.1.3 when m $\geq$ 2, associate to μ a differential form

(IV.3.17) $$F = \sum_{|J|=q} F_J \wedge dt_J \in \mathcal{C}^\infty(\Omega_0\backslash U;\Lambda^{m,m-1,q}),$$

with $F_J \in \mathcal{C}^\infty(\Omega_0\backslash U;\Lambda^{m,m-1})$, $\bar\partial_z F_J = 0$. We are using the notation $\Lambda^{m,p,q}$ to mean the vector subbundle of $\Lambda^{m+p+q}\mathbb{C}T^*(\mathbb{C}^m\times\mathbb{R}^n)$ spanned by the ex– terior products $dz_I\wedge d\bar z_J\wedge dt_K$, $|I| = m$, $|J| = p$, $|K| = q$. To say that the coefficients of $d_t\mu$ belong to $\mathfrak{G}(\mathring{U})$ is the same as saying that there are

$$A = \sum_{|J|=q+1} A_J\wedge dt_J \in \mathcal{C}^\infty(\Omega_0;\Lambda^{m,m-1,q+1}),$$
$$G = \sum_{|J|=q+1} G_J\wedge dt_J \in \mathcal{C}^\infty(\Omega_0\backslash U;\Lambda^{m,m-2,q+1}),$$
$$A_J \in \mathcal{C}^\infty(\Omega_0;\Lambda^{m,m-1}),\ G_J \in \mathcal{C}^\infty(\Omega_0\backslash U;\Lambda^{m,m-2}),$$

such that

(IV.3.18) $$d_tF = A - \bar\partial_z G \ \ in\ \Omega_0\backslash U,$$

(IV.3.19) $$\overline{\partial}_z A = 0 \quad in\ \Omega_0.$$

We make use of the long exact sequence for relative cohomology, specifically of the following segment of that sequence:

$$H^{m,m+q-1}(\Omega_0) \xrightarrow{\rho^{q-1}} H^{m,m+q-1}(\Omega_0\backslash U) \xrightarrow{\delta^q} H_U^{m,m+q}(\Omega_0) \xrightarrow{\iota^q} H^{m,m+q}(\Omega_0)$$
$$\xrightarrow{\rho^q} H^{m,m+q}(\Omega_0\backslash U).$$

The map ρ^q is defined by the restriction from Ω_0 to $\Omega_0\backslash U$; ι^q assigns to the form $f \in \mathcal{B}(\Omega_0;\Lambda^{m,m+q})$ such that $(\overline{\partial}_z+d_t)f = 0$ and supp $f \subset U$, its cohomology class mod $(\overline{\partial}_z+d_t)\mathcal{B}(\Omega_0;\Lambda^{m,m+q-1})$. The map δ^q is defined as follows: if $u \in \mathcal{B}(\Omega_0\backslash U;\Lambda^{m,m+q-1})$, $(\overline{\partial}_z+d_t)u = 0$, and if $\tilde{u}$ is an arbitrary extension of u to Ω_0, then $v = (\overline{\partial}_z+d_t)\tilde{u} \in \mathcal{B}(\Omega_0;\Lambda^{m,m+q})$, $(\overline{\partial}_z+d_t)v = 0$ and supp $v \subset U$. Set $[v] = \delta^q([u])$. From the (obvious) exactness of the above sequence we derive the following short exact sequence

$$0 \to H^{m,m+q-1}(\Omega_0\backslash U)/\mathrm{Im}\ \rho^{q-1} \xrightarrow{\delta^q} H_U^{m,m+q}(\Omega_0) \xrightarrow{\iota^q} \mathrm{Ker}\ \rho^q \to 0.$$

This is equivalent to saying that we have the direct sum decomposition:

(IV.3.20) $$H_U^{m,m+q}(\Omega_0) \cong (\mathrm{Im}\ \delta^q) \oplus (\mathrm{Ker}\ \rho^q).$$

The "factor" Im δ^q will be shown to be the image under the isomorphism i_U^q of the subspace $\mathfrak{Sol}_c^{(q)}(U,\mathcal{C}_t^\infty)$ of $\mathfrak{Sol}^{(q)}(U,\mathcal{C}_t^\infty)$ consisting of the cohomology classes of currents (IV.3.10) that have a representative (IV.3.16) such that $d_t\mu = 0$. Obviously $\mathfrak{Sol}_c^{(n)}(U,\mathcal{C}_t^\infty) = \mathfrak{Sol}^{(n)}(U,\mathcal{C}_t^\infty)$.

Let μ be a current (IV.3.16) representing a class in $\mathfrak{Sol}_c^{(q)}(U,\mathcal{C}_t^\infty)$

such that $d_t\mu = 0$, and let F be the corresponding differential form, given by (IV.3.17). Then $d_tF \equiv 0$ if $m = 1$ or, if $m \geq 2$, d_tF is $\bar\partial_z$—exact in $\Omega_0\backslash U$. In both cases this is equivalent to saying that, in (IV.3.18), $A \equiv 0$.

1. *Construction of the classes* $i_U^q([f])$

First suppose $m = 1$. In (IV.3.18) $G \equiv 0$, hence $d_tA = 0$ in $\Omega_0\backslash U$. But the coefficients of A are holomorphic with respect to z in the set Ω_0, and $\Omega_0\backslash U$ is a dense subset of Ω_0. It follows that $d_tA = 0$ in Ω_0, which is the same as the equation

$$(\bar\partial_z + d_t)A = 0 \quad \textit{in } \Omega_0.$$

Since $\bar\partial_z F = 0$ in $\Omega_0\backslash U$ we derive from (IV.3.19):

$$A = (\bar\partial_z + d_t)F \quad \textit{in } \Omega_0\backslash U.$$

Let E be an arbitrary hyperfunction extension of F to Ω_0. What precedes shows that $A - (\bar\partial_z + d_t)E \in \mathcal{B}(\Omega_0;\Lambda^{1,q+1})$ is $(\bar\partial_z + d_t)$–closed and vanishes identically in $\Omega_0\backslash U$. It defines a class in $H_U^{1,1+q}(\Omega_0)$ which we take to be $i_U^q([f])$.

When $q = n$ we repeat the argument in Part II*a* of section III.3; when $q = n - 1$ we repeat Part II*b*. Lemmas IV.3.4 and IV.3.5 are used in place of Lemmas III.3.1 and III.3.2. In the remainder of this subsection we assume $M = \text{Min}(m,n-q) \geq 2$.

We return to (IV.3.18) and (IV.3.19). We shall write $A^{(1)} = A$. By applying Lemmas IV.3.4 and IV.3.5 we can solve, inductively, the string of equations in Ω_0,

(IV.3.21) $$\bar{\partial}_z A^{(k+1)} = d_t A^{(k)},\ k = 1,\dots,M-1,$$

with $A^{(k)} \in C^\infty(\Omega_0;\Lambda^{m,m-k,q+k})$. It follows from (IV.3.18) that $d_t A^{(k)} = -\bar{\partial}_z d_t G$ in $\Omega_0\backslash U$ for $k = 1$, and for $2 \le k < k_0$ if we assume that (IV.3.21) is valid for those values of k. Then Lemma IV.3.3 when $k = 1$ and Lemma IV.3.4 for $k > 1$ enable us to find $A^{(k_0)}$.

We must distinguish two cases: either $M = n-q$, or $M = m < n-q$. In the first case we duplicate exactly the argument in Part II*c* of section III.3 (again, after substituting t for z').

Assume now $m < n-q$. Equations (IV.3.21) entail

$$(\bar{\partial}_z + d_t)\left[\sum_{k=1}^{m} (-1)^{k-1} A^{(k)}\right] = (-1)^{m-1} d_t A^{(m)}.$$

Recall that $A^{(m)} \in C^\infty(\Omega_0;\Lambda^{m,0,m+q})$. Since $\bar{\partial}_z d_t A^{(m)} = 0$ the coefficients of $d_t A^{(m)}$ are regular functions in Ω_0; and by the first part of Lemma IV.3.4 they extend as regular functions in $\Pi\times\Pi_0'$, defining a form $H \in C^\infty(\Pi\times\Pi_0';\Lambda^{m,0,m+q+1})$ such that

$$\bar{\partial}_z H = 0 \ \textit{in}\ \Pi\times\Pi_0',\quad H = (-1)^{m-1} d_t A^{(m)}\ \textit{in}\ \Omega_0.$$

As Ω_0 is dense in $\Pi\times\Pi_0'$ we must have $d_t H = 0$ in $\Pi\times\Pi_0'$. Consequently there is a form $R \in C^\infty(\Pi\times\Pi_0';\Lambda^{m,0,m+q})$ such that $\bar{\partial}_z R = 0$, $d_t R = H$ *in* $\Pi\times\Pi_0'$. For reference below we note that we have

(IV.3.22) $$d_t R = (\bar{\partial}_z + d_t) R = (-1)^{m-1} d_t A^{(m)}\ \textit{in}\ \Omega_0.$$

Setting

(IV.3.23) $$A' = \sum_{k=1}^{m} (-1)^{k-1} A^{(k)} - R,$$

[thus $A' \in C^{\infty}(\Omega_0;\Lambda^{m,m+q})$] we obtain

(IV.3.24) $$(\bar\partial_z + d_t)A' = 0 \ \textit{in}\ \Omega_0;$$

(IV.3.25) $$A' = (\bar\partial_z + d_t)(F+G) - d_t G - \sum_{k=2}^{m} (-1)^k A^{(k)} - R \quad \textit{in}\ \Omega_0\backslash U.$$

Consider first the case m = 2. By (IV.3.21) we have, in $\Omega_0\backslash U$,

$$\bar\partial_z(d_t G + A^{(2)} + R) = - d_t(\bar\partial_z G - A) = 0,$$

hence $d_t G + A^{(2)} + R$ extends as a form $K \in C^{\infty}(\Pi\times\Pi_0';\Lambda^{2,0,q+2})$ whose coefficients are regular functions. Moreover, $d_t(d_t G + A^{(2)} + R) = d_t(A^{(2)} + R) = 0$ by (IV.3.22). It follows that we can find a form $S \in C^{\infty}(\Pi\times\Pi_0';\Lambda^{2,0,q+1})$ such that $\bar\partial_z S = 0$, $d_t S = d_t G + A^{(2)} + R$ in $\Omega_0\backslash U$. In this case we derive from (IV.3.25):

(IV.3.26) $$A' = (\bar\partial_z + d_t)(F+G-S) \ \textit{in}\ \Omega_0\backslash U.$$

If E is any hyperfunction extension of $F+G-S$ to Ω_0, $A' - (\bar\partial_z + d_t)E$ is $(\bar\partial_z + d_t)$—closed in Ω_0 and vanishes in $\Omega_0\backslash U$. It defines the element $i_U^q([f])$ of $H_U^{2,q+2}(\Omega_0)$.

Now suppose m ≥ 3. We set $G^{(1)} = G$ and we solve the following string of equations in $\Omega_0\backslash U$:

$$\text{(IV.3.27)} \qquad \bar\partial_z G^{(k)} = A^{(k)} + d_t G^{(k-1)} \quad (2 \leq k \leq m-1)$$

with $G^{(k)} \in C^\infty(\Omega_0\backslash U;\Lambda^{m,m-k-1,q+k})$. Observe that $\bar\partial_z(A^{(2)}+d_t G^{(1)}) = 0$ in $\Omega_0\backslash U$, by (IV.3.18) and (IV.3.21); and

$$\bar\partial_z(A^{(k)}+d_t G^{(k-1)}) = d_t(A^{(k-1)} - \bar\partial_z G^{(k-1)}) = 0$$

by induction on $k \geq 3$, and by (IV.3.21). Lemma IV.3.4 enables us to solve recursively the equations (IV.3.27). If we write $G^{(1)} = G$ and set

$$G' = \sum_{k=1}^{m-1} (-1)^{k-1} G^{(k)},$$

we get

$$A' = (\bar\partial_z + d_t)(F+G') + (-1)^{m-1}[A^{(m)} + d_t\, G^{(m-1)}] - R.$$

We have, by (IV.3.22),

$$\text{(IV.3.28)} \qquad d_t\{(-1)^m[A^{(m)} + d_t\, G^{(m-1)}] + R\} = 0 \ \textit{in}\ \Omega_0\backslash U$$

and, by (IV.3.21) and (IV.3.27),

$$\bar\partial_z\{(-1)^m[A^{(m)} + d_t G^{(m-1)}] + R\} = (-1)^m[\bar\partial_z A^{(m)} - d_t \bar\partial_z G^{(m-1)}] =$$
$$(-1)^m[\bar\partial_z A^{(m)} - d_t A^{(m-1)}] = 0.$$

The coefficients of the differential form

$$(-1)^m[A^{(m)} + d_t\, G^{(m-1)}] + R \in C^\infty(\Omega_0\backslash U;\Lambda^{m,0,m+q}),$$

are regular functions in $\Omega_0\backslash U$. Once again we apply the first part of Lemma IV.3.4: those coefficients extend as regular functions in $\Pi\times\Pi_0'$; and (IV.3.28) remains valid after extension, since $\Omega_0\backslash U$ is dense in $\Pi\times\Pi_0'$. It is therefore possible to solve, in $\Pi\times\Pi_0'$,

$$d_t S = (-1)^m[A^{(m)} + d_t G^{(m-1)}] + R$$

in such a way that the coefficients of the form S be regular in $\Pi\times\Pi_0'$. We obtain

$$A' = (\bar\partial_z + d_t)(F+G'-S) \text{ in } \Omega_0\backslash U. \tag{IV.3.29}$$

If E is a hyperfunction extension to Ω_0 of $F+G'-S$, it follows from (IV.3.24) that $A'-(\bar\partial_z+d_t)E$ is a (m,m+q)—cocycle in the differential complex (IV.3.8) (where $\mathcal{A} = \Omega_0$). According to (IV.3.29) it vanishes in $\Omega_0\backslash U$. Thus it defines a class in the relative cohomology space $H_U^{m,m+q}(\Omega_0)$ which we set to be $i_U^q([f])$.

2. *The map i_U^q is well–defined*

Assume that $f \in \mathcal{B}(U,\mathcal{C}_t^\infty;\Lambda^q)$ "cobounds" in the differential complex (IV.3.11). This has different meanings, depending on whether q = 0 or q $\geq$ 1. When q = 0 it means that the differential form (IV.3.17), $F \in \mathcal{C}^\infty(\Omega_0\backslash U;\Lambda^{m,m-1})$, is given by

$$F = A_0 - \bar\partial_z G_0, \tag{IV.3.30}$$

where $A_0 \in \mathcal{C}^\infty(\Omega_0;\Lambda^{m,m-1})$, $G_0 \in \mathcal{C}^\infty(\Omega_0\backslash U;\Lambda^{m,m-2})$. When q $\geq$ 1, there is a form

$$\Phi = \sum_{|J|=q-1} \Phi_J \wedge dt_J \in C^{\infty}(\Omega_0 \backslash U; \Lambda^{m,m-1,q-1})$$

such that $\bar{\partial}_z \Phi = 0$ and that

$$\text{(IV.3.31)} \qquad F - d_t \Phi = A_0 - \bar{\partial}_z G_0,$$

where now $A_0 \in C^{\infty}(\Omega_0; \Lambda^{m,m-1,q})$, $G_0 \in C^{\infty}(\Omega_0 \backslash U; \Lambda^{m,m-2,q})$. The form Φ defines an element $\phi \in \mathcal{B}(U, C_t^{\infty}; \Lambda^{q-1})$ such that $d_t \phi = f$.

In any case $q \geq 0$, we have $\bar{\partial}_z A_0 = 0$ in Ω_0 and we derive from (IV.3.18) and (IV.3.31),

$$\text{(IV.3.32)} \qquad A - d_t A_0 = \bar{\partial}_z (G + d_t G_0) \quad \textit{in } \Omega_0 \backslash U.$$

If $m = 1$, both G and G_0 vanish identically; A_0 extends $F - d_t \Phi$ to Ω_0 (with the understanding that $\Phi \equiv 0$ when $q = 0$). We have $A = d_t A_0$ in $\Omega_0 \backslash U$ and since the coefficients of both A and A_0 are regular functions in Ω_0, we have $A = (\bar{\partial}_z + d_t) A_0$ in Ω_0. Recalling that E is an extension of F to Ω_0 (see IIa) we see that, in $\Omega_0 \backslash U$, $A - (\bar{\partial}_z + d_t) E = (\bar{\partial}_z + d_t)(A_0 - E) = (\bar{\partial}_z + d_t)(F - E)$. Since $\text{supp}(F - E) \subset U$ we conclude that the class of $A - (\bar{\partial}_z + d_t) E$ in $H_U^{1,q+1}(\Omega_0)$, ie., $i_U^q([f])$, vanishes.

In the remainder of this subsection we assume $m \geq 2$. When $n - q \leq m$ we duplicate the reasoning in Part III, section III.3, substituting everywhere t for z'. Let us look at the case $m < n - q$. We solve the string of equations in Ω_0,

(IV.3.33) $$\bar\partial_z B^{(1)} = A - d_t A_0,$$

and if m $\geq$ 3,

(IV.3.34) $$\bar\partial_z B^{(k)} = A^{(k)} + d_t B^{(k-1)} \quad (2 \leq k \leq m-1),$$

with $B^{(k)} \in C^\infty(\Omega_0;\Lambda^{m,m-k-1,q+k})$. In order to solve (IV.3.33) it suffices to apply Lemma IV.3.3 by taking advantage of (IV.3.32). To solve the equations (IV.3.34) we avail ourselves of (IV.3.21), from which we get directly

$$\bar\partial_z[A^{(2)} + d_t B^{(1)}] = \bar\partial_z A^{(2)} - d_t(A - d_t A_0) = 0;$$

if k $\geq$ 3, we also use (IV.3.34) with k—1 substituted for k, to get

$$\bar\partial_z[A^{(k)} + d_t B^{(k-1)}] = \bar\partial_z A^{(k)} - d_t A^{(k-1)} = 0.$$

It suffices then to apply the second part of Lemma IV.3.4.

It follows from (IV.3.33) and (IV.3.34) that

$$(\bar\partial_z + d_t)\left[\sum_{k=1}^{m-1}(-1)^{k-1}B^{(k)}\right] =$$

$$\sum_{k=1}^{m}(-1)^{k-1}A^{(k)} - d_t A_0 - (-1)^m[d_t B^{(m-1)} - A^{(m)}].$$

Recalling that $\bar\partial_z A_0 \equiv 0$ and taking (IV.3.23) into account we get, in Ω_0,

$$A' = (\bar\partial_z + d_t)\left[A_0 + \sum_{k=1}^{m-1}(-1)^{k-1}B^{(k)}\right] - (-1)^m[A^{(m)} - d_t B^{(m-1)}] - R.$$

It follows from (IV.3.24) that

$$R' = (-1)^m[A^{(m)} - d_t B^{(m-1)}] + R \in C^\infty(\Omega_0;\Lambda^{m,0,q+m})$$

has regular coefficients and satisfies $d_t R' \equiv 0$. Since m $\geq$ 2 these coefficients extend as regular functions in $\Pi \times \Pi_0'$ and the extended form R' is

d_t—closed. There is a form $S' \in C^{\infty}(\Pi\times\Pi_0';\Lambda^{m,0,q+m-1})$ such that $\bar\partial_z S' = 0$, $\mathrm{d}_t S' = R'$ in $\Pi\times\Pi_0'$. We reach the conclusion that

$$\text{(IV.3.35)}\quad A' = (\bar\partial_z + \mathrm{d}_t)\left[A_0 + \sum_{k=1}^{m-1}(-1)^{k-1}B^{(k)} - S'\right] \quad \textit{in } \Omega_0.$$

If m = 2 the argument stops here. If m $\geq$ 3 we restrict the equations (IV.3.33) and (IV.3.34) to $\Omega_0\backslash U$ and compare with (IV.3.27). We solve the following of equations in $\Omega_0\backslash U$:

$$\text{(IV.3.36)}\qquad \bar\partial_z v^{(1)} = B^{(1)} - G^{(1)} - \mathrm{d}_t G_0,$$

and if m $\geq$ 4,

$$\text{(IV.3.37)}\quad \bar\partial_z v^{(k)} = B^{(k)} - G^{(k)} + \mathrm{d}_t v^{(k-1)} \quad (k = 2,\dots,m-2),$$

with $v^{(k)} \in C^{\infty}(\Omega_0\backslash U;\Lambda^{m,m-k-2,q+k})$. First of all, by (IV.3.32) and (IV.3.33), we have $\bar\partial_z(B^{(1)} - G^{(1)} - \mathrm{d}_t G_0) = 0$ in $\Omega_0\backslash U$. It suffices there—fore to apply Lemma IV.3.4. From (IV.3.27) and (IV.3.34) we get, in $\Omega_0\backslash U$,

$$\bar\partial_z[B^{(k)} - G^{(k)} + \mathrm{d}_t v^{(k-1)}] =$$
$$\bar\partial_z[B^{(k)} - G^{(k)}] - \mathrm{d}_t[B^{(k-1)} - G^{(k-1)}] = 0.$$

Again we may apply Lemma IV.3.4, until we reach the value k = m—2.

We derive from (IV.3.36)—(IV.3.37):

$$A_0 + \sum_{k=1}^{m-1}(-1)^{k-1}B^{(k)} =$$

$$A_0 + G + d_t G_0 + \bar\partial_z v^{(1)} +$$

$$\sum_{k=2}^{m-2}(-1)^{k-1}[G^{(k)} + \bar\partial_z v^{(k)} - d_t v^{(k-1)}] + (-1)^m B^{(m-1)} =$$

$$A_0 - \bar\partial_z G_0 + G' + (\bar\partial_z + d_t)\left[G_0 + \sum_{k=1}^{m-2}(-1)^{k-1} v^{(k)}\right] +$$

$$(-1)^m[B^{(m-1)} - G^{(m-1)} + d_t v^{(m-2)}],$$

We derive from (IV.3.27), (IV.3.34) and (IV.3.37) (where k = m—2) that

$$B^{(m-1)} - G^{(m-1)} + d_t v^{(m-2)} \in \mathcal{C}^\infty(\Omega_0\backslash U;\Lambda^{m,0,q+m-1})$$

has coefficients that are regular functions in $\Omega_0\backslash U$, and therefore extend as regular functions to $\Pi\times\Pi_0'$. When $m \geq 3$ we call S'' the extension to $\Pi\times\Pi_0'$ of $(-1)^m[B^{(m-1)} - G^{(m-1)} + d_t v^{(m-2)}]$. When m = 2, we call S'' the extension (with regular coefficients) to $\Pi\times\Pi_0'$ of $B^{(1)} - G^{(1)} - d_t G_0$. In both cases,

$$A_0 + \sum_{k=1}^{m-1}(-1)^{k-1}B^{(k)} - S'' = F + G' + (d_t + \bar\partial_z)\psi \quad \textit{in } \Omega_0\backslash U, \tag{IV.3.38}$$

where $\psi = G_0 + \sum_{k=1}^{m-2}(-1)^{k-1}v^{(k)} - \Phi \in \mathcal{C}^\infty(\Omega_0\backslash U;\Lambda^{m,m+q-2})$. We extend ψ as an element of $\mathcal{B}(\Omega_0;\Lambda^{m,m+q-2})$. Combining (IV.3.35) and (IV.3.38) shows that

$$A' - (\bar\partial_z + d_t)(F + G' - S) = d_t(S - S' + S'').$$

We derive from this, and from (IV.3.29), that $d_t(S-S'+S'') \equiv 0$ in $\Omega_0\backslash U$. But $d_t(S-S'+S'') \equiv 0$ in $\Pi\times\Pi_0'$ since the coefficients of $S-S'+S''$ are regular functions in $\Pi\times\Pi_0'$; and there is a form $\varphi \in \mathcal{C}^\infty(\Pi\times\Pi_0';\Lambda^{m,0,q+m-2})$ with regular functions as coefficients, such that

$$S-S'+S'' = d_t\varphi \quad in\ \Pi\times\Pi_0'.$$

According to this and to (IV.3.38) we have

$$\text{(IV.3.39)} \qquad F+G' - S = E_0 - (\bar\partial_z+d_t)(\varphi+\psi) \quad in\ \Omega_0\backslash U,$$

where

$$E_0 = A_0 + \sum_{k=1}^{m-1}(-1)^{k-1}B^{(k)} - S' \in \mathcal{C}^\infty(\Omega_0;\Lambda^{m,m+q-1}).$$

On the other hand, according to (IV.3.35), $A' = (\bar\partial_z+d_t)E_0$. If then E is the hyperfunction extension of $F+G' - S$ used in Part 1 we see that

$$A' - (\bar\partial_z+d_t)E = (\bar\partial_z+d_t)(E_0-E) = (\bar\partial_z+d_t)\Big[E_0-E-(\bar\partial_z+d_t)(\varphi+\psi)\Big].$$

Since according to (IV.3.39) $E_0 = E + (\bar\partial_z+d_t)(\varphi+\psi)$ in $\Omega_0\backslash U$ this proves that the class $i_U^q([f])$ in $H_U^{m,m+q}(\Omega_0)$ vanishes.

3. *The map i_U^q is injective*

The argument is exactly the same as in Part IV, section III.3, after substitution of t for z' and deletion of $\bar\partial_{z'}$. We use the fact that the cohomology (in $\mathbb{C}^m\times\mathbb{R}^n$) with coefficients in the sheaf ${}_m\mathcal{O}^{(m)}$ is the same whether we use the "hyperfunction resolution" (IV.3.4) or the "smooth resolution"

$$0 \to {}_m\mathcal{O}^{(m)} \xrightarrow{i} \mathcal{C}^\infty\Lambda^{m,0} \xrightarrow{\bar\partial_z+d_t} \mathcal{C}^\infty\Lambda^{m,1} \xrightarrow{\bar\partial_z+d_t} \cdots$$
$$\xrightarrow{\bar\partial_z+d_t} \mathcal{C}^\infty\Lambda^{m,m} \to 0.$$

[$\mathscr{C}^{\infty}\Lambda^{p,q}$ stands for the sheaf of germs of differential forms (IV.3.5) with C^{∞} coefficients $f_{I,J,K}$]. The role of Lemma III.3.3 is played by the following consequence of Theorem IV.2.1:

LEMMA IV.3.6.— *Let* F *be a closed subset of* $\Pi\times\Pi_0'$ *satisfying* (IV.2.1). *Given any form*

$$h = \sum_{k=0}^{m} (-1)^{k-1} h^{(k)} \in C^{\infty}(\Pi\times\Pi_0'\backslash F;\Lambda^{m,m+q}),$$

with $h^{(k)} \in C^{\infty}(\Pi\times\Pi_0'\backslash F;\Lambda^{m,m-k,q+k})$, *there is* $v \in C^{\infty}(\Pi\times\Pi_0'\backslash F;\Lambda^{m,m-1,q})$ *such that*

$$h - (\bar\partial_z + d_t)v = \sum_{k=1}^{m} (-1)^{k-1}\tilde h^{(k)}$$

with $\tilde h^{(k)} \in C^{\infty}(\Pi\times\Pi_0'\backslash F;\Lambda^{m,m-k,q+k})$.

Proof of Lemma IV.3.6: If $h^{(0)}$ does not vanish identically we write $h^{(0)} = \sum_{|J|=q} h_J^{(0)}\wedge dt_J$ with $h_J^{(0)} \in C^{\infty}(\Pi\times\Pi_0'\backslash F;\Lambda^{m,m})$. For each J, $|J| = q$, we have $\bar\partial_z h_J^{(0)} \equiv 0$. We apply Theorem IV.2.1: there is $v_J \in C^{\infty}(\Pi\times\Pi_0'\backslash F;\Lambda^{m,m-1})$ such that $\bar\partial_z v_J = h_J^{(0)}$ in $\Pi\times\Pi_0'\backslash F$. Then

$$v = \sum_{|J|=q} v_J\wedge dt_J \in C^{\infty}(\Pi\times\Pi_0'\backslash F;\Lambda^{m,m-1,q})$$

solves $\bar\partial_z v = h^{(0)}$ in $\Pi\times\Pi_0'\backslash F$; $h - (\bar\partial_z + d_t)v$ has the desired property. □

IV.4 PARTIAL REGULARITY OF ALL HYPERFUNCTION SOLUTIONS

Our purpose in the present section is to show that there is a natural isomorphism $\mathfrak{Sol}^{(q)}(U) \cong \mathfrak{Sol}^{(q)}(U,\mathcal{C}^{\infty}_{t})$ for every $q = 0,1,\ldots,n$. Throughout the section we assume $n \geq 1$, for there is nothing to prove when $n = 0$ (ie., when there are no variables t); and also $m \geq 1$, for when $m = 0$ (ie., when there are no variables z) the statements reduce to well known facts about the De Rham complex.

We continue to make use of the notation of the previous sections, and of the polyhedrons

$$\Pi_0 = \{ x \in \mathbb{R}^m;\ |x_j| < a_j,\ j = 1,\ldots,m \},$$

$$\Pi_1 = \{ y \in \mathbb{R}^m;\ |y_j| < b_j,\ j = 1,\ldots,m \},$$

$$\Pi_0' = \{ x' \in \mathbb{R}^n;\ |x_j'| < a_j',\ j = 1,\ldots,n \},$$

$$\Pi_1' = \{ y' \in \mathbb{R}^n;\ |y_j'| < b_j',\ j = 1,\ldots,n \},$$

and $\Pi = \Pi_0 + i\Pi_1$, $\Pi' = \Pi_0' + i\Pi_1'$ (see Theorem III.2.1).

We continue to deal with the coarse local embedding $(x,t) \to (Z(x,t),t)$; but now we regard it as valued in $\mathbb{C}^m \times \mathbb{C}^n$ rather than in $\mathbb{C}^m \times \mathbb{R}^n$ (it maps $\Pi_0 \times \Pi_0'$ onto $\Sigma_0 \subset \Pi \times \Pi'$). In other words, we regard it as a special kind of embedding (III.2.1), in which $Z'(x,t) \equiv t$. What is crucial for our purpose here is the fact that the submanifold Σ_0 is contained in the subspace $y' = 0$, ie., in the product set $\Pi \times \Pi_0'$.

The meaning of the relative cohomology space

$$H_U^{m,m+n+q}(\mathbb{C}^{m+n} \backslash \partial U)$$

remains the same as in Theorem III.2.2; ∂U denotes the boundary of U in the image Σ of the domain $\mathcal{U}' \subset \mathcal{U}$. The assumption in section III.2 was that U is a relatively compact subset of $\Sigma_0 = \Sigma \cap (\Pi \times \Pi')$. We shall also make use of the relative cohomology space $H_U^{m,m+q}(\Pi\times\Pi_0'\backslash\dot{U})$ introduced in section IV.3. But in the latter U is the image of $U_0\times\Pi_0'$ under the map $(x,t)\to(Z(x,t),t)$ with $U_0\subset\subset\Pi_0$; U is *not* a relatively compact subset of Σ_0 ($\dot{U}$ denotes the boundary of U in Σ_0).

This discrepancy is easily remedied. Continue to assume that U is the image of $U_0\times\Pi_0'$ under the map $(x,t)\to(Z(x,t),t)$; but now consider a polyhedron $\Pi_0'' = \{ x' \in \mathbb{R}^n;\ |x_j'| < a_j'',\ j = 1,\dots,n \}$ with $a_j'' > a_j'$ for each j and with radii a_j'' small enough that the polyhedron in (x,t)—space, $\Pi_0\times\Pi_0''$, can be identified to a relatively compact subset of $\mathcal{U}'$. Note that U is a closed subset, and $\Pi\times\Pi_0'\backslash\dot{U}$ an open subset, of $\Pi\times\Pi_0''\backslash\partial U$. According to the excision axiom (cf. proof of Lemma III.2.4) there is a natural isomorphism

$$\text{(IV.4.1)}\qquad H_U^{m,m+q}(\Pi\times\Pi_0'\backslash\dot{U}) \cong H_U^{m,m+q}(\Pi\times\Pi_0''\backslash\partial U).$$

But U is a relatively compact subset of the image of $\Pi_0\times\Pi_0''$ under the map $(x,t)\to(Z(x,t),t)$ and one is allowed to deal with the spaces $H_U^{m,m+n+q}(\Pi\times(\Pi_0''+\imath\Pi_1')\backslash\partial U)$. In other words we may as well assume U to be the image of $U_0\times\Theta_0$ with Θ_0 a relatively compact subset of Π_0' (that Θ_0 be a polyhedron is not really needed). Under these circumstances we are going to construct a natural isomorphism

$$\text{(IV.4.2)}\qquad H_U^{m,m+q}(\Pi\times\Pi_0'\backslash\partial U) \cong H_U^{m,m+n+q}(\Pi\times\Pi'\backslash\partial U)$$

$$[\cong H_U^{m,m+n+q}(\mathbb{C}^{m+n}\backslash\partial U);$$

cf. Remark III.2.4] which, when composed with the analogue of (IV.4.1),

$$\text{(IV.4.3)}\qquad H_U^{m,m+q}(\Pi\times\Theta_0\backslash\mathring{U}) \cong H_U^{m,m+q}(\Pi\times\Pi_0'\backslash\partial U),$$

will yield the sought isomorphism,

$$\text{(IV.4.4)}\qquad H_U^{m,m+q}(\Pi\times\Theta_0\backslash\mathring{U}) \cong H_U^{m,m+n+q}(\mathbb{C}^{m+n}\backslash\partial U).$$

To construct the isomorphism (IV.4.2) we are going to avail our-selves of the following segments of the exact sequences for relative cohomology:

$$H^{m,m+q-1}(\Pi\times\Pi_0'\backslash\partial U) \xrightarrow{\rho_0^{q-1}} H^{m,m+q-1}(\Pi\times\Pi_0'\backslash \mathscr{C}\ell U) \xrightarrow{\delta_0^q}$$

$$H_U^{m,m+q}(\Pi\Pi_0'\backslash\partial U) \xrightarrow{\iota_0^q} H^{m,m+q}(\Pi\times\Pi_0'\backslash\partial U) \xrightarrow{\rho_0^q} H^{m,m+q}(\Pi\times\Pi_0'\backslash \mathscr{C}\ell U)$$

[cf. remarks preceding (IV.3.20); here we write ρ_0^q, δ_0^q and ι_0^q to avoid confusion with the analogous maps in the exact sequence with $\Pi\times\Pi'$ in the place of $\Pi\times\Pi_0'$];

$$H^{m,m+n+q-1}(\Pi\times\Pi'\backslash\partial U) \xrightarrow{\rho^{q-1}} H^{m,m+n+q-1}(\Pi\times\Pi'\backslash \mathscr{C}\ell U) \xrightarrow{\delta^q}$$

$$H_U^{m,m+n+q}(\Pi\times\Pi'\backslash\partial U) \xrightarrow{\iota^q} H^{m,m+n+q}(\Pi\times\Pi'\backslash\partial U)$$

$$\xrightarrow{\rho^q} H^{m,m+n+q}(\Pi\times\Pi'\backslash \mathscr{C}\ell U)$$

[cf. remarks preceding (III.3.5)]. We shall construct natural isomorphisms

$$\text{(IV.4.5)}\qquad H^{m,m+n+q}(\Pi\times\Pi'\backslash\mathcal{K}) \cong H^{m,m+q}(\Pi\times\Pi_0'\backslash\mathcal{K})$$

for all $q = -1,0,1,\ldots,n$ (recall that $m \geq 1$) and $\mathcal{K} = \partial U$ or $\mathcal{K} = \mathscr{C}\ell U$. With the vertical arrows representing the isomorphisms (IV.4.5) the diagram

$$\begin{array}{ccc} H^{m,m+n+q}(\Pi\times\Pi'\backslash\partial U) & \xrightarrow{\rho^q} & H^{m,m+n+q}(\Pi\times\Pi'\backslash\mathscr{C}\ell U) \\ \downarrow & & \downarrow \\ H^{m,m+q}(\Pi\times\Pi_0'\backslash\partial U) & \xrightarrow{\rho_0^q} & H^{m,m+q}(\Pi\times\Pi_0'\backslash\mathscr{C}\ell U) \end{array}$$

will be commutative. As a consequence we will get natural isomorphisms

$$\mathrm{Ker}\ \rho^q \cong \mathrm{Ker}\ \rho_0^q,$$

$$H^{m,m+n+q-1}(\Pi\times\Pi'\backslash\mathscr{C}\ell U)/\mathrm{Im}\ \rho^{q-1} \cong H^{m,m+n+q-1}(\Pi\times\Pi_0'\backslash\mathscr{C}\ell U)/\mathrm{Im}\ \rho_0^{q-1},$$

and then the isomorphism (IV.4.2) will follow directly from the isomorphisms (III.3. 5) and (IV.3.20). We recall that the latter isomorphisms can be described quite explicitely, by means of the right inverse $i_U^q \circ \kappa^q$ of the map ι^q produced in the proofs of Theorems III.2.2 and IV.3.2.

We may and shall use the smooth resolution of the sheaf ${}_m\mathcal{O}^{m)}$, both in (z,t)—space $\Pi\times\Pi_0'$ where the differential operator is $\bar\partial_z + d_t$, and in (z,z')—space $\Pi\times\Pi'$, where the differential operator is $\bar\partial_z + d_{z'}$. Thus a class $[f]$ in $H^{m,n+q}(\Pi\times\Pi'\backslash\mathcal{K})$ will have a representative

$$\text{(IV.4.6)}\ f = \sum_{|I|+|J|+|K|=n+q} f_{I,J,K}\, dz_1\wedge\cdots\wedge dz_m \wedge d\bar z_I \wedge dx'_J \wedge dy'_K$$

with coefficients $f_{I,J,K} \in C^\infty(\Pi\times\Pi'\backslash\mathcal{K})$, satisfying $(\bar\partial_z + d_{z'})f \equiv 0$. However small the number $\epsilon > 0$ we can find a smooth form u of type (m,n+q—1) satisfying $(\bar\partial_z + d_{z'})u = f$ in the region $\{\ (z,z') \in \Pi\times\Pi';\ |y'| > \frac{1}{4}\epsilon\ \}$, keeping in mind that $y' \equiv 0$ on $\mathcal{K}$. [An easy way to construct such a C^∞ form u is by decomposing the given form f as we are going to do in a moment, and then by solving a string of equations of the kind (IV.4.8), (IV.4.9) below.] Select $\psi \in C^\infty(\mathbb{R}^n)$, $\psi(y') \equiv 0$ if $|y'| < \frac{1}{2}\epsilon$, $\psi(y') \equiv 1$ if

$|y'| > \epsilon$; $f - (\bar\partial_z + d_{z'})(\psi u)$ defines the same cohomology class as f and vanishes identically if $|y'| > \epsilon$. In other words we may assume supp $f \subset \Pi\times\Pi'^{\epsilon}$ with $\Pi'^{\epsilon} = \{ z' \in \Pi';\ |y'| \leq \epsilon \} \subset \Pi'$; if this is so we write

$$f \in \mathcal{C}^{\infty}_{\star}(\Pi\times\Pi'^{\epsilon}\backslash\mathcal{K};\tilde\Lambda^{m,n+q}).$$

If $f \in \mathcal{C}^{\infty}_{\star}(\Pi\times\Pi'^{\epsilon}\backslash\mathcal{K};\tilde\Lambda^{m,n+q})$ and if (z,x') stays in $\Pi\times\Pi'_0\backslash\mathcal{K}$ one can integrate f with respect to the variables y'_j ($j = 1,\dots,n$) alone. The integral of f with respect to y', $\int_{\Pi'_1} f$, is a differential form in $\Pi\times\Pi'_0\backslash\mathcal{K}$, of type (m,q) in the sense of the differential complex $(\bar\partial_z + d_{x'})$.

THEOREM IV.4.1.— *Let $\mathcal{K}$ be ∂U or $\mathcal{C}\ell$U. The "fibre integral"*

$$\text{(IV.4.7)}\quad \mathcal{C}^{\infty}_{\star}(\Pi\times\Pi'^{\epsilon}\backslash\mathcal{K};\tilde\Lambda^{m,n+q}) \ni f \to \int_{\Pi'_1} f \in \mathcal{C}^{\infty}(\Pi\times\Pi'_0\backslash\mathcal{K};\Lambda^{m,q})$$

induces an isomorphism of $H^{m,n+q}(\Pi\times\Pi'\backslash\mathcal{K})$ *onto* $H^{m,q}(\Pi\times\Pi'_0\backslash\mathcal{K})$.

Proof: The form $\int_{\Pi'_1} f \in \mathcal{C}^{\infty}(\Pi\times\Pi'_0\backslash\mathcal{K};\Lambda^{m,q})$ is closed since in $\Pi\times\Pi'_0\backslash\mathcal{K}$,

$$(\bar\partial_z + d_{x'})\int_{\Pi'_1} f = \int_{\Pi'_1} (\bar\partial_z + d_{z'}) f - \int_{\Pi'_1} d_{y'} f = 0.$$

Indeed the first term vanishes due to the fact that f is closed, and so does the second term, due to the fact that $|y'| \leq \epsilon$ on supp f.

On the other hand, if

$$f = (\bar\partial_z + d_{z'})v \text{ with } v \in \mathcal{C}^{\infty}(\Pi\times\Pi'\backslash\mathcal{K};\tilde\Lambda^{m,n+q-1})$$

we have the following relations between integrals in y'—space, for each $(z,x') \in \Pi\times\Pi'_0\backslash\mathcal{K}$,

$$\int_{\Pi_1'} f = \int_{|y'|<\epsilon} f = \int_{|y'|<\epsilon} (\bar\partial_z + d_{x'})v = (\bar\partial_z + d_{x'})\int_{|y'|<\epsilon} v,$$

which shows that the cohomology class of $\int_{\Pi_1'} f$ in $H^{m,q}(\Pi\times\Pi_0'\backslash\mathcal{K})$ vanishes, and thus that the map (IV.4.7) induces a linear map

$$H^{m,n+q}(\Pi\times\Pi) \to H^{m,q}(\Pi\times\Pi_0'\backslash\mathcal{K}).$$

We rewrite the form (IV.4.6) in the following manner:

$$f = \sum_{r=0}^{n} (-1)^{r-1} f^{(r)}$$

where

$$f^{(r)} = (-1)^{r-1} \sum_{|I|+|J|=n+q-r} \sum_{|K|=r} f_{I,J,K} dz_1\wedge\cdots\wedge dz_m \wedge d\bar z_I \wedge dx'_J \wedge dy'_K$$

(cf. section III.3). The equation $(\bar\partial_z + d_{z'})f = 0$ is equivalent to the string of equations

(IV.4.8)
$$(\bar\partial_z + d_{x'})f^{(0)} = 0,$$

(IV.4.9)
$$(\bar\partial_z + d_{x'})f^{(r)} = d_{y'} f^{(r-1)} \quad (r = 1,\dots,n).$$

Recalling that $f \equiv 0$ if $|y'| > \epsilon$ we are going to use the following homotopy operator for $d_{y'}$. Consider a smooth form in $\mathbb{R}^n\backslash\{0\}$,

$$g = \sum_{|K|=r} g_K dy'_K$$

such that $g \equiv 0$ if $|y'| > \epsilon$. If $K = \{k_1,\dots,k_r\}$ set

$$\omega'_K = \sum_{\alpha=1}^{r} (-1)^{\alpha-1} y'_{k_\alpha} dy'_{k_1}\wedge\cdots\wedge dy'_{k_{\alpha-1}} \wedge dy'_{k_{\alpha+1}}\wedge\cdots\wedge dy'_{k_r}$$

and

$$\text{(IV.4.10)} \qquad \mathfrak{J}g(y') = \sum_{|K|=r} \left[-\int_1^{+\infty} g_K(\lambda y')\lambda^{r-1}d\lambda \right] \omega'_K .$$

It is clear that $\mathfrak{J}g$ is an (r—1)—form in $\mathbb{R}^n \backslash \{0\}$ that vanishes identically if $|y'| > \epsilon$. And it is checked at once that

$$\text{(IV.4.11)} \qquad g = \mathfrak{J}d_{y'}g \quad \textit{if } r = 0,$$

$$\text{(IV.4.12)} \qquad g = d_{y'}\mathfrak{J}g + \mathfrak{J}d_{y'}g \quad \textit{if } 1 \leq r \leq n.$$

Let us go back to a form

$$f^{(r)} = \sum_{|I|+|J|=n+q-r} F_{I,J} \wedge dz_1 \wedge \cdots \wedge dz_m \wedge d\bar{z}_I \wedge dx'_J$$

where

$$F_{I,J} = (-1)^{r(m+n+q)-1} \sum_{|K|=r} f_{I,J,K} dy'_K .$$

We let $\mathfrak{J}$ act on $f^{(r)}$ coefficientwise:

$$\mathfrak{J}f^{(r)} = \sum_{|I|+|J|=n+q-r} (\mathfrak{J}F_{I,J}) \wedge dz_1 \wedge \cdots \wedge dz_m \wedge d\bar{z}_I \wedge dx'_J .$$

In defining $\mathfrak{J}F_{I,J}$ we regard (z,x') as parameters. It follows from the preceding expression that

$$(\bar{\partial}_z + d_{x'})\mathfrak{J}f^{(r)} = -\mathfrak{J}(\bar{\partial}_z + d_{x'})f^{(r)} .$$

We derive from (IV.4.9) where $r = 1$, and from (IV.4.11):

$$f^{(0)} = -(\bar{\partial}_z + d_{x'})\mathfrak{J}f^{(1)},$$

and from (IV.4.9) and (IV.4.12), when $1 \leq r \leq n$,

$$f^{(r)} = d_{y'}\mathfrak{J}f^{(r)} - (\bar{\partial}_z + d_{x'})\mathfrak{J}f^{(r+1)}$$

with the understanding that $f^{(n+1)} \equiv 0$. We reach the conclusion that, in the region $y' \neq 0$,

(IV.4.13) $$f = (\bar\partial_z + \mathrm{d}_{z'})u, \quad u = \sum_{r=1}^{n} (-1)^{r-1}\mathfrak{J}f^{(r)}.$$

We return to the definition (IV.4.10) of $\mathfrak{J}g$. If $\dot{y}' = y'/|y'|$ we see that

$$\mathfrak{J}g(y') = \sum_{|K|=r} \left[-\int_{|y'|}^{+\infty} g_K(s\dot{y}')s^{r-1}\mathrm{d}s \right] \omega'_K/|y'|^r.$$

Recalling the expression of ω'_K we see that, if the coefficients of g are bounded in a full neighborhood of $y' = 0$, those of $\mathfrak{J}g(y')$ will be bounded, in absolute value, by $const.|y'|^{1-r}$; and they vanish identically for $|y'| > \epsilon$. This remark shows that every partial derivative with respect to (z,x') of each coefficient of the form $|y'|^{r-1}u$ is locally bounded in $\Pi\times\Pi'\backslash\mathcal{K}$. Thus $u = \sum_{r=1}^{n} (-1)^{r-1}\mathfrak{J}f^{(r)}$ is a current of finite order in $\Pi\times\Pi'\backslash\mathcal{K}$ which depends smoothly on (z,x'). Thanks to (IV.4.13), $f - (\bar\partial_z+\mathrm{d}_{z'})u$ is a current carried by the linear subspace $y' = 0$, ie., by $\Pi\times\Pi'_0\backslash\mathcal{K}$. This means that

(IV.4.14) $$f - (\bar\partial_z+\mathrm{d}_{z'})u = \sum G_{\alpha,K}\otimes\delta^{(\alpha)}(y')\wedge \mathrm{d}y'_K,$$

where the sum is finite and $G_{\alpha,K} \in C^\infty(\Pi\times\Pi'_0\backslash\mathcal{K};\Lambda^{m,n+q-r})$ if $|K| = r$. If G denotes the right–hand side in (IV.4.14) of course $(\bar\partial_z+\mathrm{d}_{z'})G \equiv 0$.

Consider the differential complex:

(IV.4.15) $$\mathrm{d}_{y'}: \Lambda^p\mathcal{D}'_0(\mathbb{R}^n) \to \Lambda^{p+1}\mathcal{D}'_0(\mathbb{R}^n), \quad p = 0,1,\ldots,$$

where $\mathcal{D}'_0(\mathbb{R}^n)$ is the linear space of distributions supported at the origin, ie., of linear combination of partial derivatives of the Dirac measure $\delta(y')$. By using the duality between $\mathcal{D}_0'(\mathbb{R}^n)$ and the space $\mathbb{C}[[X]]$ of formal power series in n indeterminates $X_1, \ldots, X_n$, one can establish a duality between the differential complex (IV.4.15) and the differential complex

$$\text{(IV.4.16)} \qquad d\colon \Lambda^p\mathbb{C}[[X]] \to \Lambda^{p+1}\mathbb{C}[[X]], \quad p = 0,1,\ldots;$$

and also between their cohomology spaces of complementary degrees. The cohomology of (IV.4.16) vanishes in all degrees > 0, and is one–dimensional in degree zero. As a consequence the cohomology of (IV.4.15) vanishes in all degrees $< n$ and is generated, in degree n, by the class of $\delta(y')dy_1'\wedge\cdots\wedge dy_n'$. The same remains true if we allow the coefficients to be elements of the spaces $C^\infty(\Pi\times\Pi_0'\backslash\mathcal{K};\Lambda^{m,p})$. It follows that

$$\sum_\alpha G_{\alpha,[1,\ldots,n]}\otimes\delta^{(\alpha)}(y')\wedge dy_1'\wedge\cdots\wedge dy_n' = f_0\otimes\delta(y')\wedge dy_1'\wedge\cdots\wedge dy_n' +$$
$$d_{y'}\left[\sum_\beta \sum_{|K|=n-1} u_{\beta,K}^{(1)}\otimes\delta^{(\beta)}(y')\wedge dy_K'\right]$$

for certain forms f_0, $u_{\beta,K}^{(1)} \in C^\infty(\Pi\times\Pi_0'\backslash\mathcal{K};\Lambda^{m,q})$. We note that

$$G - f_0\otimes\delta(y')\wedge dy_1'\wedge\cdots\wedge dy_n' -$$
$$(\bar\partial_z + d_{z'})\left[\sum_\beta \sum_{|K|=n-1} u_{\beta,K}^{(1)}\otimes\delta^{(\beta)}(y')\wedge dy_K'\right] =$$
$$\sum_\alpha \sum_{|K|\le n-1} G_{\alpha,K}^{(1)}\otimes\delta^{(\alpha)}(y')\wedge dy_K',$$

where $G_{\alpha,K}^{(1)} \in C^\infty(\Pi\times\Pi_0'\backslash\mathcal{K};\Lambda^{m,q+1})$. Then perforce,

$$d_{y'}\left[\sum_\alpha \sum_{|K|=n-1} G_{\alpha,K}^{(1)}\otimes\delta^{(\alpha)}(y')\wedge dy_K'\right] \equiv 0.$$

If $n = 1$ this means that all the coefficients $G_{\alpha,K}^{(1)}$ must vanish identically. If $n \geq 2$ we have

$$\sum_{\alpha}\sum_{|K|=n-1} G^{(1)}_{\alpha,K}\otimes\delta^{(\alpha)}(y')\wedge dy'_K = d_{y'}\left[\sum_{\beta}\sum_{|K|=n-2} u^{(2)}_{\beta,K}\otimes\delta^{(\beta)}(y')\wedge dy'_K\right]$$

for some forms $u^{(2)}_{\beta,K} \in \mathcal{C}^{\infty}(\Pi\times\Pi'_0\backslash\mathcal{K};\Lambda^{m,q+1})$, and therefore

$$\sum_{\alpha}\sum_{|K|\leq n-1} G^{(1)}_{\alpha,K}\otimes\delta^{(\alpha)}(y')\wedge dy'_K = \sum_{\gamma}\sum_{|K|\leq n-2} G^{(2)}_{\gamma,K}\otimes\delta^{(\gamma)}(y')\wedge dy'_K + (\bar\partial_z+d_{z'})\left[\sum_{\beta}\sum_{|K|=n-2} u^{(2)}_{\beta,K}\otimes\delta^{(\beta)}(y')\wedge dy'_K\right],$$

where $G^{(2)}_{\gamma,K} \in \mathcal{C}^{\infty}(\Pi\times\Pi'_0\backslash\mathcal{K};\Lambda^{m,q+2})$. Etc. After n steps we shall conclude that

$$G - f_0\otimes\delta(y')\wedge dy'_1\wedge\cdots\wedge dy'_n = (\bar\partial_z+d_{z'})\left[\sum_{\beta}\sum_{j=1}^{n}\sum_{|K|=n-j} u^{(j)}_{\beta,K}\otimes\delta^{(\beta)}(y')\wedge dy'_K\right]. \tag{IV.4.17}$$

Combining (IV.4.14) and (IV.4.17) shows that

$$f \cong f_0\otimes\delta(y')\wedge dy'_1\wedge\cdots\wedge dy'_n, \tag{IV.4.18}$$

whence we conclude that $\int_{\Pi'_1} f = f_0$.

This entails all we need: The map (IV.4.7) is *surjective* since it has an inverse, namely the map that assigns the class

$$[f_0\otimes\delta(y')\wedge dy'_1\wedge\cdots\wedge dy'_n] \in H^{m,n+q}(\Pi\times\Pi'\backslash\mathcal{K})$$

to the class $[f_0] \in H^{m,q}(\Pi\times\Pi'_0\backslash\mathcal{K})$. The map (IV.4.7) is *injective*: When q $\geq$ 1, if there is $u_0 \in \mathcal{C}^{\infty}(\Pi\times\Pi'_0\backslash\mathcal{K};\Lambda^{m,q-1})$ satisfying $(\bar\partial_z+d_{x'})u_0 = f_0$ then

$$f_0 \otimes \delta(y')\wedge dy'_1 \wedge \cdots \wedge dy'_n = (\bar{\partial}_z + d_{z'})[u_0 \otimes \delta(y')\wedge dy'_1 \wedge \cdots \wedge dy'_n],$$

which implies that the class of f vanishes, by (IV.4.18). When q = 0, $f_0 \equiv 0 \Rightarrow [f] = 0$, again by (IV.4.18). □

REMARK IV.4.1.— In the last part of the proof we have made use of the distribution resolution of the sheaf ${}_m\mathcal{O}^{(m)}$ to compute the $(\bar{\partial}_z + d_{x'})$ and $(\bar{\partial}_z + d_{z'})$ cohomologies. □

Theorem IV.4.1 tells us that in dealing with an arbitrary element $[f]$ of $\mathfrak{Sol}(U)$ we have the right to represent it by an analytic functional $\mu_0(t) \in \mathfrak{B}(\bar{U})$ such that $\partial\mu_0/\partial t_j \in \mathfrak{B}(\dot{U})$, j = 1,...,n. It does not tell us, however, how to go directly from an analytic functional $\mu \in \mathcal{O}'(\mathcal{C}\ell U)$ such that $\partial\mu/\partial z'_j \in \mathcal{O}'(\partial U)$, j = 1,...,n, to an element $\mu_0(t) \in \mathfrak{B}(\bar{U})$ representing in U the same solution as μ. In order to go from μ to $\mu_0(t)$ one must follow the path of the constructions in the proofs of Theorems III.2.2 and IV.3.2 (these constructions precisely deal with the integral transforms of analytic functionals representating the hyperfunction solutions under consideration); and insert between the two the isomorphism of (IV.4.4),

$$H_U^{m,m+n}(\mathbb{C}^{m+n}\backslash\partial U) \to H_U^{m,m}(\Pi\times\Theta_0\backslash\dot{U}).$$

The analoguous remark is valid for any element $[f] \in \mathfrak{Sol}^{(q)}(U)$ when q $\geq$ 1. At any rate we may state:

THEOREM IV.4.2.— *For each* q = 0,1,...,n, *integration with respect to* $\mathcal{I}m\, z'$ *induces an isomorphism* $\mathfrak{Sol}^{(q)}(U) \cong \mathfrak{Sol}^{(q)}(U, C_t^\infty)$.

IV.5 CONSEQUENCES AND ADDITIONAL REMARKS

Uniqueness and supports of hyperfunction solutions

We continue to deal with the coarse local embedding

$$\mathcal{M} \supset \mathcal{M}' \ni (x,t) \to (Z(x,t),t) \in \Sigma \subset \mathbb{C}^m \times \mathbb{R}^n. \tag{IV.5.1}$$

Let $\mathcal{X}$ denote a maximally real submanifold of $\mathcal{M}$ whose intersection with the domain $\mathcal{M}'$ is defined by the equations $t = 0$. The map (IV.5.1) transforms the submanifold $\mathcal{X} \cap \mathcal{M}'$ into the slice $\Sigma \cap (\mathbb{C}^m \times \{0\})$. Let O be the point $x = 0$, $t = 0$ of $\mathcal{X} \cap \mathcal{M}'$.

As before we identify the polytope $\Pi_0 \times \Pi_0'$ to a subset of $\mathcal{M}'$. Let U_0 be a relatively compact open subset of Π_0, U the image of $U_0 \times \Pi_0'$ under the map (IV.5.1), $\overline{U}$ (resp., $\dot{U}$) the closure (resp., boundary) of U in $\Sigma_0 = \Sigma \cap (\Pi \times \Pi_0')$. We know by Theorem IV.4.2 that any hyperfunction solution u in $\mathcal{M}'$ depends smoothly on t and therefore, that the trace of u on an arbitrary slice U_{t_0} ($t_0 \in \Pi_0'$) is well defined. The invariant definition of the sheaf of hyperfunction solutions (see end of section III.2) allows us to state the analogue of Corollary I.4.1 in [TREVES, 1992]:

THEOREM IV.5.1.— *Let $\mathcal{X}$ be a closed, maximally real submanifold of $\mathcal{M}$, u a hyperfunction solution defined in an open neighborhood of $\mathcal{X}$ in $\mathcal{M}$. The trace of u on $\mathcal{X}$ is a well defined hyperfunction in $\mathcal{X}$.*

We can also generalize the uniqueness results about distribution solutions in Section II.3 of [TREVES, 1992]. For this we are going to need a lemma about the carriers of analytic functionals, in a situation where there are no variables t_j, ie., $n = 0$. Let $\mathcal{X}$ be a maximally real submanifold of $\mathbb{C}^m$ and Ω an open subset of $\mathcal{X}$. As usual assume that the origin lies in Ω and that $\mathcal{X}$ is tangent to $\mathbb{R}^m$ at 0. We are going to assume that Ω is sufficiently small to be strongly polynomially convex (Definition I.6.1 and Lemma I.6.1), and that the following condition is satisfied, for some number $c_0 > 0$,

$$\text{(IV.5.2)} \qquad \mathcal{R}e\langle z-z'\rangle^2 \geq c_0|z-z'|^2, \ \forall\, z, z' \in \Omega.$$

LEMMA IV.5.1.— *Let* U *be a relatively compact open subset of* Ω. *For every* $\delta > 0$ *set*

$$\mathcal{T}_\delta = \{ z \in \mathbb{C}^m;\ \mathrm{dist}(z,\partial U) \leq \delta \},\ U_\delta = \{ z \in U;\ \mathrm{dist}(z,\partial U) > \delta \},$$

and call $\hat{K}_\delta$ *the polynomial convex hull of* $U\cup\mathcal{T}_\delta$. *If* $\epsilon > 0$ *is sufficiently small then*

$$\text{(IV.5.3)} \qquad \hat{K}_\epsilon \subset U\cup\mathcal{T}_\delta,$$

$$\text{(IV.5.4)} \qquad \hat{K}_\epsilon\backslash U_\delta \text{ is polynomially convex.}$$

Proof: Let z_0 be any point in U_δ. If $z_1 \in \mathcal{T}_\epsilon$ we can write $z_1 = z_* + \zeta$ with $z_* \in \partial U$, $\zeta \in \mathbb{C}^m$, $|\zeta| \leq \epsilon$. We have, for a suitably large (universal) constant $C > 0$,

$$\mathcal{R}e\langle z_1-z_0\rangle^2 \geq \tfrac{1}{2}\, \mathcal{R}e\langle z_*-z_0\rangle^2 - C\epsilon^2.$$

Property (IV.5.2) entails $\mathcal{R}e\langle z_1-z_0\rangle^2 \geq \frac{1}{2}c_0\delta^2 - C\epsilon^2$; and thus, if $\epsilon/\delta < \sqrt{c_0/2C}$ we shall have

$$\text{(IV.5.5)} \qquad \mathcal{R}e\langle z_1-z_0\rangle^2 \geq c_1 > 0,\ \forall\, z_0 \in U_\delta,\ z_1 \in \mathcal{T}_\epsilon.$$

This will remain true if we decrease both δ and ϵ (while preserving $\epsilon/\delta < \sqrt{c_0/2C}$).

Set $\mathcal{E}(z) = \exp(-\langle z\rangle^2)$. If $z_0 \in U_\delta$, $|\mathcal{E}(z-z_0)| < 1$ when $z \in \Omega\backslash\{z_0\}$ thanks to (IV.5.2); also when $z \in \mathcal{T}_\epsilon$ thanks to (IV.5.5). But for any $z = z_0+\imath\eta$, $\eta \in \mathbb{R}^m$, $|\mathcal{E}(z-z_0)| \geq 1$; and $|\mathcal{E}(z-z_0)| > 1$ if $\eta \neq 0$. It follows that $U_\delta+ \imath(\mathbb{R}^m\backslash\{0\})$ cannot intersect $\hat{K}_\epsilon$ and therefore

$$\hat{K}_\epsilon \cap (U_\delta+\imath\mathbb{R}^m) = U_\delta. \tag{IV.5.6}$$

Since $\mathscr{C}\ell U$ is polynomially convex, to each δ', $0 < \delta' < \delta$, there is $\epsilon > 0$ such that $\hat{K}_\epsilon \subset \{ z \in \mathbb{C}^m;\ \mathrm{dist}(z,U) < \delta' \}$. By (IV.5.6) in which δ' is substituted for δ, this implies

$$\hat{K}_\epsilon \subset U \cup \{ z \in \mathbb{C}^m\backslash(U_{\delta'}+\imath\mathbb{R}^m);\ \mathrm{dist}(z,U) < \delta' \}.$$

Assume $\mathrm{dist}(z,U) < \delta'$. If δ' is sufficiently small we can write $z = z_*+\imath y$ with $z_* \in \Omega$ and $y \in \mathbb{R}^m$. But then $z \notin U_{\delta'}+\imath\mathbb{R}^m \Rightarrow z_* \notin U_{\delta'}$. We are left with two possibilities: $z_* \in U$ and $\mathrm{dist}(z_*,\partial U) \leq \delta'$; or else $z_* \in \Omega\backslash U$. There is $C' > 0$ (depending only on Ω) such that $|z-z_*| < C'\delta'$; if $z_* \in \Omega\backslash U$ this implies $\mathrm{dist}(z_*,\partial U) < (C'+1)\delta'$, which is then true in any case and implies $\mathrm{dist}(z,\partial U) < (2C'+1)\delta'$. Selecting $(2C'+1)\delta' < \delta$ ensures that $\{ z \in \mathbb{C}^m\backslash(U_{\delta'}+\imath\mathbb{R}^m);\ \mathrm{dist}(z,U) < \delta' \} \subset \mathcal{T}_\delta$ whence (IV.5.3).

By the earlier argument applied with $\eta = 0$ we see that, if $\epsilon' > 0$ is sufficiently small, $z_0 \in U_\delta$ cannot belong to the polynomial convex hull of $(\mathcal{T}_{\epsilon'} \cup U)\backslash U_\delta$. By (IV.5.3) we can select ϵ so small that $\hat{K}_\epsilon \subset \mathcal{T}_{\epsilon'} \cup U$. We conclude that U_δ does not intersect the polynomial convex hull of

$\hat{K}_\epsilon \backslash U_\delta$. Since this hull is contained in $\hat{K}_\epsilon$ it must be equal to $\hat{K}_\epsilon \backslash U_\delta$. This proves (IV.5.4). □

LEMMA IV.5.2.— *Let* U *be as in Lemma* IV.5.1. *If* δ *and* ϵ/δ *are sufficiently small any analytic functional* μ *carried by both* $\mathcal{Cl}\,U$ *and* $\mathcal{T}_\epsilon$ *is carried by* $\mathcal{Cl}\,U \backslash U_\delta$.

Proof: Select δ and ϵ so as to satisfy the requirements in Lemma IV.5.1; $\mathcal{Cl}\,U_\delta$ is polynomially convex and by Lemma IV.5.1, so is $\hat{K}_\epsilon \backslash U_\delta$. Equate μ to a sum $\mu_1 + \mu_2$ with $\mu_1 \in \mathcal{O}'(\mathcal{Cl}\,U_\delta)$, $\mu_2 \in \mathcal{O}'(\mathcal{Cl}\,U \backslash U_\delta)$. By hypothesis μ is carried by $\mathcal{T}_\epsilon \subset \hat{K}_\epsilon \backslash U_\delta$; so is μ_2 and therefore so is also μ_1. Since $(\mathcal{Cl}\,U_\delta) \cup (\hat{K}_\epsilon \backslash U_\delta) = \hat{K}_\epsilon$ is holomorphically convex we can apply Proposition I.6.4: μ_1 is carried by $(\mathcal{Cl}\,U_\delta) \cap (\hat{K}_\epsilon \backslash U_\delta) \subset \mathcal{Cl}\,U \backslash U_\delta$, whence the result. □

THEOREM IV.5.2.— *Let* $\mathcal{X}$ *be a closed, maximally real submanifold of* $\mathcal{M}$, *u a hyperfunction solution defined in an open neighborhood of* $\mathcal{X}$ *in* $\mathcal{M}$. *If the trace of u on* $\mathcal{X}$ *vanishes identically, then* $u \equiv 0$ *in an open neighborhood of* $\mathcal{X}$ *in* $\mathcal{M}$.

Proof: We return to the notation introduced at the beginning of this section. Select arbitrarily a representative $\mu \in \mathfrak{G}(\overline{U})$ of u and use the mean value theorem

$$\mu(t) - \mu(0) = \int_0^t d_s\mu(s)$$

where the integration is carried out on the straight–line segment $[0,t] \subset \Pi_0'$. Since $d_t\mu$ is carried by $\mathring{U}$, $\mu(t) - \mu(0)$ is carried by the union of the sets

∂U_s, $s \in [0,t]$. Given any $\epsilon > 0$ there is $\eta > 0$ such that, whatever $t \in \Pi'_0$, $|t| \leq \eta$, the union of all the sets ∂U_s, $s \in \Pi'_0$, $|s| \leq \eta$, is contained in a tubular neighborhood

$$\mathcal{T}_\epsilon(t) = \{ z \in \mathbb{C}^m;\ \mathrm{dist}(z, \partial U_t) < \epsilon \}.$$

Then, if $\mu(0)$ is carried by ∂U_0 we get that $\mu(t)$ is carried by $\mathcal{T}_\epsilon(t)$ for all $t \in \Pi'_0$, $|t| \leq \eta$. On the other hand, given any $\delta > 0$, we can select ϵ so small as to satisfy the requirements in Lemma IV.5.2. It will follow that $\mu(t)$ is carried by the set $\{ z \in \mathcal{C}\ell U_t;\ \mathrm{dist}(z, \partial U_t) \leq \delta \}$. By selecting δ sufficiently small we conclude that the hyperfunction solution u vanishes in a full neighborhood of O in $\mathcal{M}$. □

Since the specialization of Theorem IV.5.2 to distribution solutions is the only property needed to prove Theorem II.3.3 in [TREVES, 1992] the analogous result is valid for hyperfunction solutions. That result is based on the concept of *orbit* in a hypo–analytic manifold $\mathcal{M}$ (see Section I.11, [TREVES, 1992]), which we now recall. Let $\mathcal{V}$ be the tangent structure bundle of $\mathcal{M}$ (see section III.1 in the present text). An orbit (of $\mathcal{V}$) in an open subset Ω of $\mathcal{M}$ is an equivalence class $\mathcal{L}$ of points in Ω for the equivalence relation $\wp \approx \wp_*$ defined as follows: there is a sequence of points $\wp = \wp_0, \wp_1, \ldots, \wp_r = \wp_*$ such that, for every $i = 1, \ldots, r$, $\wp_{i-1}$ and $\wp_i$ lie on one and the same integral curve of some vector field $\mathcal{R}e\,L$, $L \in C^\infty(\mathcal{M};\mathcal{V})$ (ie., L a C^∞ section of $\mathcal{V}$ over $\mathcal{M}$). A theorem in [SUSSMAN, 1973] states that every orbit $\mathcal{L}$ is an immersed submanifold of Ω. Denote by $\mathfrak{g}(V)$ the Lie algebra, for the commutation bracket, generated by all the vector fields $\mathcal{R}e\,L$, $L \in C^\infty(\mathcal{M};\mathcal{V})$. Given any point $\wp \in \mathcal{L}$, the freezing of $\mathfrak{g}(V)$ at $\wp$, is contained in $T_\wp(\mathcal{L})$; and no closed submanifold of $\mathcal{L}$ whose dimension is $< \dim \mathcal{L}$ has that property. The open set Ω is foliated by the orbits.

THEOREM IV.5.3.— *Let h be a hyperfunction solution in an open subset Ω of $\mathcal{M}$. Any orbit of $\mathcal{V}$ in Ω that intersects* supp h *is entirely contained in* supp h.

The subspaces $\mathfrak{Sol}_c^{(q)}(U, \mathcal{C}_t^\infty)$

First we look at the case q = 0. As before, let U_0 be a relatively compact open subset of Π_0 and U be the image of $U_0 \times \Pi_0'$ under the map (IV.5.1). An element $f \in \mathfrak{Sol}_c^{(0)}(U, \mathcal{C}_t^\infty)$ has a representative $\mu \in \mathfrak{G}(\overline{U})$ such that $d_t\mu = 0$, ie., μ is independent of $t \in \Pi_0'$. Let U_0' be an open subset of $\mathbb{R}^m$, with $U_0 \subset U_0' \subset\subset \Pi_0$, and call U′ the image of $U_0' \times \Pi_0'$ under the map (IV.5.1). Then μ can be identified to an element of $\mathfrak{G}(\overline{U}')$ and thus defines a hyperfunction solution in U′ which extends f; this extension vanishes identically in $U' \setminus (\text{supp } f)$. As U_0' expands to fill Π_0 we get a unique extension of f to $\Sigma_0 = \Sigma \cap (\Pi \times \Pi_0')$ that vanishes identically off the support of f; we denote it also by f. The restriction of f to each slice $\Sigma \cap (\Pi \times \{t\})$ is compactly supported (in $\overline{U}_t$); one can say that f is *compactly supported with respect to x*.

It will often be the case that $\mathfrak{Sol}_c^{(0)}(U, \mathcal{C}_t^\infty) = \{0\}$; but not always, as now shown.

EXAMPLE IV.5.1.— Define a hypo—analytic structure on $\mathbb{R}^2$ by means of the function $Z = xe^{it}$; the associated vector field is $L = \partial/\partial t - ix\partial/\partial x$.

Take as manifold Σ the image of $\mathbb{R}^2$ under the map $(x,t) \to (xe^{\imath t},t)$.

For any $k = 0,1,2,\ldots$, the distribution $h_k(x,t) = e^{-\imath(k+1)t}\delta^{(k)}(x)$ is a solution of the homogeneous equation $Lh = 0$ (see Example II.5.1, [TREVES, 1992]). Consider then a series

$$\tilde{\mu}(t) = \sum_{k=0}^{+\infty} \frac{c_k}{k!} e^{-\imath(k+1)t}\delta^{(k)}$$

in which $\delta^{(k)}$ is the kth derivative (with respect to z) of the Dirac distribution, and the complex numbers c_k satisfy an inequality

$$\sup_{k\in\mathbb{Z}_+} (A^k|c_k|) < +\infty, \ \forall\ A > 0.$$

It is clear that $\tilde{\mu}$ is a C^∞ function of $t \in \mathbb{R}$ valued in $\mathcal{O}'(\mathbb{C})$; for each t, $\tilde{\mu}(t)$ is carried by $\{0\}$. We regard $\tilde{\mu}(t)$ as an element of the dual of $\mathcal{O}^1(\{0\})$, ie., of the space of the (germs of) one—forms $h(z)dz$, with h holomorphic in an open neighborhood of 0 in $\mathbb{C}$. This is in agreement with the invariant meaning that must be given to each distribution $\delta^{(k)}$: they are elements of the dual of the one forms $\varphi(x)dx$, $\varphi \in C^\infty(\mathbb{R})$. It is also in agreement with the pairing of hdz with the Cauchy transform of $\tilde{\mu}$ [see (I.2.13)]. We transfer $\tilde{\mu}(t)$ as an analytic functional $\mu(t)$ carried by $\{0\}\times\mathbb{R} \subset \Sigma$, by setting

$$<\mu_z(t),h(z)dz> = <\tilde{\mu}_w(t),e^{\imath t}h(we^{\imath t})dw>.$$

We get

$$<\mu_z(t),h(z)dz> = \sum_{k=0}^{+\infty} (-1)^k\frac{c_k}{k!}h^{(k)}(0).$$

This shows that $\mu(t)$ is independent of t; $\mu(t)$ defines an element of $\mathfrak{Sol}^{(0)}(U,C_t^\infty)$. □

Example IV.5.1 suggests a condition for the vanishing of $\mathfrak{Sol}^{(0)}(U,C_t^\infty)$.

THEOREM IV.5.4.— *Assume that* Π_0 *and* Π_0' *are sufficiently small and that* U *is the image of* $U_0 \times \Pi_0'$ *under the map* $(x,t) \to (Z(x,t),t)$, *with* U_0 *a relatively compact open subset of* Π_0. *The following two conditions are equivalent:*

(IV.5.7) $$\mathfrak{Sol}_c^{(0)}(U,\mathcal{C}_t^\infty) \neq 0.$$

(IV.5.8) *There is an orbit* $\mathcal{L}$ *in* Σ_0 *whose projection in* Π_0 *under the map* $(Z(x,t),t) \to x$ *is contained in* $\overline{U}_0$.

Proof: If U is as in (IV.5.7) then there is a hyperfunction solution u in Σ whose support is not empty and is mapped into $\mathcal{C}\ell U_0$ under the map $(Z(x,t),t) \to x$. By Theorem IV.5.3 supp u contains an orbit in Σ_0.

If (IV.5.8) holds $\mathcal{L}$ is a closed and connected submanifold of U. Provided Π_0 and Π_0' are sufficiently small, it follows from a result of Baouendi & Rothschild (Theorem II.3.4, [TREVES, 1992]) that there is a distribution solution h in U with supp $h = \mathcal{L}$. By Proposition I.4.3, *loc. cit.*, h is a $\mathcal{C}^\infty$ function of t valued in the space of distributions of x. Thus h defines an element μ of $\mathfrak{G}(\overline{U})$. It is readily checked that $d_t\mu \equiv 0$. □

When q = n, $d_t f \equiv 0$ whatever $f \in \mathcal{B}(U,\mathcal{C}_t^\infty;\Lambda^q)$; and every such a current f extends as an element of $\mathcal{B}(\Sigma;\Lambda^n)$. But even if $1 \leq q < n$ a class $[f] \in \mathfrak{Sol}_c^{(q)}(U,\mathcal{C}_t^\infty)$ has a representative of the kind (IV.3.10), $f \in \mathcal{B}(U,\mathcal{C}_t^\infty;\Lambda^q)$, that extends as an element $\tilde{f}$ of $\mathcal{B}(\Sigma_0,\mathcal{C}_t^\infty;\Lambda^q)$. We may arrange things to have $\tilde{f} \equiv 0$ in $\Sigma_0\backslash\overline{U}$ and $d_t\tilde{f} \equiv 0$. Indeed, represent f in

U by a current μ as in (IV.3.16) with coefficients $\mu_J \in \mathfrak{G}(\overline{U})$ such that $d_t\mu \equiv 0$. By regarding each $\mu_J(t)$ (for each t) as an analytic functional carried by any compact subset of $(\Sigma_0)_t$ that contains $\overline{U}_t$ we see that μ defines (uniquely) the extenension $\tilde{f}$. Thus the nonvanishing of $\mathfrak{Sol}_c^{(q)}(U,\mathcal{C}_t^\infty)$ is equivalent to the nonsolvability of an inhomogeneous equation $d_t u = f$ for certain (d_t–closed) right–hand sides $f \in \mathcal{B}(U,\mathcal{C}_t^\infty;\Lambda^q)$ whose support pro– jects in a subset of $\overline{U}_0$ under the map $(Z(x,t),t) \to x$.

Hyperfunction solutions defined by boundary values of holomorphic functions

Let $V_0 \subset\subset U_0$ be another open subset of the ball $\Pi_0 \subset \mathbb{R}^m$ and V denote the image of $V_0 \times \Pi_0'$ under the map $(x,t) \to (Z(x,t),t)$; assume the boundary ∂V_0 of V_0 is smooth. Let g be a regular function in the wedge $\mathcal{W}_\delta(U,\Gamma)$ [see (IV.2.7)]. Recall the definition (IV.2.9) of the analytic functional $\mu^\gamma_{g,V}(t)$. We have, for all $\gamma \in \Gamma$, $|\gamma| < \delta$, and all $h \in \mathcal{O}(\mathbb{C}^m)$,

$$<\mu^\gamma_{g,V}(t),h> = \int_{V_0} g(Z(x,t)+\imath\gamma,t)h(Z(x,t)+\imath\gamma)dZ(x,t).$$

The argument that lead us to Formula (III.1.18) leads us here to the formula

$$(\partial/\partial t_j)<\mu^\gamma_{g,V}(t),h> = \int_{V_t+\imath\gamma} (\partial g/\partial t_j)h dz +$$

$$\sum_{i=1}^m \pm \int_{\partial V_0} [(gh)(Z(x,t)+\imath\gamma,t)](\partial Z_i/\partial t_j)dZ_1\wedge\cdots\wedge dZ_{i-1}\wedge dZ_{i+1}\wedge\cdots\wedge dZ_m.$$

This shows that if $\nu_j(t)$ represents the boundary value of $\partial g/\partial t_j$ in V to every open neighborhood $\mathcal{N}$ of the boundary $\dot{V}$ of V in $\Pi\times\Pi_0'$ and to every open subset W_0 of Π_0' there is $\epsilon > 0$, $0 < \epsilon < \delta$, such that

$$\nu_j(t) - (\partial/\partial t_j)\mu^{\gamma}_{g,V}(t) \in \mathfrak{G}(\mathcal{N}_0),\ \forall\ \gamma \in \Gamma,\ |\gamma| < \epsilon,$$

where $\mathcal{N}_0 = \{ (\zeta,t) \in \mathcal{N};\ t \in W_0\}$. On the other hand, the proof of Theorem IV.2.3 implies that if ϵ is sufficiently small, for the same γ's,

$$(\partial/\partial t_j)\mu^{\gamma}_{g,V}(t) - (\partial/\partial t_j)\mu_{g,V}(t)$$

will also belong to $\mathfrak{G}(\mathcal{N}_0)$. Since $\mathcal{N}$ and W_0 are arbitrary this means that

$$b_V(\partial g/\partial t_j) = (\partial/\partial t_j)b_V g$$

and, therefore, by letting V expand to fill U,

$$b_U(\partial g/\partial t_j) = (\partial/\partial t_j)b_U g,\ \ j = 1,\dots,n. \tag{IV.5.9}$$

THEOREM IV.5.5.— *For the the boundary value of the regular function g in the wedge $\mathcal{W}_\delta(U,\Gamma)$ to be a solution in U it is necessary and sufficient that g be holomorphic with respect to z and locally constant with respect to t.*

Proof: For $b_U g$ to be a solution in U it is necessary that the right–hand sides in Equations (IV.5.9) vanish identically. By Theorem IV.2.4 this is equivalent to the vanishing of the regular functions $\partial g/\partial t_j$ throughout the wedge $\mathcal{W}_\delta(U,\Gamma)$. □

IV.6 FBI TRANSFORM OF HYPERFUNCTIONS DEPENDING SMOOTHLY ON t

Same set—up as in the previous sections; in particular we continue to make use of the coarse embedding (IV.5.1). It is convenient to carry out the analysis on the manifold $\Sigma \subset \mathbb{C}^m \times \mathbb{R}^n$, the image of $\mathcal{M}' \subset \mathcal{M}$ under the map (IV.5.1). In order to reformulate in $\mathcal{M}$ the coming definitions and results it suffices to recall that (IV.5.1) establishes an isomorphism between the hypo—analytic manifold $\mathcal{M}'$ (inheriting its structure from $\mathcal{M}$) and the hypo—analytic manifold Σ whose structure is defined by the restrictions of the complex coordinates $z_1,\dots,z_m$.

We are going to make use of two concepts introduced at the beginning of section II.2: that of the *real structure bundle* $\mathbb{R}T'_{\mathcal{X}}$ of a maximally real submanifold $\mathcal{X}$ of $\mathbb{C}^m$ and that of such a submanifold being *well—positioned* at one of its points. We are going to apply this notion to the slices $\Sigma_t = \{ z \in \mathbb{C}^m;\ (z,t) \in \Sigma \}$: we need that, for all t sufficiently close to 0, the maximally real submanifolds $\Sigma_t \subset \mathbb{C}^m$ be well—positioned at the points $Z(0,t)$ [cf. (II.2.2), (II.2.3)]. Actually it is convenient to also require the slice Σ_0 through the origin to be *very well—positioned at* 0, meaning that, given *any* number κ, $0 < \kappa < 1$, there is an open neighbor—hood Ω of 0 in Σ_0 such that (II.2.2) and (II.2.3) are valid.

All this can be achieved by applying Proposition IX.5.1 in [TRE—VES, 1992], possibly after carrying out a substitution $Z \to c_1 Z + c_2 Z^2$ ($c_1 \neq 0$) to ensure the validity of

$$\text{(IV.6.1)} \qquad |\Phi(x,t)| \leq const.(|x|^3+|t|),\ \forall\, x \in \Pi_0,\ t \in \Pi'_0,$$

where Π_0 and Π'_0 are suitably small polyhedra in x–space $\mathbb{R}^m$ and t–space $\mathbb{R}^n$ respectively, both centered at the origin (see section IV.1).

We recall that $\mathbb{R}T'_{\Sigma_t}$ is a subspace of the structure bundle of $\mathbb{C}^m$ restricted to Σ_t, $T'^{1,0}\big|_{\Sigma_t}$. We identify to $\mathbb{C}^m$ each fibre $T'^{1,0}_z$ of $T'^{1,0}$ by means of the coordinates $\zeta_1,\dots,\zeta_m$ with respect to the basis $dz_1,\dots,dz_m$. Letting the base point z vary allows us to identify $T'^{1,0}$ to $\mathbb{C}^m\times\mathbb{C}^m$ (viewed as a vector bundle over $\mathbb{C}^m$). Then $\mathbb{R}T'_{\Sigma_t}\big|_{Z(x,t)}$ is identified to the subspace of $\mathbb{C}^m\times\mathbb{C}^m$ consisting of the points $(Z(x,t),{}^tZ_x(x,t)^{-1}\xi)$, $\xi\in\mathbb{R}^m$. A conic subset of $T'^{1,0}$ will be regarded as a set of points $(z,\zeta)\in\mathbb{C}^m\times\mathbb{C}^m$ preserved by every dilation $(z,\zeta)\to(z,\rho\zeta)$, $\rho>0$.

We shall use the notation of sections IV.1 and IV.2: Π_0 and Π'_0 are suitably small polyhedra in x–space $\mathbb{R}^m$ and t–space $\mathbb{R}^n$ respectively; U will be an open neighborhood of 0 in Σ satisfying (IV.2.1), ie. U is the image under the map (IV.5.1) of a product set $U_0\times\Pi'_0$ with $U_0\subset\subset\Pi_0$ an open neighborhood of 0 in $\mathbb{R}^m$ [cf. (IV.2.8)]. We denote by $\overline{U}$ (resp., $\dot{U}$) the image of $(\mathcal{C}\ell U_0)\times\Pi'_0$ (resp., $\partial U_0\times\Pi'_0$) under (IV.5.1).

Consider then an analytic functional carried by $\overline{U}$ depending smoothly on $t\in\Pi'_0$, ie., an element $\mu(t)\in\mathfrak{G}(\overline{U})$ (Definition IV.1.1). We can define its FBI transform with respect to z [cf. (II.2.4)]:

$$\text{(IV.6.2)}\qquad \mathcal{F}\mu(t;z,\zeta)=\langle\mu_{z'}(t),e^{i\zeta\cdot(z-z')-\langle\zeta\rangle\langle z-z'\rangle^2}\Delta(z-z',\zeta)\rangle.$$

It is a $\mathcal{C}^\infty$ function in the region

$$\{\,(t,z,\zeta)\in\mathbb{R}^n\times\mathbb{C}^m\times\mathbb{C}^m;\ t\in\Pi'_0,\ |\mathcal{Im}\,\zeta|<|\mathcal{Re}\,\zeta|\,\},$$

holomorphic with respect to (z,ζ); $\Delta(z,\zeta)$ is the Jacobian determinant of the map $\zeta \to \zeta + \imath\langle\zeta\rangle z$.

In what follows we must deal with triples $(t,z,\zeta) \in \mathbb{R}^n\times\mathbb{C}^m\times\mathbb{C}^m$. A conic subset A of $\mathbb{R}^n\times\mathbb{C}^m\times\mathbb{C}^m$ will be a set stable under every dilation $(t,z,\zeta) \to (t,z,\rho\zeta)$, $\rho > 0$. We shall also use the notation $A_t = \{ (z,\zeta) \in \mathbb{C}^m\times\mathbb{C}^m;\ (t,z,\zeta) \in A \}$. If S is a subset of Σ we define

$$\mathbb{R}T'_\Sigma\backslash 0\Big|_S = \{ (t,z,\zeta) \in \mathbb{R}^n\times\mathbb{C}^m\times\mathbb{C}^m;\ (z,t) \in S,\ (z,\zeta) \in \mathbb{R}T'_{\Sigma_t},\ \zeta \neq 0 \}.$$

A routine modification of the proof of Theorem II.2.1 shows the following.

THEOREM IV.6.1.— *Provided* U *is sufficiently small, to every* $\epsilon > 0$ *there is an open and conic neighborhood* $\mathcal{U}^\epsilon$ *of* $\mathbb{R}T'_\Sigma\backslash 0\big|_U$ *in* $\mathbb{R}^n\times\mathbb{C}^m\times\mathbb{C}^m$ *such that the following is true:*

(IV.6.3) *If* $\mu \in \mathfrak{G}(\overline{U})$, *to each* $\beta \in \mathbb{Z}^n_+$ *there is* $C_\beta > 0$ *such that*

$$|\partial_t^\beta \mathcal{F}\mu(t;z,\zeta)| \leq C_\beta e^{\epsilon|\zeta|},\ \forall\ (t,z,\zeta) \in \mathcal{U}^\epsilon.$$

There is also a parallel to Theorem II.2.2:

THEOREM IV.6.2.— *Provided* U *is sufficiently small, to every compact set* $K_0 \subset U_0$ *there are an an open and conic neighborhood* $\mathcal{U}$ *of the set* $\mathbb{R}T'_\Sigma\backslash 0\big|_F$ *in* $\mathbb{R}^n\times\mathbb{C}^m\times\mathbb{C}^m$, *with* F *the image of* $K_0\times\Pi'_0$ *under the map* (IV.5.1), *and a constant* $c > 0$ *such that the following is true:*

(IV.6.4) *If* $\mu \in \mathfrak{G}(\dot{U})$, *to each* $\beta \in \mathbb{Z}^n_+$ *there is* $C_\beta > 0$ *such that*

$$|\partial_t^\beta \mathcal{F}\mu(t;z,\zeta)| \leq C_\beta e^{-c|\zeta|},\ \forall\ (t,z,\zeta) \in \mathcal{U}.$$

The various inversion formulas proved in section II.2 extend to hyperfunctions depending smoothly on t. We shall content ourselves with stating the generalization of (II.2.15) and (II.2.21). Below $\hat{\Omega}_t$ stands for the image of $\mathbb{R}^m$ under the map $x \to \hat{Z}(x,t) = x + i\chi(x)\Phi(x,t)$; $\chi \in C_c^{\infty}(\Pi_0)$, $\chi(x) \equiv 1$ if $|x| < r$, $|\chi^{(\alpha)}(x)| \leq C_\alpha r^{-|\alpha|}$ for all $\alpha \in \mathbb{Z}_+^m$, $x \in \mathbb{R}^m$. The number $r > 0$ is selected suitably small. We assume $|x| < r$ in U_0; recall that U, is the image of $U_0 \times \Pi_0'$ under the map (IV.5.1).

THEOREM IV.6.3.— *Provided* U *is sufficiently small, the following is true. Let* F *be the image of* $K_0 \times \Pi_0'$ *under the map* (IV.5.1), *with* K_0 *a compact subset of* U_0, *and let* V *be the image of* $V_0 \times \Pi_0'$, *with* V_0 *open*, $K_0 \subset V_0 \subset\subset U_0$. *Then, whatever* $\mu(t) \in \mathfrak{B}(F)$ *and* $h \in \mathcal{O}(\mathbb{C}^m)$,

(IV.6.5)
$$<\mu(t),h> =$$

$$\lim_{\epsilon \to +0} (2\pi)^{-m} \int_{z \in V_t} \int_{\zeta \in \mathbb{R}T'_{\Sigma_t}|_z} \mathcal{F}\mu(t;z,\zeta) e^{-\epsilon<\zeta>^2} h(z) dz d\zeta;$$

(IV.6.6)
$$<\mu(t),h> =$$

$$\lim_{\epsilon \to +0} \int_{z \in V_t} \int_{z' \in \hat{\Omega}_t} \int_{\zeta \in \mathbb{R}T'_{\Sigma_t}|_{z'}} \mathcal{F}\mu(t;z',\zeta) h(z) \cdot$$

$$e^{i\zeta\cdot(z-z') - <\zeta><z-z'>^2 - \epsilon<\zeta>^2} <\zeta>^{m/2} dz dz' d\zeta / (2\pi^3)^{m/2}.$$

It is then natural to introduce the analogues of the spaces $\mathfrak{E}(U)$

and $\mathfrak{E}_0(U)$ of section II.2. We shall call them $\mathfrak{E}(U,\mathcal{C}^\infty_t)$ and $\mathfrak{E}_0(U,\mathcal{C}^\infty_t)$. Their definition is based on the evident generalizations of (II.2.8) and (II.2.9). Then the FBI transform (IV.6.2) induces a linear map

$$\mathcal{F}\colon \mathcal{B}(U,\mathcal{C}^\infty_t) \to \mathfrak{E}(U,\mathcal{C}^\infty_t)/\mathfrak{E}_0(U,\mathcal{C}^\infty_t),$$

which we shall call the FBI transform of hyperfunctions in U depending smoothly on t. The reader will easily check that, for every $u \in \mathcal{B}(U,\mathcal{C}^\infty_t)$,

$$\partial_z^\alpha \partial_t^\beta \mathcal{F} u = \mathcal{F} \partial_z^\alpha \partial_t^\beta u, \quad \forall\ \alpha \in \mathbb{Z}^m_+,\ \beta \in \mathbb{Z}^n_+. \tag{IV.6.7}$$

It is possible to make use of the "inversion formulas" (IV.6.5) and (IV.6.6) exactly as in section II.3 we made use of (II.2.15) and (II.2.21), to obtain a decomposition theorem for the elements of $\mathcal{B}(U,\mathcal{C}^\infty_t)$. On the sub—ject of the boundary value of a regular function in a wedge $\mathcal{W}_\delta(V,\Gamma)$ with edge $V \subset \Sigma$ we refer the reader to Example IV.2.1. We consider a finite subfamily $\Gamma_1,\ldots,\Gamma_r \in \mathfrak{C}_m$ to which it is possible to associate an equal number of open cones in $\mathbb{R}^m\backslash\{0\}$, $\mathscr{C}_1,\ldots,\mathscr{C}_r$, satisfying Conditions (II.3.1) and (II.3.2).

THEOREM IV.6.4.— *If the open neighborhood* U *of* 0 *in* Σ *is sufficiently small, to each open subset of* Σ, $V \subset\subset U$, *there is a number* $\delta > 0$ *such that, if* $u \in \mathcal{B}(U,\mathcal{C}^\infty_t)$ *then, for each* $j = 1,\ldots,r$, *there is a regular function* f_j *in the wedge* $\mathcal{W}_\delta(V,\Gamma_j)$ *such that*

$$u\big|_V = \sum_{j=1}^{r} b_V f_j.$$

Next we look at the FBI transform of a *hyperfunction solution* $u \in \mathfrak{Sol}(U,\mathcal{C}^\infty_t)$, ie. $u \in \mathcal{B}(U,\mathcal{C}^\infty_t)$ and $d_t u \equiv 0$. In view of (IV.6.7) this implies

$$d_t \mathcal{F}u \equiv 0. \tag{IV.6.8}$$

We underline the meaning of (IV.6.8): if $\mu \in \mathfrak{G}(\overline{U})$ represents u then every partial derivative $\partial u/\partial t_j$ belongs to $\mathfrak{G}(\dot{U})$ $(j = 1,...,n)$ and therefore

$$\mathcal{F}(\partial u/\partial t_j) = \partial(\mathcal{F}u)/\partial t_j \in \mathfrak{E}_0(U,\mathcal{C}^\infty_t).$$

Roughly speaking, (IV.6.8) says that the FBI transform of a hyperfunction solution u in a conic neighborhood of $\mathbb{R}T'_{\Sigma_0}\big|_0$ in $\mathbb{R}^n \times \mathbb{C}^m \times \mathbb{C}^m$ is equivalent (modulo functions that decay exponentially as $|\zeta| \to +\infty$) to the trace of that same FBI transform at $t = 0$. This is consistent with the interpretation of a solution as a hyperfunction that is (locally) constant with respect to t. From this the criterion of hypo-analyticity generalizing Theorem II.2.5 is immediate. Recall that a hyperfunction solution u in Σ is said to be hypo-analytic at 0 if there is a holomorphic function $\tilde{h}$ in some open neighborhood $\tilde{U}_0$ of 0 in $\mathbb{C}^m$ (which we regard as a regular function in $\tilde{U}_0 \times \mathbb{R}^n$ — independent of t) such that $u = \tilde{h}$ in a neighborhood of 0 in Σ. Of course this concept can be transferred to a neighborhood of O in $\mathcal{M}'$ by means of the map (IV.5.1).

THEOREM IV.6.5.— *For a hyperfunction solution u in Σ to be hypo-analytic at 0 it is necessary and sufficient that $\mathcal{F}u$ be exponentially decaying in some open conic neighborhood $\tilde{\mathcal{U}}$ of $\mathbb{R}T'_{\Sigma_0}\big|_0$ in $\mathbb{R}^n \times \mathbb{C}^m \times \mathbb{C}^m$.*

The following remark should underline the coherence of our picture. If $\mathcal{F}u$ decays exponentially in some open conic neighborhood of

$\mathbb{R}T'_\Sigma\backslash 0\big|_0$ in $\mathbb{R}^n\times\mathbb{C}^m\times\mathbb{C}^m$ Theorem II.2.5 asserts that the trace of u on the submanifold $t = 0$ is equal, in a neighborhood of 0 in Σ, to the restriction of a holomorphic function $h(z)$ in an open neighborhood of 0 in $\mathbb{C}^m$. But the restriction of $h(z)$ regarded as a function of (z,t) is itself a solution. It follows from Theorem IV.5.2 that $u \equiv h$ in a full neighborhood of 0 in Σ.

REMARK IV.6.1.— In general there is no jump theorem analogous to Theorem IV.6.4 for solutions. Such a theorem would require that every hyperfunction $b_V f_j$ be a solution. According to Theorem IV.5.5, the latter would mean that f_j be holomorphic with respect to z and independent of t [assuming that the geometry of the wedge $\mathcal{W}_\delta(V,\Gamma_j)$ is sufficiently simple] in brief, $f_j = f_j(z)$ holomorphic.

Actually such a theorem is valid in hypo—analytic structures of the *hypersurface type*, as will be shown in section IV.8. But an (unpublished) article of J.—M. Trepreau gives the example of a generic submanifold $\mathcal{M}$ of $\mathbb{C}^3$ of real codimension two, passing through the origin, in which the following happens: if u is any hyperfunction solution [ie., any CR hyperfunction (see section IV.7)] in an open neighborhood U of 0 in $\mathcal{M}$, and if u_0 denotes its trace on some maximally real submanifold $\mathcal{X}$ of $\mathbb{C}^m$, $0 \in \mathcal{X} \subset U$, then the hypo—analytic wave—front set of u_0 at the origin is either empty or equal to $\mathbb{R}T'_{\mathcal{X}}\big|_0$. Comparing this with Definition II.3.2 one can easily see that, unless u is hypo—analytic at 0 in $\mathcal{M}$ ($\Leftrightarrow$ u_0 is hypo—analytic at 0 in $\mathcal{X}$) u cannot be represented, in any neighborhood of the origin, as the sum of the boundary values of (finitely many) holomorphic functions in wedges with edge in $\mathcal{M}$. □

Below $T^*_{\wp}\mathcal{M}$ denotes the real cotangent space to $\mathcal{M}$ at a point $\wp$, $T'_{\wp}$ the fibre at $\wp$ of the cotangent structure bundle. The intersection $T^0_{\wp} = T^*_{\wp}\mathcal{M} \cap T'_{\wp}$ is the *characteristic set* of the structure at $\wp$: it is made up of all the covectors $\xi \in T^*_{\wp}$ such that $< \xi, L > = 0$ whatever $L \in \mathcal{V}_{\wp}$. If $T^0_{\wp} = 0$ the locally integrable structure of $\mathcal{M}$ is said to be *elliptic*. Notice that $T^0_{\wp} = 0 \Rightarrow T^0_{\wp'} = 0$ for all points $\wp'$ in a neighborhood of $\wp$.

To say that $\mathcal{X}$ is a maximally real submanifold of $\mathcal{M}$ is equivalent to saying that the pullback map $\mathbb{C}T^*\mathcal{M}\big|_{\mathcal{X}} \to \mathbb{C}T^*\mathcal{X}$ induces an isomorphism $T'\big|_{\mathcal{X}} \cong \mathbb{C}T^*\mathcal{X}$. In particular it induces an injection of $T^0\big|_{\mathcal{X}} = T' \cap T^*\mathcal{M}\big|_{\mathcal{X}}$ into $T^*\mathcal{X}$. Since the pre–image of $T^*\mathcal{X}$ is, by definition, the real structure bundle of $\mathcal{X}$, $\mathbb{R}T'_{\mathcal{X}}$, we see that the latter contains $T^0\big|_{\mathcal{X}}$.

Instead of applying this to $\mathcal{M}$ we apply it to the submanifold Σ of $\mathbb{C}^m \times \mathbb{R}^n$ and, as the maximally real submanifolds, to the slices Σ_t. We use the coordinates x_i and t_j on Σ. A generic point of T'_{Σ} is a pair

$$(Z(x,t), \zeta \cdot dZ(x,t)),\ x \in \Pi_0,\ t \in \Pi'_0,\ \zeta \in \mathbb{C}^m.$$

It belongs to T^0 if and only if it can be equated to a point

$$(Z(x,t), \xi \cdot x + \tau \cdot dt) \in T^*\Sigma, \text{ with } \xi \in \mathbb{R}^m,\ \tau \in \mathbb{R}^n,$$

ie., $\zeta \cdot dZ = \zeta \cdot Z_x(x,t)dx + \zeta \cdot Z_t(x,t)dt = \xi \cdot dx + \tau \cdot dt$. We must therefore have

$$\zeta = {}^tZ_x(x,t)^{-1}\xi, \quad \mathcal{Im}\, \zeta \cdot Z_t(x,t) = 0.$$

In passing notice that the natural isomorphism of $T^*\Sigma_t$ onto $\mathbb{R}T'_{\Sigma_t}$ is given by

$$(Z(x,t), \xi \cdot dx) \to (Z(x,t), \zeta \cdot dZ(x,t)), \text{ with } \zeta = {}^tZ_x(x,t)^{-1}\xi,$$

and therefore we have indeed $T^0\big|_{\Sigma_t} \subset \mathbb{R}T'_{\Sigma_t}$ as stated above. Since $Z = x+i\Phi(x,t)$, $T^0\big|_{\Sigma_t} \subset T'\cap T^*\Sigma_t$ is the set of points $(Z(x,t),{}^tZ_x(x,t)^{-1}\xi)$ with $\xi \in \mathbb{R}^m$ such that

$$\text{(IV.6.9)} \qquad \mathcal{Re}\{\xi\cdot Z_x(x,t)^{-1}(\partial\Phi/\partial t_j)(x,t)\} = 0,\ j = 1,\dots,n$$

(Z_x^{-1} is an m×m complex matrix and $\partial\Phi/\partial t_j$ is a real m—vector). Taking (IV.1.1) into account shows that T_0^0 can be identified to the linear subspace of $T^*\Sigma_0$ ($\cong \mathbb{R}T'_{\Sigma_0} \cong \mathbb{R}^m$) consisting of the covectors ξ such that

$$\text{(IV.6.10)} \qquad \xi\cdot(\partial\Phi/\partial t_j)(0,0) = 0,\ j = 1,\dots,n.$$

The next statement is the version fitting the present context of a classical theorem of M. Sato:

THEOREM IV.6.6.— *If u_0 is the trace at $t = 0$ of a hyperfunction solution u in Σ then* $WF_{ha}(u_0) \subset T^0\big|_{\Sigma_0}$.

Proof: From (IV.6.8) we derive the following. Let U be the image of $U_0\times\Pi'_0$ under the map (IV.5.1), with $U_0 \subset\subset \Pi_0$ an open neighborhood of 0 in $\mathbb{R}^m$. There is a representative $\mu(t) \in \mathfrak{G}(\overline{U})$ of u in U such that the function $(z,\zeta) \to \mathcal{F}\mu(0;z,\zeta) - \mathcal{F}\mu(t;z,\zeta)$ decays exponentially in a conic neighborhood of $\{0\}\times(\mathbb{R}^m\backslash\{0\})$ in $\mathbb{C}^m\times\mathbb{C}^m$. If there were a compact set K′ $\subset \Pi_0\backslash\{0\}$ such that $\mu(t)$ is carried by the image of $K'\times\Pi'_0$ under the map

(IV.5.1), whatever $t \in \Pi_0'$ $\mathcal{F}\mu(t;z,\zeta)$ would decay exponentially in a conic neighborhood of $\{0\}\times(\mathbb{R}^m\backslash\{0\})$ in $\mathbb{C}^m\times\mathbb{C}^m$, and so would be $\mathcal{F}\mu(0;z,\zeta)$. Since we can decompose $\mu(t)$ into a sum $\mu_1(t) + \mu_2(t)$ with $\mu_1(t)$ carried by the image F of $\mathrm{K}\times\Pi_0'$ and $\mu_2(t)$ by that of $[(\mathcal{C}\ell\mathrm{U}_0)\backslash \mathcal{I}nt\mathrm{K}]\times\Pi_0'$ whatever the compact neighborhood $\mathrm{K} \subset \Pi_0$ of 0, it follows that we can replace $\mu(t)$ by $\mu_1(t)$.

Now let $\xi \in \mathbb{R}^m$ violate Condition (IV.6.10). After an $\mathbb{R}$—linear substitution of the variables t we may assume that

$$\xi\cdot(\partial\Phi/\partial t_1)(0,0) < 0. \tag{IV.6.11}$$

Below we take $t = (\delta,0,...,0)$, with $\delta > 0$ suitably small. By (IV.6.2) to every $\epsilon > 0$ there is $C_\epsilon > 0$ such that

$$|\mathcal{F}\mu_1(t;z,\zeta)| \leq C_\epsilon \operatorname*{Max}_{z'\in \mathrm{F}_t^\epsilon} \{|\Delta(z-z',\zeta)|e^{-\mathcal{Q}(z-z',\zeta)|\zeta|}\}, \tag{IV.6.12}$$

where

$$\mathrm{F}_t^\epsilon = \{ z' \in \mathbb{C}^m;\ \mathrm{dist}(z',\mathrm{F}_t) \leq \epsilon \}$$

and

$$\mathcal{Q}(z,\zeta) = \mathcal{I}m(\zeta\cdot z + \imath\langle\zeta\rangle\langle z\rangle^2)/|\zeta|.$$

Taking $z' = Z(x',t) + z_*$ with $Z(x',t) \in \mathcal{C}\ell\mathrm{F}_t$ (hence $x' \in \mathrm{K}$) and $|z_*| \leq \epsilon$, and writing $\hat{\xi} = \xi/|\xi|$, we get

$$\begin{aligned}\mathcal{Q}(-z',\hat{\xi}) = \mathcal{I}m\{-\hat{\xi}\cdot Z(x',t) + \imath\langle Z(x',t)\rangle^2\} + \mathrm{O}(\epsilon) = \\ -\hat{\xi}\cdot\Phi(x',t) + |x'|^2 - |\Phi(x',t)|^2 - C_0\epsilon \geq \\ -\hat{\xi}\cdot\Phi(0,t) + |x'|^2 - C_1(|x'|^3+\delta|x'|+\delta^2) - C_0\epsilon,\end{aligned}$$

by virtue of (IV.6.1) and of our choice of t. At this juncture we avail ourselves of (IV.6.11). There is a constant $c_0 > 0$ that depends only on Φ, such that

$$\mathcal{Q}(-z',\xi)/|\xi| \geq [c_0 - C_1(|x'|+\delta)]\delta + |x'|^2(1-C_1|x'|) - C_0\epsilon.$$

Note that also the constants C_0 and C_1 only depend on Φ. We can choose K, δ and ϵ so small that

$$\forall\, x' \in \mathrm{K},\ |x'|+\delta < \tfrac{1}{4}c_0/C_1,\ |x'| < 1/C_1;\ C\epsilon < \tfrac{1}{4}c_0.$$

We obtain

$$\mathrm{Q}(z-z',\zeta)/|\zeta| > \tfrac{1}{2}c_0$$

when $z = 0$, $\zeta = \xi$. But it is clear that the same inequality will be true if (z,ζ) stay in a sufficiently thin conic neighborhood of $(0,\xi)$ in $\mathbb{C}^m\times\mathbb{C}^m$. Combining this fact with (IV.6.12) allows us to assert that $\mathcal{F}\mu(t;z,\zeta)$ decays exponentially in a conic neighborhood of $\{0\}\times(\mathbb{R}^m\backslash\{0\})$ in $\mathbb{C}^m\times\mathbb{C}^m$, and that so does $\mathcal{F}\mu(0;z,\zeta)$. Then, in order to reach the desired conclusion it suffices to apply Theorem II.4.1. □

Returning to the hypo–analytic manifold $\mathcal{M}$ we recall the following. Given any maximally real submanifold $\mathcal{X}$ of $\mathcal{M}$ passing through O it is possible to choose the coordinates x_i and t_j in a suitably small open neighborhood $\mathcal{M}'$ of O in such a way that $\mathcal{X}\cap\mathcal{M}' = \{\ (x,t) \in \mathcal{M}';\ t = 0\ \}$ ([TREVES, 1992], p. 76). It can then be shown (cf. [BAOUENDI–CHANG–TREVES, 1983]) that, given any hyperfunction solution u in an open neighborhood $\mathrm{U} \subset \mathcal{M}'$ of O, the set of points $\xi \in \mathrm{T}^0_O$ that belong to hypo–analytic wave front set of the trace of u on $\mathcal{X}$ is independent of the choice of $\mathcal{X}$. One can therefore define that set to be the *hypo–analytic wave front set* of the hyperfunction solution u at O.

IV.7 USE OF THE FINE EMBEDDING. CR HYPERFUNCTIONS

As before let $\mathcal{M}$ be a hypo—analytic manifold, and

$$(\mathcal{M}',Z_1,\ldots,Z_m,x_1.,x_m,t_1,\ldots,t_n)$$

a hypo—analytic chart in $\mathcal{M}$ centered at a point O. So far we have used the local embedding (IV.5.1). It is often convenient to use a finer embedding (see Sections I.7 and II.7 in [TREVES, 1992]). Assume as before that $Z_j = x_j+\imath\Phi_j(x,t)$ (j = 1,...,m) and let ν denote the rank at the origin of the map $\mathbb{R}^n \ni t \to \Phi(0,t) = (\Phi_1(0,t),\ldots,\Phi_m(0,t)) \in \mathbb{R}^m$. After a $\mathbb{C}$—linear substitution of the Z_j's we may achieve that the following holds true:

$$d_t\Phi_1\wedge\cdots\wedge d_t\Phi_\nu \neq 0,\ d\Phi_{\nu+1} =\cdots= d\Phi_m = 0 \text{ at the origin.}$$

After contracting $\mathcal{M}'$ about O we may take $y_1 = \Phi_1(x,t),\ldots,\ y_\nu = \Phi_\nu(x,t)$ as coordinates, replacing ν of the t_k's. We shall also write s_k instead $x_{\nu+k}$ if $1 \leq k \leq m-\nu$. After relabeling the t_ℓ we end up with a coordinate chart centered at O (ie., all the coordinates vanish at O),

$$\text{(IV.7.1)}\qquad (\mathcal{M}',x_1,\ldots,x_\nu,y_1,\ldots,y_\nu,s_1,\ldots,s_d,t_1,\ldots,t_{n'}),$$

and a set of hypo—analytic first integrals,

$$\text{(IV.7.2)}\quad z_j = x_j+\imath y_j\ (1 \leq j \leq \nu),\ w_k = s_k+\imath\varphi_k(z,s,t)\ (1 \leq k \leq d).$$

Here m = ν+d, n = ν+n$'$; the functions φ_k are real—valued, smooth and

$$\text{(IV.7.3)}\qquad \varphi_k\Big|_O = d\varphi_k\Big|_O = 0 \quad (k = 1,\ldots,d).$$

This yields a diffeomorphism of $\mathcal{M}'$ onto a submanifold Σ of $\mathbb{C}^m\times\mathbb{R}^{n'}$, namely the map

$$\mathcal{M}' \ni (x,y,s,t) \to (z,w,t) \tag{IV.7.4}$$

with z and w as in (IV.7.2), with (IV.7.3) holding. We shall refer to (IV.7.4) as the *fine local embedding.*

REMARK IV.7.1.— Let $\mathcal{V} \subset \mathbb{C}T\mathcal{M}$ denote the tangent structure bundle of $\mathcal{M}$, $T' \subset \mathbb{C}T^*\mathcal{M}$ the cotangent structure bundle of $\mathcal{M}$, ie., the orthogonal of $\mathcal{V}$ for the duality between tangent and cotangent vectors. We recall the following terminology ([TREVES, 1992], Definition I.2.3 and p. 42): the hypo—analytic structure of $\mathcal{M}$ is said to be

real if $\mathcal{V}$ is spanned by its intersection $\mathcal{V}_0$ with the real tangent bundle $T\mathcal{M}$ of $\mathcal{M}$;

complex if $\mathbb{C}T\mathcal{M} = \mathcal{V} \oplus \overline{\mathcal{V}}$, ie., $\mathbb{C}T^*\mathcal{M} = T' \oplus \overline{T}'$ ($\oplus$: direct sum);

elliptic if $\mathbb{C}T\mathcal{M} = \mathcal{V} + \overline{\mathcal{V}}$, ie., $T' \cap \overline{T}' = 0$;

CR if $\mathcal{V} \cap \overline{\mathcal{V}} = 0$, ie., $\mathbb{C}T^*\mathcal{M} = T' + \overline{T}'$.

In the hypo—analytic chart (IV.7.1)—(IV.7.2) this classification translates itself as follows: in $\mathcal{M}'$ the structure of $\mathcal{M}$ is

real $\Leftrightarrow$ $\nu = 0$ and $\varphi_k \equiv 0$ for every $k = 1,\dots,d$;

complex $\Leftrightarrow$ $d = n' = 0$;

elliptic $\Leftrightarrow$ $d = 0$;

CR $\Leftrightarrow$ $n' = 0$.

We also recall that there are hypo—analytic structures that do not belong to any of the above classes, such as the Mizohata structure on $\mathbb{R}^2$, defined by the function $Z = s + it^2$, or the structure on $\mathbb{R}^2$ in Example IV.5.1. □

Define the vector fields

$$N_k = \sum_{k'=1}^{d} c_{kk'}\,\partial/\partial s_{k'} \quad (k = 1,\ldots,d)$$

where the d×d matrix $(c_{kk'})_{1\leq k,k'\leq d}$ is the inverse of the Jacobian matrix $\partial w/\partial s = (\partial w_k/\partial s_{k'})_{1\leq k,k'\leq d}$. [the functions w_k are as in (IV.7.2)]. A basis of $\mathcal{V}$ in $\mathcal{M}'$ consists of the vector fields

(IV.7.5)
$$L_j = \partial/\partial\bar{z}_j - i\sum_{k=1}^{d}(\partial\varphi_k/\partial\bar{z}_j)N_k, \quad j = 1,\ldots,\nu;$$
$$L_{\nu+l} = \partial/\partial t_l - i\sum_{k=1}^{d}(\partial\varphi_k/\partial t_l)N_k, \quad l = 1,\ldots,n'.$$

The vector fields $L_1,\ldots,L_n,N_1,\ldots,N_d$ commute. Together with the vector fields

$$L'_j = \partial/\partial z_j - i\sum_{k=1}^{d}(\partial\varphi_k/\partial z_j)N_k, \quad j = 1,\ldots,\nu,$$

they form a (commuting) basis of $\mathbb{C}T\mathcal{M}$ over $\mathcal{M}'$.

The vector bundle $\Lambda_{\mathcal{M}}^{0,q}$ on $\mathcal{M}$ has been defined in (III.1.3). Here the space of C^∞ sections of $\Lambda_{\mathcal{M}}^{0,q}$ over $\mathcal{M}'$, $C^\infty(\mathcal{M}';\Lambda_{\mathcal{M}}^{0,q})$, will be identified to the space of differential forms in $\mathcal{M}'$,

(IV.7.6)
$$f = \sum_{|J|+|K|=q} f_{J,K}\,d\bar{z}_J\wedge dt_K,$$

with coefficients $f_{J,K} \in C^\infty(\mathcal{M}')$ [cf. (III.1.11)]. The differential operator L acts on a current (IV.7.6) according to the formula:

(IV.7.7) $$Lf = (\bar{\partial}_z + d_t)f - \imath \sum_{k=1}^{d} (\bar{\partial}_z + d_t)\varphi_k \wedge N_k f,$$

with the vector fields N_k acting coefficientwise on the form f. Since the L_j commute we have $L^2 = 0$. This leads to the differential complex

(IV.7.8) $$L: \mathcal{C}^\infty(\mathcal{M}';\Lambda_{\mathcal{M}}^{0,q}) \to \mathcal{C}^\infty(\mathcal{M}';\Lambda_{\mathcal{M}}^{0,q+1}), \quad q = 0,1,\dots,n.$$

We may consider the analogous complex with $\Lambda_{\mathcal{M}}^{m,q}$ substituted for $\Lambda_{\mathcal{M}}^{0,q}$; the form f given by (IV.7.6) must be replaced by the form

$$dz_1 \wedge \cdots \wedge dw_1 \wedge \cdots \wedge dw_d \wedge f.$$

And as usual we may consider the differential complexes analogous to (IV.7.8) with $\mathcal{D}'$ (or $\mathcal{C}_c^\infty$, or $\mathcal{E}'$) substituted for $\mathcal{C}^\infty$.

As the reader has undoubtedly noticed, the role of t in the coarse local embedding is played by (x,y,t) in the fine local embedding; the role of x is played by s. We let x vary in a polyhedron

$$\tilde{\Pi}_0 = \{ x \in \mathbb{R}^\nu;\ |x_i| < \tilde{a}_i,\ i = 1,\dots,\nu \},$$

y varies in

$$\tilde{\Pi}_1 = \{ y \in \mathbb{R}^\nu;\ |y_j| < \tilde{b}_j,\ j = 1,\dots,\nu \};$$

thus z varies in $\tilde{\Pi} = \tilde{\Pi}_0 + \imath\tilde{\Pi}_1$. We let s vary in

$$\Pi_0 = \{ s \in \mathbb{R}^d;\ |s_k| < a_k,\ k = 1,\dots,d \}$$

and t in

$$\Pi_0' = \{ t \in \mathbb{R}^{n'};\ |t_l| < a_l',\ l = 1,\dots,n' \}.$$

We shall call Σ the image of the map

(IV.7.9) $$\mathcal{M}' \ni (z,s,t) \to (z, s+\imath\varphi(z,s,t), t) \in \mathbb{C}^\nu \times \mathbb{C}^d \times \mathbb{R}^{n'},$$

and by Σ_0 the image of $\tilde{\Pi}\times\Pi_0\times\Pi_0'$ [we have used the notation $\varphi = (\varphi_1,\ldots,\varphi_d)$]. We select an open subset U_0 of $\mathbb{R}^d$ whose closure is contained in Π_0 and we denote by U the image of $\tilde{\Pi}\times U_0\times\Pi_0'$ under the map (IV.7.9), by $\overline{U}$ (resp., $\dot{U}$) the closure (resp., boundary) of U in Σ_0. The slices of U will now be the sets

$$U_{z,t} = \{\, w \in \mathbb{C}^d;\ (z,w,t) \in U \,\}.$$

One can develop a theory of the hyperfunction solutions that depend smoothly on (z,t), within the framework of the fine local embedding, that parallels the theory of hyperfunction solutions that depend smoothly on t, based on the coarse embedding. All it requires is to adapt all the concepts introduced in the previous sections of Chapter III. We content ourselves with pointing out the most salient ones.

If $\mathcal{A}$ is an open subset of $\mathbb{C}^\nu\times\mathbb{C}^d\times\mathbb{R}^{n'}$, we use the space $\mathfrak{F}_f(\mathcal{A})$ of C^∞ functions of (z,w,t) in $\mathcal{A}$ that are holomorphic with respect to w (the subscript f will be used systematically to indicate that the fine local embedding is being used). If K_0 is a compact subset of Π_0' and if K the image of $\tilde{\Pi}\times K_0\times\Pi_0'$ under the map (IV.7.9), $\mathfrak{G}_f(K)$ will denote the space of analytic functionals $\mu(z,t)$ in w-space $\mathbb{C}^d$ which depend in C^∞ fashion on $(z,t) \in \tilde{\Pi}\times\Pi_0'$, such that for each (z,t), $\mu(z,t)$ is carried by the slice $K_{z,t}$ [the formal definition of $\mathfrak{G}_f(K)$ must be stated in analogy with Definition IV.1.1]. The space $\mathcal{B}_f(U,C^\infty_{z,t})$ of hyperfunctions in U that depend smoothly on $(z,t) \in \tilde{\Pi}\times\Pi_0'$ is defined to be the quotient $\mathfrak{G}_f(\overline{U})/\mathfrak{G}_f(\dot{U})$.

DEFINITION IV.7.1.— *We shall denote by* $\mathfrak{Sol}_f(U)$ *the subspace of* $\mathcal{B}_f(U,C^\infty_{z,t})$ *consisting of those hyperfunctions u in* U *that have a represen–*

tative $\mu \in \mathfrak{G}_f(\overline{U})$ *such that* $\partial\mu/\partial\bar{z}_j$, $\partial\mu/\partial t_k$ *belong to* $\mathfrak{G}_f(\overline{U})$ *for all* $j = 1,\ldots,\nu$, $k = 1,\ldots,n'$ $(= n-\nu)$.

Another way of phrasing Definition IV.7.1 is by saying that $u \in \mathcal{B}_f(U,\mathcal{C}^\infty_{z,t})$ belongs to $\mathfrak{Sol}_f(U)$ if $\bar\partial_z u \equiv 0$, $d_t u \equiv 0$ in U.

Next we introduce the space $\mathcal{B}_f(U,\mathcal{C}^\infty_{z,t};\Lambda^q)$ of *currents* (IV.7.6) with coefficients $f_{J,K}$ that belong to the space $\mathcal{B}_f(U,\mathcal{C}^\infty_{z,t})$. Of course,

$$\mathcal{B}_f(U,\mathcal{C}^\infty_{z,t};\Lambda^0) = \mathcal{B}_f(U,\mathcal{C}^\infty_{z,t}) \text{ and } \mathcal{B}_f(U,\mathcal{C}^\infty_{z,t};\Lambda^q) = 0 \text{ if } q > n.$$

The role of the differential complex (IV.3.11) is played here by the differential complex

$$\text{(IV.7.10)}\quad \bar\partial_z + d_t\colon \mathcal{B}_f(U,\mathcal{C}^\infty_{z,t};\Lambda^q) \to \mathcal{B}_f(U,\mathcal{C}^\infty_{z,t};\Lambda^{q+1}),\quad q = 0,1,\ldots,n.$$

DEFINITION IV.7.2.— *We denote by* $\mathfrak{Sol}_f^{(q)}(U)$ *the* q^{th} *cohomology space of the complex* (IV.7.10).

The analogue of Theorem IV.3.2 is valid here. We avail ourselves of the same resolution (IV.3.5) as in the coarse embedding, with the understanding that the variable $z_{\nu+k}$ is now called w_k $(k = 1,\ldots,d = m-\nu)$. The open subset U of Σ is as described above; now Π_1 is a polyhedron in $\mathbb{R}^d$ of the same kind as Π_0, such that $\varphi(\tilde\Pi\times\Pi_0\times\Pi_0') \subset \Pi_1$; we write $\Pi = \Pi_0+\imath\Pi_1 \subset \mathbb{C}^d$. We can define restriction mappings $\mathfrak{r}^U_V$ and r^U_V just as for the coarse embedding [see (IV.3.12) and Lemma IV.3.3]. Below we also use smaller polyhedrons $\tilde\Pi' \subset \tilde\Pi$, $\Pi_0'' \subset \Pi_0'$.

Returning to the proof of Theorem IV.3.2 we note that the form

(IV.3.17) must here be replaced by a form

$$F = \sum_{|J|+|K|=q} F_{JK} d\bar{z}_J \wedge dt_K$$

where

$$F_{JK} = dw_1\wedge\cdots\wedge dw_d\wedge \sum_{|I|=d-1} F_{IJK} d\bar{w}_I \in C^\infty(\tilde{\Pi}\times\Pi\times\Pi_0'\setminus U;\Lambda^{d,d-1})$$

satisfies $\bar{\partial}_w F_{JK} \equiv 0$. Equation (IV.3.18) was the starting point of the definition of the isomorphism i_U^q in Theorem IV.3.2. Here it must be replaced by the equation

$$(\bar{\partial}_z + d_t)F = A - \bar{\partial}_w G \quad in\ \tilde{\Pi}\times\Pi\times\Pi_0'\setminus U$$

with $A = \sum_{|J|+|K|=q+1} A_{JK} d\bar{z}_J \wedge dt_K$, $G = \sum_{|J|+|K|=q+1} G_{JK} d\bar{z}_J \wedge dt_K$,

$$A_{JK} = dw_1\wedge\cdots\wedge dw_d\wedge \sum_{|I|=d-1} A_{IJK} d\bar{w}_I \in C^\infty(\tilde{\Pi}\times\Pi\times\Pi_0'\setminus\mathring{U};\Lambda^{d,d-1}),$$

$$G_{JK} = dw_1\wedge\cdots\wedge dw_d\wedge \sum_{|I|=d-2} G_{IJK} d\bar{w}_I \in C^\infty(\tilde{\Pi}\times\Pi\times\Pi_0'\setminus U;\Lambda^{d,d-2});$$

and $\bar{\partial}_w A_{JK} \equiv 0$. The construction proceeds essentially as in Section IV.3 and produces a (d,d+q)—cocycle in the differential complex

(IV.7.11)

$$\bar{\partial}_z + \bar{\partial}_w + d_t : C^\infty(\tilde{\Pi}\times\Pi\times\Pi_0'\setminus\mathring{U};\tilde{\Lambda}_*^{d,r}) \to C^\infty(\tilde{\Pi}\times\Pi\times\Pi_0'\setminus\mathring{U};\tilde{\Lambda}_*^{d,r+1}),$$

$$r = 0,1,\ldots,d+n,$$

where we have denoted by $\Lambda_*^{d,r}$ the bundle spanned by the exterior products $dw_1\wedge\cdots\wedge dw_d\wedge d\bar{w}_I\wedge d\bar{z}_J\wedge dt_K$, $|I|+|J|+|K| = r$. We observe that exterior multiplication of all currents by $dz_1\wedge\cdots\wedge dz_\nu$ induces an isomorphism $\Lambda_*^{d,r} \cong \Lambda^{m,r}$. Taking all this into account leads to the "correct" version of Theorem IV.3.2 for the fine embedding:

THEOREM IV.7.1.— *Let* q *be an integer,* $0 \leq q \leq n$. *There are natural isomorphisms* i_U^q *and* i_V^q *such that the diagram*

$$\text{(IV.7.12)}\qquad \begin{array}{ccc} \mathfrak{Sol}_f^{(q)}(U) & \xrightarrow{i_U^q} & H_U^{m,d+q}(\tilde{\Pi}\times\Pi\times\Pi_0'\backslash\dot{U}) \\ {\scriptstyle r_V^U}\downarrow & & \downarrow{\scriptstyle r_V^U} \\ \mathfrak{Sol}_f^{(q)}(V) & \xrightarrow[i_V^q]{} & H_V^{m,d+q}(\tilde{\Pi}'\times\Pi\times\Pi_0''\backslash\dot{V}) \end{array}$$

is commutative.

The presheaf $\{H_U^{m,d+q}(\tilde{\Pi}\times\Pi\times\Pi_0'\backslash\dot{U}),r_V^U\}$ (present notation) leads to the same sheaf as the presheaf $\{H_U^{m,m+q}(\Pi\times\Pi_0'\backslash\dot{U}),r_V^U\}$ (notation of section IV.3). The proof of this fact is based on excision and integration with respect to y. The latter is possible because the image of a set $W_0 = \tilde{\Pi}'\times U_0\times\Pi_0''$, with $\tilde{\Pi}' \subset\subset \tilde{\Pi}$, $\Pi_0'' \subset\subset \Pi_0'$, under the coarse embedding

$$(x,y,s,t) \to (z,w,y,t)$$

is the product of the image of W_0 under the fine embedding $(x,y,s,t) \to (z,t,w)$ with a polyhedron in y–space. And the argument that has led to Theorem IV.4.2, leads here to

THEOREM IV.7.2.— *Integration with respect to* $\mathcal{Im}\, z'$ *induces an isomorphism* $\mathfrak{Sol}^{(q)}(U) \cong \mathfrak{Sol}_f^{(q)}(U)$ $(0 \leq q \leq n)$.

Here z' stands for the complexification of the pair of real variables (y,t).

CR hyperfunctions

An important reason for introducing fine local embeddings is that they are especially adapted to the study of hypo—analytic CR structures. Indeed, in that case there are no t variables. The local embedding is simply the map

$$\tilde{\Pi}\times\Pi_0 \ni (z,s) \to (z,w) \in \Sigma\cap(\tilde{\Pi}\times\Pi) \subset \mathbb{C}^m$$

with $w = (w_1,...,w_d)$ given as in (IV.7.2). If the subset U of Σ is defined as usual, $\mathcal{B}(\mathrm{U},\mathcal{C}^\infty_z)$ denotes the space of hyperfunctions in U that depend in $\mathcal{C}^\infty$ fashion on z. The solutions in U, ie., the elements of $\mathfrak{Sol}_f(\mathrm{U})$, are those elements of $\mathcal{B}(\mathrm{U},\mathcal{C}^\infty_z)$ that are holomorphic with respect to z. Thus a hyperfunction solution $u \in \mathfrak{Sol}_f(\mathrm{U})$ will be represented by an analytic functional depending smoothly on z, $\mu \in \mathfrak{G}_f(\overline{\mathrm{U}})$, such that $\bar\partial_z\mu \in \mathfrak{G}_f(\dot{\mathrm{U}})$.

In the CR case the notion of a wedge with edge on Σ [see (IV.2.7)] is closely related to that of a wedge with edge on a maximally real submanifold of $\mathbb{C}^m$ [see (II.1.1)]. A strict analogy [treating the variable called here (x,y) like the variable called t in section IV.2] would have us replace the wedge (IV.2.7) by the set

$$\{ (z,w+\imath\gamma) \in \mathbb{C}^\nu\times\mathbb{C}^d;\ (z,w) \in \mathrm{U},\ \gamma \in \Gamma,\ |\gamma| < \delta \},$$

where Γ would be a cone $\gamma \subset \mathbb{R}^d\backslash\{0\}$. But if we view (y,γ) as a point in an open cone in $\mathbb{R}^m\backslash\{0\}$ we may instead deal with wedges of the following type

$$\{ (z,w)+\imath\gamma \in \mathbb{C}^m;\ (z,w) \in \mathrm{U},\ \gamma \in \Gamma,\ |\gamma| < \delta \}, \tag{IV.7.13}$$

where now Γ is an open cone in $\mathbb{R}^m\backslash\{0\}$. A regular function in the open set (IV.7.13) will then be a $\mathcal{C}^\infty$ function which is holomorphic with respect to w. Here, of course, there are also holomorphic functions, that is, holomorphic with respect to (z,w).

In the CR case Theorem IV.5.5 has an especially simple formulation:

THEOREM IV.7.3.— *In order for the boundary value of a regular function* f *in the set* (IV.7.13) *to be a CR hyperfunction in* U *it is necessary and sufficient that* f *be holomorphic.*

IV.8 FBI MINITRANSFORM. JUMP THEOREM IN HYPO–ANALYTIC STRUCTURES OF THE HYPERSURFACE TYPE

Throughout this section we shall reason within the framework of section IV.7 and of the fine local embedding (IV.7.4) or (IV.7.9). We shall use systematically the notation of section IV.7. In particular, $\tilde{\Pi}$, Π_0, Π'_0 will be the polyhedra in z–space $\mathbb{C}^\nu$, in s–space $\mathbb{R}^d$ and in t–space $\mathbb{C}^{n'}$ respectively, specified in connection with (IV. 7.9); U will denote the image of $\tilde{\Pi}\times U_0\times\Pi'_0$ under the map (IV.7.9), with $U_0 \subset\subset \Pi_0$ an open neighborhood of 0 in $\mathbb{R}^d$ and $\overline{U}$ (resp., $\dot{U}$) the image of $\tilde{\Pi}\times(\mathscr{C}\ell U_0)\times\Pi'_0$ [resp. $\tilde{\Pi}\times(\partial U_0)\times\Pi'_0$]. We need, however, to strengthen our requirements on the function $\varphi(z,s,t)$. After a contraction of $\tilde{\Pi}$, Π_0, Π'_0 about 0 and a biholomorphic substitution on w we can achieve that, in the product set $\tilde{\Pi}\times\Pi_0\times\Pi'_0$,

$$|\varphi(z,s,t)| \leq const.(|z|+|s|^3+|t|). \tag{IV.8.1}$$

If Π'_0 is small enough, (IV.8.1) ensures that each slice $\Sigma_{z,t} = \{ w \in \mathbb{C}^d;$ $(z,w,t) \in \Sigma \}$ is well–positioned at the point $(z,0,t)$; and that $\Sigma_{0,0}$ is very well positioned at the origin.

In section IX.6, [TREVES, 1992], the *FBI minitransform* of a distribution in U depending smoothly on (z,t) was defined. Here we extend this definition to an arbitrary hyperfunction in U depending smoothly on $(z,t) \in \tilde{\Pi}\times\Pi'_0$. ie., an element u of $\mathcal{B}_f(U,\mathcal{C}^\infty_{z,t})$. Let $\mu(z,t) \in \mathfrak{G}_f(\overline{U})$ represent u; we set

(IV.8.2) $\mathcal{F}\mu(z,t;w,\sigma) =$

$$< \mu_{w'}(z,t), e^{\imath\sigma\cdot(w-w')-<\sigma><w-w'>^2}\Delta(w-w',\sigma)>$$

where $\Delta(w,\sigma)$ is the Jacobian determinant of the map $\sigma \to \sigma + \imath<\sigma>w$; $\mathcal{F}\mu$ is a C^∞ function in the region $\{ (z,t,w,\sigma) \in \mathbb{C}^\nu\times\mathbb{R}^{n'}\times\mathbb{C}^d\times\mathbb{C}^d;\ (z,t) \in \tilde{\Pi}\times\Pi'_0,$ $|\mathcal{Im}\,\sigma| < |\mathcal{Re}\,\sigma| \}$, holomorphic with respect to (w,σ).

Theorems IV.6.1 and IV.6.2 have their analogues here, whose statements are left to the reader. We shall content ourselves with stating the analogue of Theorem IV.6.3:

THEOREM IV.8.1.— *Provided* U *is sufficiently small, the following is true. Let* F *be the image of* $\tilde{\Pi}\times K_0\times\Pi'_0$ *under the map* (IV.7.9), *with* K_0 *a compact subset of* U_0, *and let* V *be the image of* $\tilde{\Pi}\times V_0\times\Pi'_0$, *with* V_0 *open*, $K_0 \subset V_0 \subset\subset U_0$. *Then, whatever* $\mu(z,t) \in \mathfrak{G}_f(F)$ *and* $h \in \mathcal{O}(\mathbb{C}^d)$,

(IV.8.3) $<\mu(z,t),h> =$

$$\lim_{\epsilon\to+0} (2\pi)^{-d}\int_{w\in V_{z,t}}\int_{\sigma\in\mathbb{R}T'_{\Sigma_{z,t}}|_w} \mathcal{F}\mu(z,t;w,\sigma)e^{-\epsilon<\sigma>^2}h(w)dwd\sigma;$$

(IV.8.4) $<\mu(z,t),h> =$

$$\lim_{\epsilon\to+0}\int_{w\in V_{z,t}}\int_{w'\in\hat{\Omega}_{z,t}}\int_{\sigma\in\mathbb{R}T'_{\Sigma_{z,t}}|_w} \mathcal{F}\mu(z,t;w',\sigma)h(w)\cdot$$

$$e^{\imath\sigma\cdot(w-w')-<\sigma><w-w'>^2-\epsilon<\sigma>^2}<\sigma>^{d/2}dwdw'd\sigma/(2\pi^3)^{d/2}.$$

In (IV.8.4) $\hat{\Omega}_{z,t}$ stands for the image of $\mathbb{R}^d$ under the map

$$s \to \hat{w}(z,s,t) = s + \imath\chi(s)\varphi(z,s,t).$$

Here $\chi \in C_c^\infty(U_0)$, $\chi(s) \equiv 1$ if $|s| < r$, $|\chi^{(\alpha)}(s)| \leq C_\alpha r^{-|\alpha|}$ for all $\alpha \in \mathbb{Z}_+^d$, $s \in \mathbb{R}^d$. The number $r > 0$ is selected suitably small (keep in mind that $s \in U_0 \Rightarrow |s| < r$).

In analogy with has been done in section IV.6 we identify the structure bundle $T'^{1,0}$ of $\mathbb{C}^d$ to $\mathbb{C}^d\times\mathbb{C}^d$ viewed as a vector bundle over $\mathbb{C}^d$, via the map $(w,\sigma\cdot dw) \to (w,\sigma)$. Then the real structure bundle of $\Sigma_{z,t}$, $\mathbb{R}T'_{\Sigma_{z,t}}$, is identified to the set of points $(w(z,s,t), {}^t w_s(z,s,t)^{-1}\sigma)$, $\sigma \in \mathbb{R}^d$. A subset A of $\mathbb{C}^\nu\times\mathbb{C}^d\times\mathbb{R}^{n'}\times\mathbb{C}^d$ will be called *conic* if A is stable under every dilation $(z,w,t,\sigma) \to (z,w,t,\rho\sigma)$, $\rho > 0$. We shall also write

$$A_{z,t} = \{\ (w,\sigma) \in \mathbb{C}^d\times\mathbb{C}^d;\ (z,w,t,\sigma) \in A\ \}.$$

And if S is a subset of Σ we define

$$\mathbb{R}T'_\Sigma\backslash 0\big|_S = \{\ (z,w,t,\sigma) \in \mathbb{C}^\nu\times\mathbb{C}^d\times\mathbb{R}^{n'}\times\mathbb{C}^d;\ (z,w,t) \in S,\ (w,\sigma) \in \mathbb{R}T'_{\Sigma_{z,t}},\ \sigma \neq 0\ \}.$$

We may introduce the space $\mathfrak{E}_f(U,C^\infty_{z,t})$ of C^∞ functions in some open and conic neighborhood $\mathcal{U}$ of $\mathbb{R}T'_\Sigma\backslash 0\big|_U$ in $\mathbb{C}^\nu\times\mathbb{C}^d\times\mathbb{R}^{n'}\times\mathbb{C}^d$, holomorphic with respect to (w,σ) and having the following property:

(IV.8.5) *Given any number $\epsilon > 0$ there is an open and conic neighborhood $\mathcal{U}^\epsilon \subset \mathcal{U}$ of $\mathbb{R}T'_\Sigma\backslash 0\big|_U$ in $\mathbb{C}^\nu\times\mathbb{C}^d\times\mathbb{R}^{n'}\times\mathbb{C}^d$ such that*

$$\sup_{(z,w,t,\sigma)\in\mathcal{U}^\epsilon}\left[e^{-\epsilon|\sigma|}\,|h(z,w,t,\sigma)|\right] < +\infty.$$

A function $h \in \mathfrak{E}_f(U,C^\infty_{z,t})$ will belong to the subspace $\mathfrak{E}_{of}(U,C^\infty_{z,t})$ if it is defined in an open and conic neighborhood $\mathcal{U}$ of $\mathbb{R}T'_{\Sigma}\backslash 0\big|_{U}$ in $\mathbb{C}^{\nu}\times\mathbb{C}^{d}\times\mathbb{R}^{n'}\times\mathbb{C}^{d}$ and if the following is true:

(IV.8.6) *to each compact set* $K \subset U$ *there are an open and conic neighborhood* $\mathcal{U}_K \subset \mathcal{U}$ *of* $\mathbb{R}T'_{\Sigma}\backslash 0\big|_{K}$ *in* $\mathbb{C}^{\nu}\times\mathbb{C}^{d}\times\mathbb{R}^{n'}\times\mathbb{C}^{d}$ *and a constant* $c > 0$ *such that*

$$\sup_{(z,w,t,\sigma)\in\mathcal{U}_K}\left[e^{c|\sigma|}\,|h(z,w,t,\sigma)|\right] < +\infty.$$

The FBI minitransform of elements of $\mathfrak{G}_f(\overline{U})$ induces a linear map

$$\mathcal{F}\colon \mathcal{B}_f(U,\mathcal{C}^{\infty}_{z,t}) \to \mathfrak{E}_f(U,\mathcal{C}^{\infty}_{z,t})/\mathfrak{E}_{of}(U,\mathcal{C}^{\infty}_{z,t}),$$

which is the FBI minitransform of elements of $\mathcal{B}_f(U,\mathcal{C}^{\infty}_{z,t})$.

Formula (IV.6.7) has its parallel here:

(IV.8.7)

$$\partial^{\alpha}_{x}\partial^{\beta}_{y}\partial^{\gamma}_{w}\partial^{\kappa}_{t}\mathcal{F}u = \mathcal{F}\partial^{\alpha}_{x}\partial^{\beta}_{y}\partial^{\gamma}_{w}\partial^{\kappa}_{t}u,\ \forall\ \alpha,\beta\in\mathbb{Z}^{\nu}_{+},\ \gamma\in\mathbb{Z}^{d}_{+},\ \kappa\in\mathbb{Z}^{n'}_{+};$$

and so does Theorem IV.6.5. Since, in the fine local embedding, the hyperfunction solutions in U are the elements $u \in \mathcal{B}_f(U,\mathcal{C}^{\infty}_{z,t})$ such that $(\bar{\partial}_z+d_t)u \equiv 0$, it follows from (IV.8.7) that, for such a hyperfunction solution u,

(IV.8.8) $$(\bar{\partial}_z+d_t)\mathcal{F}u \equiv 0.$$

The FBI minitransform can be used to characterize the hypo—analyticity, and to describe the hypo—analytic wave—front set, of solutions, just like the FBI "maxitransform" was used in section IV.6.

Rather than dwelling on these rather routine generalizations we shall focus our attention on the case d = 1.

We recall that (for any value of d) $T^0_{\wp} = T^*_{\wp}\mathcal{M} \cap T'_{\pi}$ is the *characteristic set* at the point $\wp$ of the locally integrable structure of $\mathcal{M}$. In the fine local embedding (IV.7.9) the characteristic set at O is spanned over $\mathbb{R}$ by $d\varphi_1,\ldots,d\varphi_d$.

DEFINITION IV.8.1.— *The locally integrable structure of* $\mathcal{M}$ *is said to be of the* ***hypersurface type*** *at the point* $\wp$ *if* $\dim_{\mathbb{R}} T^0_{\wp} = 1$.

Thus, in the fine local embedding (IV.7.9), for the structure of $\mathcal{M}$ to be of the hypersurface type at O means that d = 1. In the remainder of the section we assume this to be the case and write w, s, σ instead of w_1, s_1, σ_1, and φ instead of φ_1. The hypersurface Σ is defined, near 0, by the equation $\mathcal{Im}\ w = \varphi(z,s,t)$, which shows that Σ is a hypersurface in $\mathbb{C}^{\nu}\times\mathbb{C}\times\mathbb{R}^{n'}$. Therein lies the reason for Definition IV.8.1.

When d = 1 one deals with wedges of the following type [cf. (IV.7.13)]. Let U be an open neighborhood of 0 in Σ, δ a number > 0; we write

$$\mathcal{W}_{\delta}(U,\mathbb{R}_{\pm}) =$$
$$\{\ (z,w,t) \in \mathbb{C}^{\nu}\times\mathbb{C}\times\mathbb{R}^{n'};\ \exists\ \tau \in \mathbb{R}_{+},\ (z,w-\imath\tau,t) \in U,\ 0 < \pm\tau < \delta\ \}.$$

By a *regular function* f in is an open subset $\tilde{\Omega}$ of $\mathbb{C}^{\nu}\times\mathbb{C}\times\mathbb{R}^{n'}$ we mean a $\mathcal{C}^{\infty}$ function in $\tilde{\Omega}$ which is holomorphic with respect to w; we write $f \in \mathfrak{F}(\tilde{\Omega})$. We can therefore talk of the boundary value on U, $b_U f$, of a function $f \in \mathfrak{F}(\mathcal{W}_{\delta}(U,\mathbb{R}_{\pm}))$. For $b_U f$ to be a solution it is necessary and sufficient that f be holomorphic with respect to z and locally constant with respect to t.

THEOREM IV.8.2.— *Suppose* Σ *is a hypersurface in* $\mathbb{C}^\nu\times\mathbb{C}\times\mathbb{R}^{n'}$ *and* U *a sufficiently small open neighborhood of* 0 *in* Σ. *To every open subset of* Σ, $V \subset\subset U$, *there is a number* $\delta > 0$ *such that the following holds:*

(IV.8.9) *To every hyperfunction* $u \in \mathcal{B}_f(U,\mathcal{C}^\infty_{z,t})$ *there are two regular functions* $f^{\pm} \in \mathfrak{F}(\mathcal{W}_\delta(V,\mathbb{R}_\pm))$ *such that*

$$u\Big|_V = b_V f^+ + b_V f^-.$$

Furthermore, if $W \subset\subset U$ *is a sufficiently small open neighborhood of* 0 *in* Σ, *the following is true:*

(IV.8.10) *To every hyperfunction solution* u *in* U *there are two regular functions* $h^{\pm} \in \mathfrak{F}(\mathcal{W}_\delta(W,\mathbb{R}_\pm))$ *such that*

$$(\bar\partial_z+\mathrm{d}_t)h^{\pm} = 0 \text{ and } u\Big|_W = b_W h^+ + b_W h^-.$$

Proof: The part that concerns (IV.8.9) is simply a restatement in Theorem IV.6.4 in the present circumstances, since here the family of cones $\Gamma_1,\dots,\Gamma_r$ is unique: it consists of the two open half—lines $\mathbb{R}_\pm$. A hyperfunction solution in U can be identified to an element u of $\mathcal{B}_f(U,\mathcal{C}^\infty_{z,t})$ such that $(\bar\partial_z+\mathrm{d}_t)u \equiv 0$. Let $V \subset\subset U$ be an open neighborhood of 0 in Σ and $\delta > 0$ a number as in (IV.8.9); we have, in V:

$$b_V(\bar\partial_z+\mathrm{d}_t)f^+ = (\bar\partial_z+\mathrm{d}_t)b_V f^+ = -(\bar\partial_z+\mathrm{d}_t)b_V f^- = -b_V(\bar\partial_z+\mathrm{d}_t)f^-$$

with $f^{\pm} \in \mathfrak{F}(\mathcal{W}_\delta(V,\mathbb{R}_\pm))$. From the elementary version of the theorem of the Edge of the Wedge we derive that there is a one—form

$$g = \sum_{j=1}^{\nu} g'_j(z,w,t)\mathrm{d}\bar z_j + \sum_{k=1}^{n'} g''_k(z,w,t)\mathrm{d}t_k$$

with regular coefficients g'_j, g''_k in $\tilde{\Omega} = \mathcal{W}_\delta(V,\mathbb{R}_+)\cup V\cup\mathcal{W}_\delta(V,\mathbb{R}_-)$, such that

$$g = (\bar\partial_z + d_t)f^+ \text{ in } \mathcal{W}_\delta(V,\mathbb{R}_+),\ g = -(\bar\partial_z + d_t)f^- \text{ in } \mathcal{W}_\delta(V,\mathbb{R}_-).$$

Obviously $(\bar\partial_z + d_t)g \equiv 0$ in $\tilde{\Omega}$. Let now the polyhedron $\tilde{\Pi}\times\tilde{\Pi}_0\times\Pi'_0$ centered at the origin in $\mathbb{C}^\nu\times\mathbb{C}\times\mathbb{R}^{n'}$ be contained in $\tilde{\Omega}$; here

$$\tilde{\Pi}_0 = \{\, s \in \mathbb{C};\ |\mathcal{R}e\, s| < a,\ |\mathcal{I}m\, s| < b \,\},\ a,\ b > 0.$$

We can solve the equation

$$(\bar\partial_z + d_t)v = g \text{ in } \tilde{\Pi}\times\tilde{\Pi}_0\times\Pi'_0.$$

Set

$$h^+ = f^+ - v \text{ in } \mathcal{W}_\delta(V,\mathbb{R}_+)\cap(\tilde{\Pi}\times\tilde{\Pi}_0\times\Pi'_0),$$

$$h^- = f^- + v \text{ in } \mathcal{W}_\delta(V,\mathbb{R}_-)\cap(\tilde{\Pi}\times\tilde{\Pi}_0\times\Pi'_0).$$

We have

$$(\bar\partial_z + d_t)h^\pm \equiv 0 \text{ in } \mathcal{W}_\delta(V,\mathbb{R}_\pm)\cap(\tilde{\Pi}\times\tilde{\Pi}_0\times\Pi'_0),$$

and if $W \subset V\cap(\tilde{\Pi}\times\tilde{\Pi}_0\times\Pi'_0)$, $b_W h^+ + b_W h^- = b_W f^+ + b_W f^-$, whence the result. □

In the CR case, ie., when there are no variables t, Σ is a hypersurface in $\mathbb{C}^m$ ($= \mathbb{C}^\nu\times\mathbb{C}$) and we obtain the expected result:

COROLLARY IV.8.1.— *In the neighborhood of every point of the real C^∞ hypersurface Σ in $\mathbb{C}^m$ every CR hyperfunction is equal to the sum of the boundary value of a holomorphic function on one side of Σ and of the boundary value of a holomorphic function on the opposite side.*

IV.9 NONVANISHING OF THE LOCAL COHOMOLOGY IN HYPO–ANALYTIC STRUCTURES OF THE HYPERSURFACE TYPE

In the present section we propose to extend to hyperfunctions the main theorem in [CORDARO–TREVES, 1991]. The result applies to a hypo–analytic structure *of the hypersurface type* (Definition IV.8.1), and shows that under an appropriate hypothesis, the stalk of the sheaf $\mathcal{S}o\ell^{(q+1)}$ at a point O of the base manifold $\mathcal{M}$ does not vanish (for a certain value of q, $0 \leq q \leq n-1$). This, of course, is another way of saying that, in neighborhoods of O, the inhomogeneous equation $Lu = f$ does not have any solution u of type (0,q) for some L–closed current f of type (0,q+1).

Our basic hypothesis will be the existence of a hypo–analytic first integral in an open neighborhood U of $\mathcal{M}$ which is *acyclic* in dimension q ([CORDARO–TREVES, 1991], Definition 2.3). We now recall the meaning of such a hypothesis. First of all, as we have said, the hypo–analytic structure of $\mathcal{M}$ is assumed to be of the hypersurface type (near O). This is the same as saying that there is a hypo–analytic function w in some open neighborhood of O, such that $dw\big|_O$ is *real–valued* and $dw\big|_O \neq 0$; and that w is essentially unique, in the sense that the differential at O of any other hypo–analytic function h in a neighborhood of O, is collinear to $dw\big|_O$ unless $d\bar{h}\big|_O \neq dh\big|_O$.

It is then natural to take w as one of the first integrals in the local representation of the hypo–analytic structure. This is why, throughout the present section, we shall reason in the framework of the fine local embedding (section IV.7). Thus we shall use the coordinates x_i, y_j, s_k, t_l in (IV.7.1) and the first integrals (IV.7.2) under Hypothesis (IV.7.3). Actually we shall transfer the whole analysis onto Σ, the image of the open subset $\mathcal{M}'$ of $\mathcal{M}$ under the map (IV.7.4). Our assumption that the structure is of the hypersurface type at O is equivalent to the condition d = 1, ie., there is only one variable s. By appropriately selecting the remaining first integrals $z_1,\dots,z_\nu$ and the variables $t_1,\dots,t_{n'}$, we can arrange things in such a way that $w = s + i\varphi(z,s,t)$ (the number of variables z and t is unrestricted). The reader will then notice that Σ is a real hypersurface in $\mathbb{C}^m\times\mathbb{R}^{n'}$, which is the reason why the structure is said to be of the hypersurface type. With the proviso that d = 1 the set–up will be the one described in section IV.7.

We shall use the following notation for the various vector bundles that enter the picture: $\Lambda^{p,q}$ will denote the subbundle of $\mathbb{C}T^*(\mathbb{C}^m\times\mathbb{R}^{n'})$ spanned by the exterior products $dz_I\wedge d\bar{z}_J\wedge dt_K$ with $|I| = p$, $|J|+|K| = q$, and $dw\wedge dz_I\wedge d\bar{z}_J\wedge dt_K$ with $|I| = p-1$, $|J|+|K| = q$; $\Lambda_0^{p,q}$ will denote the vector bundle over (z,t)–space $\mathbb{C}^\nu\times\mathbb{R}^{n'}$ spanned by the products $dz_I\wedge d\bar{z}_J\wedge dt_K$ with $|I| = p$, $|J|+|K| = q$; $\Lambda_\Sigma^{p,q}$ will denote the pullback of $\Lambda^{p,q}$ to Σ.

Call U (resp., V) the image of $\tilde{\Pi}\times\mathcal{I}\times\Pi_0'$ (resp., $\tilde{\Pi}'\times\mathcal{I}'\times\Pi_0''$) via the map (IV.7. 4); $\tilde{\Pi}' \subset\subset \tilde{\Pi}$ are polyhedrons in z–space $\mathbb{C}^\nu$, $\Pi_0'' \subset\subset \Pi_0'$ in t–space $\mathbb{R}^{n'}$; $\mathcal{I}' \subset\subset \mathcal{I}$ are open intervals in the s–line $\mathbb{R}$, centered at the origin. We consider a *regular* value w_0 of $w = s+i\varphi(z,s,t)$ in U, with $s_0 = \mathcal{R}e\, w_0 \in \mathcal{I}'$: $d_{x,y,t}\varphi$ does not vanish at any point of the *level set* $\mathcal{S}_0 = \{ (z,w,t) \in U\ ;\ s = s_0,\ \varphi(z,s_0,t) = \mathcal{I}m\, w_0 \}$. We regard $\mathcal{S}_0$ as a C^∞ sub–manifold of the subset $\tilde{\Pi}\times\Pi_0'$ of (z,t)–space $\mathbb{C}^\nu\times\mathbb{R}^{n'}$, whose intersection $\mathcal{S}_0'$ with $\tilde{\Pi}'\times\Pi_0''$ is nonempty. We make use of the sheaf $\mathscr{C}^\infty\Lambda_0^{p,q}$ of germs of differential forms $f \in C^\infty(\mathcal{A};\Lambda_0^{p,q})$ ($\mathcal{A}$: open subset of $\mathbb{C}^\nu\times\mathbb{R}^{n'}$), ie., of forms (IV.3.6) but with C^∞ coefficients. Here we make use of the resolution

(IV.9.1) $$0 \to {}_\nu\mathcal{O}^{(p)} \xrightarrow{i} \mathscr{C}^\infty\Lambda_0^{p,0} \xrightarrow{\bar\partial_z+d_t} \mathscr{C}^\infty\Lambda_0^{p,1} \xrightarrow{\bar\partial_z+d_t} \cdots \xrightarrow{\bar\partial_z+d_t} \mathscr{C}^\infty\Lambda_0^{p,n} \to 0;$$

${}_\nu\mathcal{O}^{(p)}$ is the sheaf of germs of differential forms $\sum_{|I|=p} f_I(z)dz_I$ with holomorphic coefficients f_I ($z \in \mathbb{C}^\nu$, $0 \le p \le \nu$). Consider arbitrary sections

$$\alpha \in \Gamma(\mathcal{S}_0; \mathscr{C}^\infty\Lambda_0^{0,q}),\ \beta \in \Gamma_c(\mathcal{S}'_0; \mathscr{C}^\infty\Lambda_0^{\nu,n-q-1})$$

such that $(\bar\partial_z+d_t)\alpha = 0$, $(\bar\partial_z+d_t)\beta = 0$. When $q = 0$ one must further require that $\int_{\mathcal{S}_0} h\beta = 0$ for all $h \in \mathcal{O}(\mathbb{C}^m)$. The integral $\int_{\mathcal{S}_0} \alpha\wedge\beta$ only depends on the cohomology classes

$$[\alpha] \in H^q(\mathcal{S}_0, {}_\nu\mathcal{O}),\ [\beta] \in H_c^{n-q-1}(\mathcal{S}'_0, {}_\nu\mathcal{O}^{(\nu)}).$$

The bilinear functional

$$H^q(\mathcal{S}_0, {}_\nu\mathcal{O})\times H_c^{n-q-1}(\mathcal{S}'_0, {}_\nu\mathcal{O}^{(\nu)}) \ni ([\alpha],[\beta]) \to \int_{\mathcal{S}_0} \alpha\wedge\beta$$

is called the *intersection number of* $\mathcal{S}_0$ *in dimension* q, *relative to the pair* (U,V) and is denoted by $I_{\mathcal{S}_0,U,V}$ (Definition 2.2, *loc. cit.*). Then the hypo–analytic function w is said *acyclic in dimension* q *at the point* O if to each open neighborhood U of O in Σ there is an open subneighborhood V, with U and V of the type indicated above, such that $I_{\mathcal{S}_0,U,V} \equiv 0$ whatever the regular level set of w in U that intersects V (Definition 2.3, *loc. cit.*).

The following definition is quite natural: the differential equation $Lu = f$ is said to be *locally solvable at* O *in degree* q if given any open neighborhood U of O in Σ there is an open subneighborhood V such that the following is true:

$(\text{IV.9.2})_q$ $\quad \forall\, f \in \mathcal{C}^{\infty}(\text{U};\Lambda_{\Sigma}^{0,q+1})$, $\text{L}f \equiv 0$ *in* U, $\exists\, u \in \mathcal{D}'(\text{V};\Lambda_{\Sigma}^{0,q})$ *such that* $\text{L}u = f$ *in* V.

We are allowed to choose the neighborhoods U and V in (IV.9.2) to be the images of $\tilde{\Pi}\times\mathcal{I}\times\Pi_0'$ and $\tilde{\Pi}'\times\mathcal{I}'\times\Pi_0''$ respectively, via the map (IV.7.4).

We can now recall the main result in [CORDARO–TREVES, 1991] (see also sections VIII.4, VIII.5 in [TREVES, 1992]):

THEOREM IV.9.1.— *If the equation* $\text{L}u = f$ *is locally solvable at O in degree* q *then every germ of a hypo–analytic function whose differential at O spans the characteristic set at O must be acyclic at O in dimension* q.

We propose to generalize Theorem IV.9.1 to hyperfunctions. More precisely we propose to prove:

THEOREM IV.9.2.— *Let w be a hypo–analytic function in Σ whose differential spans the characteristic set at O. Suppose there are neighborhoods* V ⊂ U *of O in Σ, respectively the image of* $\tilde{\Pi}\times\mathcal{I}\times\Pi_0'$ *and* $\tilde{\Pi}'\times\mathcal{I}'\times\Pi_0''$ *under the map* (IV.7.4), *such that the intersection number in dimension* q *of some regular level set* $\mathcal{S}_0$ *of w relative to the pair* (U,V) *does not vanish. Then the following property does not hold:*

$(\text{IV.9.3})_q$ $\quad \forall\, f \in \mathcal{B}_{\text{f}}(\text{U},\mathcal{C}^{\infty}_{z,t};\Lambda^{q+1})$, $(\bar{\partial}_z+\text{d}_t)f \equiv 0$, $\exists\, u \in \mathcal{B}_{\text{f}}(\text{V},\mathcal{C}^{\infty}_{z,t};\Lambda^{q})$ *such that* $(\bar{\partial}_z+\text{d}_t)u = f$ *in* V.

Proof: As before we suppose that $\mathcal{S}_0 = \{\, (z,t) \in \tilde{\Pi}\times\Pi_0';\ (z,w_0,t) \in \Sigma \,\}$. This means that $\mathcal{S}_0$ is defined by the equation $\varphi(z,s_0,t) = \mathcal{Im}\, w_0$ $(s_0 =$

$\mathcal{R}e\, w_0$). We are assuming that $d_{z,t}\varphi(z,s_0,t) \neq 0$ at every point of $\mathcal{S}_0$ and that $\mathcal{S}_0$ intersects $\tilde{\Pi}'\times\Pi''_0$. It is easy to see (cf. Section 3, [CORDARO–TREVES, 1991]) that, possibly after changing s into $-s$, the hypothesis about the intersection number implies the following property:

(IV.9.4)$_q$ *There exist* $f_0 \in \mathcal{C}^\infty(\tilde{\Pi}\times\Pi'_0;\Lambda_0^{0,q})$, $g \in \mathcal{C}_c^\infty(\tilde{\Pi}'\times\Pi''_0;\Lambda_0^{\nu,n-q-1})$ *such that*

$$\mathrm{supp}\,(\bar\partial_z+d_t)f_0 \subset \{\, (z,t) \in \tilde{\Pi}\times\Pi'_0;\ \varphi(z, \mathcal{R}e\, w_0,t) < \mathcal{I}m\, w_0 \,\},$$
$$\mathrm{supp}\,(\bar\partial_z+d_t)g \subset \{\, (z,t) \in \tilde{\Pi}'\times\Pi''_0;\ \varphi(z, \mathcal{R}e\, w_0,t) > \mathcal{I}m\, w_0 \,\},$$
$$\int f_0\wedge(\bar\partial_z+d_t)g \neq 0.$$

In the sequel f_0 and g will denote forms that verify the requirements in (IV.9.4)$_q$. We define an L–closed form belonging to $\mathcal{C}^\infty(U;\Lambda_\Sigma^{0,q+1})$ [with L as in (IV.7.7)] as the pullback to U of the current

$$\frac{1}{\pi}\frac{1}{w-w_0}(\bar\partial_z+d_t)f_0 \in \mathcal{C}^\infty(\tilde{\Pi}\times\mathbb{C}\times\Pi'_0;\Lambda^{0,q+1}).$$

This form can be identified to a hyperfunction current $f \in \mathcal{B}_f(U,\mathcal{C}^\infty_{z,t};\Lambda^{q+1})$ [of the kind (IV.7.6) but with coefficients in $\mathcal{B}_f(U,\mathcal{C}^\infty_{z,t})$] such that $(\bar\partial_z+d_t)f = 0$, thereby defining a class $[f] \in \mathfrak{Sol}_f^{q+1}(U)$. We are going to prove that there is no $u \in \mathcal{B}_f(V,\mathcal{C}^\infty_{z,t};\Lambda^q)$ satisfying $(\bar\partial_z+d_t)u = f$ in V.

We shall reason *a contrario* and assume that u exists. We select a representative of u in V,

$$\omega = \sum_{|J|+|K|=q-1} \omega_{J,K}d\bar{z}_J\wedge dt_K$$

with coefficients $\omega_{J,K} \in \mathfrak{G}_f(\bar{V})$; $\omega \in \mathfrak{G}_f(\bar{V};\Lambda^{q-1})$ in the notation used below. We remind the reader that the fine local embedding is being used and therefore, that $\mathfrak{G}_f(\mathcal{K})$ with $\mathcal{K} = \overline{V}$ or $\mathcal{K} = \dot{V}$, stands for the space of analytic functionals in the complex w-plane that depend smoothly on (z,t) and are carried by $\mathcal{K}$. Here $\overline{V}$ is the closure, and $\dot{V}$ the boundary, of V in $\Sigma\cap(\tilde{\Pi}'\times\Pi'\times\Pi''_0)$, where $\Pi' = \mathcal{I}'+\imath\mathcal{I}_1$ [in accordance with the nota— tion introduced in section IV.7], Let $f_{J,K}$, $|J|+|K| = q+1$, denote the coefficients of f; define $\mu_{J,K} \in \mathfrak{G}_f(\overline{V})$ by setting, for every $h \in \mathcal{O}(\mathbb{C})$,

$$<\mu_{J,K}(z,t),h> =$$

$$\int_{\mathcal{I}'} f_{J,K}(z,s+\imath\varphi(z,s,t),t)h(s+\imath\varphi(z,s,t))[1+\imath\varphi_s(z,s,t)]ds.$$

Then f is represented in V by

$$\mu = \sum_{|J|+|K|=q+1} \mu_{J,K} d\bar{z}_J \wedge dt_K.$$

We have $(\bar{\partial}_z+d_t)\mu \in \mathfrak{G}_f(\dot{V};\Lambda^{q+2})$, and there is a current

$$\nu = \sum_{|J|+|K|=q+1} \nu_{J,K} d\bar{z}_J \wedge dt_K \in \mathfrak{G}_f(\dot{V};\Lambda^{q+1}),$$

such that

$$(\bar{\partial}_z+d_t)\omega = \mu + \nu. \tag{IV.9.5}$$

We evaluate the coefficients of the currents on both sides of Equation (IV.9.5) on an entire function $h(w)$. We get two C^∞ forms of bidegree $(0,q+1)$ in (z,t)—space. We carry out the exterior multiplication of those two forms with the form g in $(IV.9.4)_q$; g has bidegree $(\nu,n-q-1)$. The resulting forms are smooth, compactly supported and have total degree $\nu+n = 2\nu+n'$, and thus they can be integrated over $\mathbb{C}^\nu\times\mathbb{R}^{n'}$:

$$\int <(\bar{\partial}_z+d_t)\omega\wedge g,h> = \int <\mu\wedge g,h> + \int <\nu\wedge g,h>. \tag{IV.9.6}$$

Hypothesis $(\mathrm{IV}.9.4)_q$, combined with the fact that $s_0 \in \mathcal{I}'$, entails the existence of two positive numbers, δ, η, such that

$$\text{(IV.9.7)} \qquad [s_0-\eta, s_0+\eta] \subset \mathcal{I}';$$

$$\text{(IV.9.8)} \qquad \varphi(z,s,t) \geq \mathcal{I}m\, w_0 + \delta \;\; \textit{if } |s-s_0| \leq \eta,\; (z,t) \in \operatorname{supp}(\bar\partial_z+\mathrm{d}_t)g.$$

Call V_1 the image of $\tilde\Pi'\times(s_0-\frac{1}{2}\eta, s_0+\frac{1}{2}\eta)\times\Pi''_0$ under the map

$$(z,s,t) \to (z, s+ \imath\varphi(z,s,t), t)$$

and consider a decomposition

$$\omega = \omega_1 + \omega_2,\; \omega_1 \in \mathfrak{G}_f(\overline{V}_1;\Lambda^q),\; \omega_2 \in \mathfrak{G}_f(\overline{V}\backslash V_1;\Lambda^q).$$

Putting this into (IV.9.6) and integrating by parts yields

$$\text{(IV.9.9)} \quad \int \langle\mu\wedge g, h\rangle = (-1)^{q-1}\int \langle\omega_1\wedge(\bar\partial_z+\mathrm{d}_t)g, h\rangle - \int \langle\nu', h\rangle,$$

where $\nu' = \nu\wedge g - (-1)^{q-1}\omega_2\wedge(\bar\partial_z+\mathrm{d}_t)g \in \mathfrak{G}_f(\overline{V}\backslash V_1;\Lambda^n)$. Consider the compact subset K' of the w-plane defined by the inequalities

$$|\mathcal{R}e\, w - s_0| \leq \eta,\quad \mathcal{I}m\, w_0 + \tfrac{1}{2}\delta \leq \mathcal{I}m\, w \leq M' \;(< +\infty).$$

If $0 < \epsilon < \frac{1}{2}\delta$ the set

$$\mathcal{N}'^{\epsilon} = \{\, (z, s+\imath\varphi(z,s,t)+\imath\gamma, t) \in \mathbb{C}^m\times\mathbb{R}^{n'};$$
$$(z,t) \in \tilde\Pi'\times\Pi''_0,\; |s-s_0| < \eta,\; |\gamma| < \epsilon \,\}$$

is an open neighborhood of $\overline{V}_1$ in $\tilde\Pi'\times\mathbb{C}\times\Pi''_0$. There is a constant $C > 0$ such that an arbitrary coefficient $(\omega_1)_{I,J}$ of ω_1 satisfies an inequality

$$|\langle(\omega_1)_{I,J}, h\rangle| \leq C \sup_{\mathcal{N}'^{\epsilon}_{z,t}} |h|$$

for all $(z,t) \in (\bar\partial_z+\mathrm{d}_t)g$. It follows from (IV.9.8) that

$$\mathcal{N}'^{\epsilon}_{z,t} = \{\ s+\imath\varphi(z,s,t)+\imath\gamma \in \mathbb{C};\ |s-s_0| < \eta,\ |\gamma| < \epsilon\ \}$$

is contained in K$'$ provided $(z,t) \in \text{supp}\,(\overline{\partial}_z+d_t)g$ and M' is sufficiently large. If this is so we have

$$\text{(IV.9.10)}\qquad \left|\int <\omega_1\wedge(\overline{\partial}_z+d_t)g,h>\right| \leq \textit{const.}\ \underset{K'}{\text{Max}}\ |h|,\ \ \forall\ h \in \mathcal{O}(\mathbb{C}).$$

In dealing with the term $\int <\nu',h>$ we reason in similar fashion. The set

$$\mathcal{N}'' = \{\ (z,s+\imath\varphi(z,s,t)+\imath\gamma,t) \in \mathbb{C}^m\times\mathbb{R}^{n'};$$
$$(z,s,t) \in \tilde{\Pi}'\times\mathcal{I}\times\Pi''_0,\ |s-s_0| > \tfrac{1}{4}\eta,\ |\gamma| < 1\ \}$$

is an open neighborhood of $\overline{V}\backslash V_1$ in $\tilde{\Pi}'\times\mathbb{C}\times\Pi''_0$. There is a constant $C > 0$ such that an arbitrary coefficient $\nu'_{I,J}$ of ν' satisfies an inequality

$$|<\nu'_{I,J},h>| \leq C \sup_{\mathcal{N}''_{z,t}} |h|$$

for all $(z,t) \in \text{supp}\ g$; and

$$\mathcal{N}''_{z,t} = \{\ s+\imath\varphi(z,s,t)+\imath\gamma \in \mathbb{C};\ s \in \mathcal{I},\ |s-s_0| > \tfrac{1}{4}\eta,\ |\gamma| < 1\ \}$$

is contained in the compact set

$$K'' = \{\ w \in \mathbb{C};\ \mathcal{R}e\,w \in \mathcal{C}\ell\mathcal{I},\ |\mathcal{R}e\,w - s_0| \geq \tfrac{1}{4}\eta,\ |\mathcal{I}m\,w| \leq M''\ \}$$

for a suitably large constant $M'' > 0$. We obtain

$$\text{(IV.9.11)}\qquad \left|\int <\nu',h>\right| \leq \textit{const.}\ \underset{K''}{\text{Max}}\ |h|,\ \ \forall\ h \in \mathcal{O}(\mathbb{C}).$$

By combining (IV.9.9), (IV.9.10) and (IV.9.11) we reach the following conclusion. There is a compact subset K of the region in the w—plane,

$$\mathcal{R} = \{ w \in \mathbb{C};\ |\, \mathcal{R}e\, w - s_0 | > \tfrac{1}{8}\eta \ \ or\ \ \mathcal{I}m\, w > \mathcal{I}m\, w_0 + \tfrac{1}{4}\delta \},$$

such that, for a suitably large positive constant C,

$$\text{(IV.9.12)} \qquad \left| \int <\mu\wedge g,h> \right| \leq C \operatorname*{Max}_{K} |h|, \ \forall\, h \in \mathcal{O}(\mathbb{C}^m).$$

We return to the definition of μ. Up to sign $\int <\mu\wedge g,h>$ is equal to the integral

$$\mathcal{T}(h) = \pi^{-1}\int_{s\in\mathcal{I}'}\int_{z\in\mathbb{C}^\nu}\int_{t\in\mathbb{R}^{n'}} h(w)\,\frac{1}{w - w_0}[(\bar\partial_z + d_t)f_0]\wedge g\wedge dw$$

where $w = s+\imath\varphi(z,s,t)$. We select $\chi_0 \in C^\infty_c(\mathcal{I}')$, $\chi_0(s) \equiv 1$ if $|s-s_0| < \tfrac{1}{2}\eta$, $\chi_0 \equiv 0$ if $|s-s_0| \geq \eta$ and we write $\chi(w) = \chi_0(\mathcal{R}e\, w)$; then

$$\mathcal{T}(h) = \mathcal{T}(\chi h) + \mathcal{T}((1-\chi)h).$$

There is a compact subset K_1 of the region $\{ w \in \mathbb{C};\ |\,\mathcal{R}e\, w - s_0| \geq \tfrac{1}{2}\eta \}$ such that, for some constant $C > 0$ and every $h \in \mathcal{O}(\mathbb{C})$,

$$\text{(IV.9.13)} \qquad |\mathcal{T}((1-\chi)h)| \leq C \operatorname*{Max}_{K_1} |h|.$$

In order to deal with the term $\mathcal{T}(\chi h)$ we shall compute the total differential in U, $d\left[\pi^{-1}(w-w_0)^{-1}\chi_0(s)h(w)f_0\wedge g\wedge dw\right]$. Recall that

$$g = g_0\wedge dz_1\wedge\cdots\wedge dz_\nu \text{ with } g_0 \in C^\infty_c(\tilde\Pi'\times\Pi''_0;\Lambda^{0,n-q-1}).$$

It follows that

$$d\left[\pi^{-1}(w-w_0)^{-1}\chi_0(s)h(w)f_0\wedge g\wedge dw\right] = h(w)L\left[\pi^{-1}(w-w_0)^{-1}\chi_0(s)f_0\wedge g\wedge dw\right] =$$

$$h(w_0)\delta(w-w_0)L\bar w\wedge f_0\wedge g\wedge dw + \pi^{-1}(w-w_0)^{-1}(\chi h)(w)[(\bar\partial_z+d_t)f_0]\wedge g\wedge dw +$$

$$(-1)^q\pi^{-1}(w-w_0)^{-1}h(w)f_0\wedge d(\chi g)\wedge dw.$$

We have availed ourselves of the property that $\chi_0(s_0) = 1$. Integration with respect to (z,s,t) over $\mathbb{C}^\nu\times\mathbb{R}\times\mathbb{R}^{n'}$ yields

$$\mathcal{T}(\chi h) = (-1)^{q-1}\pi^{-1}\int_{z\in\mathbb{C}^\nu}\int_{s\in\mathbb{R}}\int_{t\in\mathbb{R}^{n'}}(w-w_0)^{-1}h(w)f_0\wedge d(\chi g)\wedge dw -$$
$$h(w_0)\int_{z\in\mathbb{C}^\nu}\int_{s\in\mathbb{R}}\int_{t\in\mathbb{R}^{n'}}\delta(w-w_0)L\bar{w}\wedge f_0\wedge g\wedge dw.$$

On supp $d(\chi g)$ we have either $|s-s_0| \geq \frac{1}{2}\eta$ or, if not, $(z,t) \in$ supp $(\bar{\partial}_z+d_t)g$. If we take (IV.9.8) into account we see that the form $(w-w_0)^{-1}d(\chi g)$ is smooth, and that there is a compact subset K_2 of the region $\mathcal{R}$ such that, for some $C > 0$ and every $h \in \mathcal{O}(\mathbb{C})$,

(IV.9.14) $$\left|\int_{z\in\mathbb{C}^\nu}\int_{s\in\mathbb{R}}\int_{t\in\mathbb{R}^{n'}}(w-w_0)^{-1}h(w)f_0\wedge d(\chi g)\wedge dw\right| \leq C\operatorname*{Max}_{K_2}|h|.$$

Combining (IV.9.12), (IV.9.13) and (IV.9.14) yields

(IV.9.15) $$c_0|h(w_0)| \leq C\operatorname*{Max}_{K_0}|h|,\ \forall\ h \in \mathcal{O}(\mathbb{C}),$$

where $K_0 = K \cup K_1 \cup K_2 \subset\subset \mathcal{R}$ and

$$c_0 = \left|\int_{z\in\mathbb{C}^\nu}\int_{s\in\mathbb{R}}\int_{t\in\mathbb{R}^{n'}}\delta(w-w_0)L\bar{w}\wedge f_0\wedge g\wedge dw\right|.$$

We contend that $c_0 > 0$. We note that

$$L\bar{w} = L(w+\bar{w}) = -2i(\bar{\partial}_z+d_t)\varphi \neq 0$$

on the level set $\mathcal{S}_0$ and therefore

$$c_0 = c_1\left|\int_{(z,t)\in\mathcal{S}_0}f_0\wedge g\right|,\ c_1 > 0.$$

Set $V_0^+ = \{\ (z,t) \in \tilde{\Pi}'\times\Pi_0'';\ \varphi(z,s_0,t) > \mathcal{Im}\, w_0\ \}$. By Stokes' theorem and recalling that supp g is compact, we see that

$$\int_{(z,t)\in\mathcal{S}_0}f_0\wedge g = \int_{V_0^+}\left[[(\bar{\partial}_z+d_t)f_0]\wedge g + (-1)^q f_0\wedge(\bar{\partial}_z+d_t)g\right].$$

But it follows from $(IV.9.4)_q$ that

$$\int_{V_0^+} [(\bar\partial_z + d_t) f_0] \wedge g = 0, \int_{V_0^+} f_0 \wedge (\bar\partial_z + d_t) g = \int_{\mathbb{C}^\nu \times \mathbb{R}^{n'}} f_0 \wedge (\bar\partial_z + d_t) g \neq 0,$$

whence the claim.

Going back to (IV.9.15) we observe that the open set $\mathcal{R}$ is simply connected and thus K_0 is contained in a compact set $\hat{K}$ whose complement is connected and contains w_0. The inequality (IV.9.15) demands that $\hat{K}$ not be polynomially convex, whence a contradiction. This completes the proof of Theorem IV.9.2. □

The reader will find in [CORDARO–TREVES, 1991] (see also Section VIII.6 of [TREVES, 1992]) a description of a number of situations in which the hypothesis on the nonvanishing of the intersection number is valid. Here we shall content ourselves with mentioning a couple of cases that seem noteworthy.

EXAMPLE IV.9.1.— *Case* m = 1. In this case there are no variables z_i's. There is a single first integral, $w = s + i\varphi(s,t)$. The sets $w = const.$ are referred to as the *fibres of the structure* (in the vicinity of O). The germs of the fibres at points near O are invariants of the locally integrable structure of $\mathcal{M}$ (underlying its hypo—analytic structure, see p. 87, [TREVES, 1992]). The vanishing of the intersection number in dimension q is equiv—alent to the triviality of the homology in the same dimension. More pre—cisely, the validity of either Condition $(IV.9.4)_q$ or its counterpart [obtained by reversing the inequalities in $(IV.9.4)_q$] is equivalent to the following property:

$(IV.9.16)_q$ *There is a fibre $\mathcal{F}$ in* U *such that a singular* q*—cycle in* $\mathcal{F} \cap V$ *does not bound in* $\mathcal{F}$.

When q = 0 this means that *there are two distinct connected components of a fibre $\mathcal{F}$ in* U *which intersect* V. The prototype of structures in which this occurs is the Mizohata structure in $\mathbb{R}^2$ defined by the first integral $Z = x+\imath t^2$: every fibre in the complement of the x–axis $t = 0$ consists of two points.

Always in the case m = 1, it is proved in [CORDARO—TREVES, 1994] that the negation of $(\mathrm{IV}.9.16)_q$ implies, and therefore is equivalent to, the validity of Property $(\mathrm{IV}.9.3)_q$. □

EXAMPLE IV.9.2.— Suppose m ≥ 1 (but always d = 1) and that there is a decomposition of the coordinates,

$$z' = (z_1,\dots,z_\mu),\ z'' = (z_{\mu+1},\dots,z_\nu),$$

$$t' = (t_1,\dots,t_{\mu'}),\ t'' = (t_{\mu'+1},\dots,t_{n'}),$$

such that, for some numbers suitably close to zero, $s_0 \in \mathbb{R}$, $b_0 > 0$, the following holds:

(IV.9.17) $$\forall\ (z,t) \in \mathscr{C}\ell(\tilde{\Pi}\times\Pi_0'),\ \varphi(z,s_0,t) = b_0 \Rightarrow |z'|+|t'| \neq 0;$$

(IV.9.18)
$$\{\ (z,t) \in \tilde{\Pi}\times\Pi_0';\ z'' = 0,\ t'' = 0,\ \varphi(z,s_0,t) \leq b_0\ \} \subset\subset \tilde{\Pi}'\times\Pi_0''.$$

Then, if $s_0 = \mathscr{R}e\, w_0$ and $\mathscr{I}m\, w_0$ is sufficiently close to b_0, Property $(\mathrm{IV}.9.2)_q$ will hold for $q = \mu+\mu'$.

Special examples of situations in which (IV.9.17)—(IV.9.18) are valid are discussed on pp. 386—388 in [TREVES, 1992]. Typical of such examples are the CR structures of the hypersurface type in which the Levi form is nondegenerate and the number of its positive, or of its negative, eigenvalues is exactly equal to q+1. Among these are the *Lewy structures* (see [TREVES, 1992], Sections VII.5, VII.6), in particular the prototype of Lewy structures, ie., the CR structure on $\mathbb{R}^3$ defined by the first integrals $z = x+\imath y$, $w = s+\imath|z|^2$. In it the tangent structure bundle is spanned by the Lewy vector field $\partial/\partial\bar{z} - \imath z\partial/\partial s$. There is no solvability in degree zero, whether in the distribution or in the hyperfunction sense. □

HISTORICAL NOTES

Hyperfunctions made their formal appearance in two articles of Mikio Sato at the end of the Nineteen Fifties. Hyperfunctions on the real line were introduced in [SATO, 1959] as boundary values of holomorphic functions in $\mathbb{C}\setminus\mathbb{R}$. In [SATO, 1960] the theory was extended to $\mathbb{R}^n$ $(n \geq 2)$ through relative cohomology spaces. If Ω is an open subset of $\mathbb{R}^n$ and V an open subset of $\mathbb{C}^n$ such that $V \cap \mathbb{R}^n = \Omega$, the space of hyperfunctions in Ω was defined as

$$(1) \qquad \mathcal{B}(\Omega) = H^n_\Omega(V, {}_n\mathcal{O})$$

[${}_n\mathcal{O}$ is the sheaf of germs of holomorphic functions in $\mathbb{C}^n$]. The relation between the definition in the present text and Sato's is clarified by our Theorem III.2.2 and Remark III.2.1. Equating hyperfunctions to sums of boundary values of holomorphic functions in wedges (with edge on $\mathbb{R}^n$) ensues by reinterpreting $H^n_\Omega(V, {}_n\mathcal{O})$ in the language of Čech cohomology.

Soon after the appearance of Sato's hyperfunctions A. Martineau saw the possibility of basing their definition on the theory of *analytic functionals.* The latter had attracted scant attention before the Nineteen Fifties, and few publications were devoted to them: one exception is [FANTAPPIÈ, 1943]. Of course, the duality theory of the spaces of holomorphic functions had to bring them in (as in [GROTHENDIECK, 1953]). The isomorphism (see section I.2) via the Laplace—Borel transform between the space of analytic functionals in $\mathbb{C}^n$ and the space of entire functions of exponential type (also in $\mathbb{C}^n$) is a variant of the Paley—Wiener theorem, useful in the study of PDE with constant coefficients (or of convolution equations) in the complex domain (see [MALGRANGE, 1955/56]).

But at the end of the Fifties Martineau was in a position to widen the role of the analytic functionals. Large tracts of several complex variables theory had been recently illuminated by the Oka–Cartan methods and the language of sheaves. It had been shown in [GRAUERT, 1958] that every compact subset of $\mathbb{R}^n$ possesses both the Runge and the Cousin properties, a fact which is the key to the analytic functionals approach to hyperfunctions, through the following implication:

(2) *the intersection of two compact subsets of* $\mathbb{R}^n$*, each of which carries a given analytic functional* μ*, also carries* μ*,*

a property patently untrue for arbitrary compact subsets of complex space. The proof of (2) in [MARTINEAU, 1959/60] is based on the cohomology with coefficients in the sheaf ${}_n\mathcal{O}$. Thus can one talk of the *support* of an analytic functional in real space. From then on analytic functionals could be identified to the hyperfunctions in $\mathbb{R}^n$ that are *compactly supported.*

The monographs [SCHAPIRA, 1970] and [HÖRMANDER, 1983] adopt Martineau's point of view. In both, Property (2) is derived by potential theory methods: Schapira applies the Cauchy–Kowalewski theorem to the Laplace equation, Hörmander represents any analytic functional carried by $\mathbb{R}^n$ as the jump of an appropriate harmonic function in $\mathbb{R}^{n+1}$. The link between hyperfunctions and harmonic functions had been explored earlier, in [KOMATSU, 1969], [SCHAPIRA, 1969]; see also [KAWAI–KOMATSU, 1971/72]. This trend culminates in [KOMATSU, 1991], where hyperfunctions are defined as classes of harmonic functions (following a suggestion in [SATO, 1959]).

But a trend in a different direction was started by the observation, in [MARTINEAU, 1960/61], that Serre's duality ([SERRE, 1955]) could be used to prove a natural isomorphism

(3) $$H^{n,n-1}(\mathbb{C}^n\backslash K) \cong \mathcal{O}'(K)$$

for any *polynomially convex* compact subset K of $\mathbb{C}^n$ and $n \geq 2$ [$\mathcal{O}'(K)$ is the space of analytic functionals in $\mathbb{C}^n$ carried by K]. In tribute to its elegance and directness we have devoted an entire section, section I.6, to retracing the same route, with minor departures. This, despite the fact that in the prior sections, our own presentation (admittedly restricted to small compact subsets of a maximally real submanifold of $\mathbb{C}^n$) is based on an explicit realization of the isomorphism (3), without recourse to Serre's duality.

An isomorphism similar to (3) for n = 1 is a classical feature of the theory of analytic functionals in the plane or on the Riemann sphere, antedating Sato's theory (see [GROTHENDIECK, 1953]): $H^{1,0}(\mathbb{C}^n\backslash K)$ must be replaced by the space of one–forms $h(z)dz$ with h holomorphic in $\mathbb{C}\backslash K$ and $|h(z)| \to 0$ as $z \to \infty$ (if n = 1, K polynomially convex $\Leftrightarrow$ $\mathbb{C}\backslash K$ connected). The isomorphism is realized by the Cauchy transform (section I.2). The feasibility of an explicit integral representation of the iso–morphism (3) for $n \geq 2$ was demonstrated in [TSUNO, 1982], by way of the Ramirez–Henkin formulas (see also [YE, 1993]). This is essentially the path followed in the present book, but through integral representations simpler than Tsuno's (Theorem I.4.2), built out of reproducing kernels similar to those considered in [HARVEY–WELLS, 1972]. Harvey and Wells prove that the totally real submanifolds of $\mathbb{C}^n$ of class C^1 are *totally real subsets* of $\mathbb{C}^n$ in the sense of [HARVEY, 1969], sets on which, accord–ing to the latter work, the sheaf of hyperfunctions can be defined.

Our presentation (in sections I.3, I.4, I.5) of hyperfunction theory on a maximally real (ie., totally real, of maximum dimension, n) sub–manifold $\mathcal{X}$ of $\mathbb{C}^n$ can now easily be situated: we combine the Martineau approach via analytic functionals with *Gaussian approximation* as in [BAOUENDI—TREVES, 1981] and integral representations of the isomorphism (3) for small compact subsets K of $\mathcal{X}$. [When $\mathcal{X} = \mathbb{R}^n$ the results are valid for compact sets of any size.] Another noteworthy feature, borrowed from [SCHAPIRA, 1970], is the so—called *completeness* of the presheaf of hyperfunctions (Theorem I.5.1). And in order to define the boundary value of a holomorphic function in a wedge, with edge $U \subset \mathcal{X}$, as a hyperfunction in U (Theorem II.1.1), we follow [YE, 1993] (Ye adapts an idea in [HÖRMANDER, 1983]).

The holomorphic extension of hyperfunctions to wedges leads directly to the notion of *essential support* (or *singular spectrum*) of a hyperfunction, as shown first in [SATO, 1970]. Shortly later Hörmander defined the $\mathcal{C}^\infty$ singular spectrum by means of cutoff functions and Fourier tranform, and gave it the name of *wave front set*. A $\mathcal{C}^\omega$ version of the $\mathcal{C}^\infty$ definition can then be devised by a special choice (introduced earlier by L. Ehrenpreis) of the cutoff functions. For this version the name *analytic wave front set* was coined. A third definition of the $\mathcal{C}^\omega$ wave front set of a distribution u can be derived through the variant of the Fourier transform introduced in [BRÓS—IAGOLNITZER, 1973, also 1974/75] and called here, and elsewhere, the Fourier—Brós—Iagolnitzer (abbreviated to FBI) transform:

(4) *the complement of the wave front set of u in an open subset Ω of $\mathbb{R}^n$ is the conic region of the cotangent bundle over Ω in which the FBI transform of u decays exponentially as the frequency tends to infinity.*

The equivalence for distributions of the various notions of analytic wave front set was established in [BONY, 1976/77]. A very general theory of the FBI transform and its use in the microlocal analysis of $\mathcal{C}^\omega$ singularities was presented in [SJÖSTRAND, 1982].

Because the base manifold $\mathcal{X}$ is "curved", one is forced to modify the FBI transform (originally defined on the flat base $\mathbb{R}^n$); and because in general $\mathcal{X}$ will not be real—analytic, the concept of analytic wave front set must also be generalized. This motivates the definition of a *hypo—analytic structure* and of the *hypo—analytic wave front set* of a distribution, pro—posed in [BAOUENDI—CHANG—TREVES, 1983] (see also [SJÖSTRAND, 1982(2)]). The analytic functionals approach to hyperfunctions makes the extension of the FBI transform and of the hypo—analytic wave front set to the latter very easy, as the reader can see in sections II.2, II.3, II.4. The extension is further facilitated by the systematic use of the *real structure bundle* of the maximally real submanifold $\mathcal{X}$ (introduced in [TREVES, 1992]). The characterization (4) remains valid for a hyperfunction u, as shown when $\mathcal{X} = \mathbb{R}^n$ in [HÖRMANDER, 1983] and when $\mathcal{X}$ is a totally real submanifold of $\mathbb{C}^n$, in [BAOUENDI—ROTHSCHILD, 1988] where the objects of study are sums of boundary values of holomorphic functions in wedges with edge on $\mathcal{X}$ (and arbitrary growth at the edge).

From its inception hyperfunction theory has been intimately linked with the Theorem of the Edge of the Wedge (see [MARTINEAU, 1964, 1967/68, 1970]). This theorem has its origin in problems of analytic continuation encountered in Quantum Mechanics (see [BOGOLJOUBOV—MEDVEDEV—POLIVANOV, 1958], [BREMERMANN—OEHME—TAYLOR, 1958]; a survey of the subject can be found in [MORIMOTO, 1973]). For wedges with nonflat edges, particular cases of Martineau's Edge of the Wedge Theorem (Theorem II.5.1) can be found in [PINČUK, 1976] and [YE, 1993]. We prove the result in full generality, for systems of wedges with edge on the maximally real submanifold $\mathcal{X}$, by the FBI transform method, in the spirit of [BAOUENDI—CHANG—TREVES, 1983] (complemented by a simple argument of exterior algebra).

Microfunctions (section II.7), the microlocal version of hyper—

functions, were introduced by Sato and provide the foundation for microlocal analysis in the analytic category ([SATO–KAWAI–KASHIWARA, 1970]), and more generally, in the hypo–analytic category.

Since the foundations (hypo–analytic structures) and the material (hyperfunction solutions and hyperfunction cohomology classes) in Chapters III and IV of the present monograph are of a greater novelty than those in Chapters I and II, their historical precedents are fewer. As a rule, the older results, such as they are, deal with analytic dependence on parameters, and real–analytic *generic* submanifolds of complex space (whose hypo–analytic structure is thus the CR structure inherited from the complex structure of the ambient space). We shall limit ourselves to a list of results that are related to our own:

In [KANEKO, 1988] it is proved that if the projection into x–space of the support of a hyperfunction $u(x,t)$ is contained in a compact set K (independent of t), then for $u(x,t)$ to depend analytically on t (according to the definition in [SATO, 1960]) it is necessary and sufficient that $< u(\cdot,t),h >$ be an analytic function of t whatever the holomorphic function h in some neighborhood of K [$u(x,t)$ is regarded as an analytic functional in x–space]. This can be compared with the material in Chapter IV where we deal with analytic functionals depending on parameters.

The notion of CR hyperfunctions on a real–analytic generic submanifold $\mathcal{M}$ of $\mathbb{C}^N$ is self–evident: they are the hyperfunctions annihilated by the CR vector fields (whose coefficients are analytic). The articles [KASHIWARA–KAWAI, 1972, 1973] show that, if d = $\operatorname{codim}_{\mathbb{R}} \mathcal{M}$, the sheaf of CR–hyperfunctions is isomorphic to the sheaf of relative cohomology of degree d with coefficients in the sheaf ${}_N\mathcal{O}$, $\mathcal{H}_{\mathcal{M}}^{d}({}_N\mathcal{O})$. This can be compared with our main theorem (Theorem IV.3.2) or the version of it (Theorem IV.6.1) that applies to CR structures (among others). In the same set–up [TAJIMA, 1988] shows that if the support $\mathcal{N}$ of a CR hyper–

function in $\mathcal{M}$ is a real analytic submanifold of $\mathcal{M}$ then it is totally characteristic with respect to the CR vector fields (this is a direct consequence of our Theorem IV.5.3). A special case of the uniqueness Theorem IV.5.2, still when $\mathcal{M}$ is real–analytic but now $\text{codim}_{\mathbb{R}} \mathcal{M} = 1$, is obtained in [TAJIMA, 1990].

A microlocal extension of Martineau's theorem of the Edge of the Wedge was recently described in [SCHAPIRA–TRÉPREAU, 1990]: the base manifold is a generic submanifold $\mathcal{X}$ of a complex manifold $\mathcal{M}$ and the wedges lie in an open subset Ω of the complement of the zero section in the conormal bundle to $\mathcal{X}$ in $\mathcal{M}$. The theorem of the Edge of the Wedge is valid under a certain hypothesis on the Levi form of $\partial\Omega$ and provided $\mathcal{X}$ does not contain any complex curve.

In connection with section IV.9 one must mention the result in [SCHAPIRA, 1969], according to which the local solvability, in hyper–functions, of a first–order, complex linear PDE with analytic coefficients is equivalent to the validity of the so–called condition $(\mathcal{P})$.

BIBLIOGRAPHICAL REFERENCES

Baouendi, M. S., Chang, C. H. and Treves, F.— *Microlocal hypo—analyticity and extension of CR functions*, J. Diff. Geom. **18** (1983), 331—391.

Baouendi, M. S. and Rothschild, L. P.— *Extension of holomorphic functions in generic wedges and their wave–front sets*, Comm. in PDE **13** (1988), 1441—1466.

Baouendi, M. S. and Treves, F.— *A property of the functions and distributions annihilated by a locally integrable system of complex vector fields*, Ann. of Math. **113** (1981), 387—421.

Bogoljoubov, N., Medvedev, B. V. and Polivanov, M. K.— *Voprosy teorii dispersionnykh sootnosheniy,* Fitzmatgiz 1958.

Bony, J.—M.— *Equivalence de diverses notions de spectre singulier analytique*, Séminaire Goulaouic—Schwartz 1976/1977, Exposé n° III.

Bredon, G.— Sheaf Theory, McGraw—Hill, New York 1967.

Bremermann, H. J., Oehme, R. and Taylor, J. G.— *A proof of dispersion relations in quantized field theories*, Phys. Rev. **109** (1958), 2178—2190.

Bros, J. and Iagolnitzer, D.— *Causality and local analyticity; mathematical study.* Proceed. Conf. sur la théorie de la renormalisation, Ann. Inst. Poincare **18** (1973), 147—184.

.— *Support essentiel et structure analytique des distributions*, Séminaire Goulaouic—Lions—Schwartz, 1974—75, Exp. 18.

Cartan, H.— Séminaire, Ecole Norm. Sup. Paris 1950/51.

Cordaro, P. D. and Treves, F.— *Homology and cohomology in hypo—analytic structures of the hypersurface type*, J. Geometric Analysis **1** (1991), 39—70.

Cordaro, P. D. and Treves, F.— *Necessary and sufficient conditions for the local solvability in hyperfunctions of a class of systems of complex vector fields,* 1994, preprint.

De Wilde, M.— Closed graph theorems and webbed spaces, Pitman London 1978.

Eilenberg, S. and Steenrod, N.— Foundations of Algebraic Topology, Princeton Univ. Press, Princeton N. J. 1952.

Fantappiè, L.— *Teoría de los funcionales analíticos y sus applicaciones,* Consejo Supr. de Investig. Cientif. Barcelona 1943.

Godement, R.— Théorie des faisceaux, Hermann Paris 1958.

Grauert, H.— *On Levi's problem and the imbedding of real–analytic manifolds*, Ann. of Math., Series 2, **68** (1958), 460—472.

Grothendieck, A.— *Sur certains espaces de fonctions holomorphes I*, J. für reine und angew. Math. **192** (1953), 34—64.

Gunning, R. C. and Rossi, H.— Analytic functions of several complex variables, Prentice—Hall. Englewood Cliffs N. J. 1965

Harvey, F. R.— *Hyperfunctions and partial differential equations*, Thesis Stanford Univ. 1966, published in part in Proceed. Nat. Acad. U.S.A. **55** (1966), 1041—1046.

.— *The theory of hyperfunctions on totally real subsets of a com— plex manifold with applications to extension problems*, Amer. J. of Math. **91** (1969), 853—873.

Harvey R. and Wells Jr, R. O.— *Compact holomorphically convex subsets of a Stein manifold*, Trans. Amer. Math. Soc. **136** (1969), 509—516.

.— *Holomorphic Approximation and Hyperfunction Theory on a* C^1 *totally real submanifold of a complex manifold*, Math. Ann. **197** (1972), 287—318.

Hörmander, L.— An introduction to complex analysis in several variables, Van Nostrand Princeton. N.J. 1966.

.— The analysis of linear partial differential operators I & II, Springer—Verlag 1983.

Kaneko, A.— *A topological characterization of hyperfunctions with real analytic parameters.* Scientific Papers College Arts & Sci., University of Tokyo **38** (1988), 1—6.

Kashiwara, M. and Kaway, T.— *On the boundary value problem for elliptic systems of linear differential equations, Part I,* Proceed. Japan Acad. **48** (1972), 712—715; *Part II,* ibid **49** (1973), 164—168.

Kawai, T. and Komatsu, H.— *Boundary values of hyperfunction solutions of linear partial differential equations,* Public. RIMS Kyoto University, **7** (1971—72), 95—104.

Komatsu, H.— *Resolution by hyperfunctions of sheaves of solutions of differential equations with constant coefficients,* Math. Ann. **176** (1968), 77—86.

.— *Boundary values for solutions of elliptic equations,* Proceed. Intern. Conf. on Funct. Anal. and related topics 1969, Univ. of Tokyo Press, Tokyo 1970, 107—121.

.— *Microlocal analysis in Gevrey and in complex domains,* in Microlocal Analysis and Applications, Cattabriga, L., and Rodino, L., edit., Lecture Notes in Math. # 1495, Springer Veralg 1991.

Malgrange, B.— *Existence et approximation des solutions des équations aux dérivées partielles et des équations de convolution,* Ann. Inst. Fourier Grenoble **6** (1955/56), 271—355.

Martineau, A.— *Fonctions analytiques et distributions; support des fonc—tionelles analytiques,* Séminaire Schwartz, 4e année (1959/60), n° 19.

.— *Les hyperfonctions de M. Sato,* Séminaire Bourbaki, 13e année, 1960/61, n° 214.

.— *Distributions et valeurs au bord des fonctions holomorphes,* Proceed. Intern. Summer Institute Lisbon 1964, 195—326.

.— *Théorèmes sur le prolongement analytique du type "Edge of the Wedge Theorem",* Séminaire Bourbaki, 20e année, 1967/68, n°

340.

.— *Le "edge of the wedge theorem" en théorie des hyperfonctions de Sato*, Proceed. Intern. Confer. Funct. Anal. and Related Topics, Univ. of Tokyo Press, Tokyo 1970, 95—106.

Morimoto, M.— *Une remarque sur le théorème du "Edge of the wedge" de A. Martineau*, Proceed. Japan Acd. **45** (1969), 446—448.

.— *Edge of the Wedge Theorem and Hyperfunctions,* Springer Lecture Notes in Math. **287** (1973), 41—81.

Pinčuk, S. I.— *Bogoljubov's theorem on the "edge of the wedge" for generic manifolds*, Math. USSR Sbornik **23** (1974), 441—455.

Sato, M.— *Theory of hyperfunctions*, J. Fac. Sci., Univ. Tokyo, Part I in **I** (1959), 139—193; Part II in **I** (1960), 387—437.

.— *Hyperfunctions and partial differential equations*, Proceed. Intern. Confer. Funct. Anal. and Related Topics, Univ. of Tokyo Press, Tokyo 1969, 91—94.

Sato, M., Kawai, T. and Kashiwara, M.— *Hyperfunctions and pseudo—differential operators*, Lecture Notes in Math. #287 (1973), Springer—Verlag, 265—529.

Schapira, P.— *Une équation aux dérivées partielles sans solutions dans l'espace des hyperfonctions*, C. R. Acad. Sci. Paris **265** (1967), 665—667.

.— *Solutions hyperfonctions des équations aux dérivées partielles du premier ordre,* Bull. Soc. Math France **97** (1969), 243—255.

.— Théorie des hyperfonctions, Lecture Notes in Math. **126**, Springer—Verlag 1970.

.— *Problème de Dirichlet et solutions hyperfonctions des équations elliptiques,* Boll. UMI **3** (1969), 367—372.

Schapira, P. and Trépreau, J.—M.— *Microlocal pseudoconvexity and "Edge of the Wedge" theorem.* Duke Math. J. **61** (1990), 105—118.

Schwartz, L.— Théorie des Distributions, Hermann Paris 1966.

Serre, J.–P.– *Un théorème de dualité*, Comm. Math. Helv. **29** (1955), 9–26.

Sjöstrand, J.– Singularités analytiques microlocales, Astérisque **95**, Soc. Math. France 1982.

.– *The FBI–transform for CR submanifolds of* $\mathbb{C}^n$, Prépublications Départ. Math. Université de Paris–Sud (1982).

Sussman, H.– *Orbits of families of vector fields and integrability of distributions*, Trans. Amer. Math. Soc. **180** (1973), 171–188.

Tajima, S.– *Support of CR–hyperfunctions*, Proceed. Japan Acad. **64** (1988), 239–240.

.– *Unique continuation theorem for CR–hyperfunctions*, Nihonkai Math. J. **1** (1990), 1–9.

Treves, F.– Topological Vector Spaces, Distributions and Kernels, Academic Press New York 1967.

.– Basic Linear PDE, Academic Press New York 1975.

.– Hypo–analytic structures. Local Theory, Princeton Univ. Press. Princeton N. J. 1992.

Tsuno, Y.– *Integral representations of an analytic functional*, J. Math. Soc. Japan **34** (1982), 379–391.

Ye Zaifei.– *Holomorphic extension and decomposition from a totally real manifold*, Trans. Amer. Math. Soc. **339** (1993), 1–33.

INDEX OF TERMS

Note: Boldface indicates where definition of term can be found